Green Energy and Technology

For further volumes:
http://www.springer.com/series/8059

Giuliano Dall'O'

Green Energy Audit of Buildings

A Guide for a Sustainable Energy Audit of Buildings

Springer

Giuliano Dall'O'
Department Architecture
Built environment and Construction
engineering (ABC)
Politecnico di Milano
Milan
Italy

Additional material to this book can be downloaded from http://extras.springer.com/8059.

ISSN 1865-3529 ISSN 1865-3537 (electronic)
ISBN 978-1-4471-5063-3 ISBN 978-1-4471-5064-0 (eBook)
DOI 10.1007/978-1-4471-5064-0
Springer London Heidelberg New York Dordrecht

Library of Congress Control Number: 2013933593

Printed on acid-free paper

Springer is part of Springer Science+Business Media (www.springer.com)

Preface

The *energy issue* has been in recent years, and will be in the years to come, the most critical factor for a sustainable development of our society. The energy demand in the entire world, driven by expectations of a better quality of life in emerging countries, as well as by population increase, has reached levels that cannot be sustained in the future.

Two elements make the situation even more critical for the coming years: *fossil fuels are limited* and their widespread use causes significant adverse *environmental impacts* such as the production of carbon dioxide, one of the causes of *global warming*, but also acid rain, higher ozone concentration, or particulates, common emergencies in urban areas.

There is a widespread awareness of how the energy problem must be dealt with, being the implementation of *long-term energy policies*, very different from those adopted so far. A drastic improvement in *energy efficiency* at all processing levels and in all sectors (e.g. industry, transports, construction, services), and a significant use of *alternative energy sources* (e.g. solar, wind, biomass), through a planned process of replacing the fossil origin fuels, are necessary. A significant and necessary challenge supported by new strategies but, above all, by a new cultural approach, oriented towards environmental sustainability.

These changes in energy policies have concretely been initiated in the last ten years: almost all countries are committed to achieving the goals of reducing carbon dioxide emissions on the basis of international agreements (e.g. the Kyoto protocol). The European Union is defining medium- and long-term targets and programmes through to 2050, by introducing verification steps in order to correct the strategies adopted by the Member States.

This international challenge should not only be seen as an emergency, but as an interesting opportunity, as it can generate *positive effects in the economy*. The implementation of new energy policies requires the use of new technologies for energy conversion and end-uses: this stimulates the research in the various fields involved and obviously new skills for the actors (e.g. engineers, architects, energy managers, specialists, building managers, etc.). The whole of these new technologies, and the related services, is the basis of the *green economy*: an interesting opportunity to develop the economy towards greater sustainability targets.

The European Union is promoting among the Member States a common strategy to face the challenge.

Policies for energy efficiency have been developed not in Europe alone but also in other parts of the world, often with the aim of stimulating the economy in recession: in 2009 the U.S. Government proposed a series of economic measures for public and private employers to give a clear impetus to the development of the *green economy*.

The rapid evolution of rules for the energy efficiency of new buildings, results in a large energy performance gap between new buildings and existing ones that constitute the real estate. The improvement in the energy performance of existing buildings, through energy retrofit measures, is indeed a great chance for operators and, more generally, for the *green building economy*. A correct definition of the retrofit measures, that should be cost-effective, is however not so easy: it requires a detailed analysis of the causes of energy wastes, through a methodological and a professional approach. Accurate and complete energy audits are essential as a means to assess and verify a project's success at meeting performance goals.

The book *Green Energy Audit* fits into the framework discussed above, as a tool to support, from the technical and economical points of view, all the actions aimed at improving the performance of existing buildings and their facilities. The energy audit methodology proposed in this book has led us to define an acronym different from the traditional one: *Green Energy Audit*.

The added value lies in the "*green*", a word that refers to and summarizes a widespread concept; that of environmental sustainability. In fact, the *Green Energy Audit* is not limited to providing tools and methods to reduce energy consumption, but it poses a more ambitious objective: to contribute to an overall improvement in the sustainability of the building under consideration. The book also provides the elements to understand what possible benefit will be obtained, once implemented measures to improve the sustainability, to start a process of certification under LEED® protocols.

The technical literature offers many rigorous books and handbooks dealing with the matter related to energy audits, however, the publications are often intended for experts with in-depth, baseline knowledge of the subject.

Indeed this book is designed to give the reader, in a simple and accessible way, and without requiring in-depth skills, a comprehensive method to analyze the building and the facilities. It shows how to properly use the audit instrumentation for field surveys and monitoring, to define the baseline energy balance, to identify the retrofit measures assuming different scenarios, to make economic evaluations, hence evaluating the improvement of global sustainability, and to prepare a clear and effective technical report.

The book *Green Energy Audit of Buildings* is aimed at many types of audience:

- engineers and architects who are already operating at various levels in the energy fields, and who may have experience in the field of energy audit. It will assist in completing their methodological approach and provide ideas to improve their expertise;

- energy assessors who wish to undertake the profession of auditor;
- lecturers and gradate/postgraduate students of science, engineering, architecture, and construction who want to improve their knowledge on the issues of sustainability of buildings and energy audit;
- not only those responsible for maintenance of buildings, but also for all non-technical energy operators, such as real estate managers or building managers.

The purpose of this book is, above all, to spread knowledge and interest on the topics of energy and environmental quality of buildings, so contributing, pragmatically, to the change in real estate.

I am grateful to many people, beginning with the colleagues who have provided written contributions, that have been included in the book in specific topics. A special thanks goes to my colleague and friend Luca Sarto, who also helped me in the review of the entire book. I am also grateful to the colleagues and the students who provided valuable feedback and encouraging comments on most of the material of this book. I am grateful to Mark Izard who helped me in finding the correct way to express the concepts, but at the same time his help was for me a useful comparison in viewpoints on the technical aspects.

Finally I express my thanks to Mr. Anthony Doyle for agreeing to include this book in the prestigious series "*Green Energy and Technology*" of Springer, giving me the opportunity to contribute to this important project.

Milan, April 2013 Giuliano Dall'O'

Contents

Chapter 1
Introduction

This opening chapter describes the objectives and the purposes of this book, giving the reader a general framework for a first understanding of the audit methodology. Green Energy Audit proposes a different approach to the analysis of the existing buildings, the objective is to improve sustainability by applying retrofit measures not only in the area of energy efficiency.

Finally, this chapter examines the structure of the book, defining the various parts that constitute it. The book is conceived to form a useful instrument for learning about and acting on the subject matter.

1.1 Awareness, as Starting Point for Performance Improvement

The development of our society is definitely oriented towards an improvement in environmental sustainability. This change, almost unanimously considered necessary, covers all areas responsible for consumption of energy, and more generally speaking, of resources: the main ones are transport, industry and construction (residential and the tertiary sector).

As regard the European Union, the Green Paper "Towards a European strategy for the security of energy supply" [1] estimated that the residential and tertiary sector, the major part of which is buildings, accounts for more than 40 % of final energy consumption in the Community and is expanding, a trend which is bound to increase its energy consumption and hence also its carbon dioxide emissions.

Energy policies will cover both new buildings and existing buildings. The European Directive 2002/91/EC [2], named energy performance building directive (EPBD) highlights the fact that buildings will have an impact on long-term energy consumption and new buildings should therefore meet minimum energy performance requirements, tailored to the local climate. The transposition of the above-mentioned Directive in the Member States, generated new regulations and laws aimed to significantly increase the energy performances of new constructions. The more recent Directive 2010/31/EC [3] goes further, in terms of energy

G. Dall'O', *Green Energy Audit of Buildings*, Green Energy and Technology,
DOI: 10.1007/978-1-4471-5064-0_1,

performances for buildings: by 31 December 2020, all new buildings must be nearly zero-energy buildings; and after 31 December 2018, new buildings occupied and owned by public authorities must be nearly zero-energy buildings.

The real problem, however, concerns the existing building stock. The energy performance of existing buildings is very low and an energy policy that focuses on turn-over (i.e., replacement of existing buildings with new buildings) is not realistic since this process would take too many years.

As regard existing buildings, EPBD states that "Major renovations of existing buildings above a certain size should be regarded as an opportunity to take cost-effective measures to enhance energy performance. Major renovations are cases such as those where the total cost of the renovation related to the building shell and/or energy installations such as heating, hot water supply, air-conditioning, ventilation and lighting is higher than 25 % of the value of the building, excluding the value of the land upon which the building is situated, or those where more than 25 % of the building shell undergoes renovation; the improvement of the overall energy performance of an existing building does not necessarily mean a total renovation of the building but could be confined to those parts that are most relevant for the energy performance of the building and are cost-effective".

Also if the potential in terms of primary energy reduction in existing buildings is relatively high, more than 20 %, problems to be faced in carrying out actions on a large-scale are many:

- each building has different characteristics and therefore requires a careful analysis of the retrofit measures that can be taken in relation to the construction technologies, plant characteristics, methods of use, but also to the climatic conditions;
- it is not so easy to gain a good understanding of facility energy use;
- the market offers a wide range of possible technologies to improve the energy efficiency of buildings and related facilities, to take advantage of renewable energy sources and to improve the management: the choice may be difficult owing to lack of skills;
- the economic aspect should not be underestimated, since the retrofit, without financial support, may require considerable investments which may moreover have long payback times.

Actions to improve the energy performance of buildings cannot be improvised, but must be defined considering awareness, professional skills and, above all, an effective methodological approach.

1.2 A Comprehensive Approach

The *energy audit of buildings* is the most efficient tool to promote concrete energy retrofit actions, overcoming the critical aspects listed above.

There are many definitions of energy audit, since there are several operating modes with which to apply it. Generally we can define energy audit as an inspection, survey and analysis of energy flows in a building, in a process or in a system, aimed to reduce the amount of energy input without negatively affecting the outputs.

The reduction of primary energy use in buildings is an important goal and correlation between the use of primary energy from fossil fuels and the environmental impact is evident. However, if our goal is to improve the overall sustainability of buildings, we must go beyond this, considering also other aspects not always related to energy uses.

The concepts that underlie the new methodology approach developed in this book, start with a simple consideration: when a building is analysed in order to improve its energy performance, is it possible to make an additional effort to improve its level of sustainability? The answer is obviously affirmative.

A comprehensive approach to improving the sustainability of buildings is not only possible but also convenient. Environmental certification of buildings, also if promoted on a voluntary basis, is attracting ever greater interest. An interest which is justified by the widespread awareness of the importance of considering not only the energy aspects, which are also prevalent, but also environmental ones.

A growing number of buildings are certified according to international protocols, such as leadership in energy and environmental design (LEED)® developed in the United States or building research establishment environmental assessment method (BREEAM)®, developed in the UK, but also many others.

Rating systems have been developed to measure the sustainability of buildings and provide best-practice experiences in their highest certification level. Bauer et al. [4] define a comparison of different Rating Systems for Sustainable Buildings, considering the main international protocols. The key aspects of assessment are different, however, some items are commons in all protocols:

- Energy and atmospheric pollution;
- Site quality, transport and land consumption;
- Health and indoor air quality;
- Resources and materials.

The Green Energy Audit, the audit procedure discussed in this book, is not limited to providing tools and methods to reduce energy consumption, but it poses a more ambitious objective: to contribute to an overall improvement in the sustainability of the building. For this reason, the procedure considers and evaluates all retrofit measures that can contribute to improving the sustainability of the building. The preliminary assessment of the rating of global sustainability is made with reference to the LEED® Protocols new construction (NC) and major renovations and existing buildings: operation and maintenance (EBOM).

1.3 Structure of the Book

The field of application of this book is the construction sector: residential buildings, commercial and tertiary buildings. The energy and the resources that are considered are related to the building utilities and facilities provided to the user (e.g., thermal comfort, internal air quality, visual comfort, water management etc.).

The book does not consider the uses of resources related to the production cycles in the industrial field, although the overall methodological approach may not be different and many techniques described here are also relevant in this field, in particular those which concern field surveys and monitoring. Three operational levels are described and discussed: Walkthrough Audit, Standard Audit and Simulation Audit: for each of these three levels, tools and methods are provided. The book is structured into two parts: the first part defines the methods, the second part concerns the tools.

1.3.1 Methodologies

This part of the book gives the reader a way to apply the Green Energy Audit in a systematic way: it is structured in eight chapters which are summarised below.

- Chapter 2 presents the *general aspects of Green Energy Audit*. Aims and objectives of the work are highlighted and particularly those aspects that make the process convenient.
- Chapter 3 is devoted to an overview of the *application of the methodology*. The whole activity is divided into several phases, each of which is defined in a synthetic way within the various sections.
- Chapter 4 is concerned with the phase of *acquisition of basic information*, with the objective of providing criteria and methods to effectively manage data and information in an orderly and efficient way.
- Chapter 5 deals with the *field survey and monitoring issues*. It describes the characteristics of the instrumentation needed to carry out a proper survey for a better understanding of facility energy use.
- Chapter 6 introduces the issue of the *infrared audit*, giving the reader an overview on the infrared audit process, with not only the potential of the instrument as a support for the energy audit but also warnings about pitfalls to avoid in order to ensure a correct use of it.
- Chapter 7 focuses on the approach for the *definition of the baseline* and the *improvement evaluation*.
- Chapter 8, which is devoted to the *definition of the green energy plan*, gives general criteria for the definition of the retrofit measures to be adopted. This chapter provides the reader with a framework for structuring the technical report.

1.3.2 Tools

This part of the book contains four chapters that deepen topics of interest.

- Chapter 9 concerns the correlation between *Green Energy Audit* and *LEED®* Protocols NC and EBOM.
- Chapter 10 is devoted to the *economic evaluation of retrofit actions*. The most common economic evaluation methods are presented and discussed in this chapter.
- Chapter 11 deals with the *contractual schemes*. It analyzes and discusses the typical structure of a third party financing contract.
- Chapter 12 contains insights regarding *comfort and well-being of the occupants* (thermal comfort, internal air quality and lighting comfort);
- Chapter 13 contains a set of 97 *retrofit measure sheets* that define the main retrofit measures that can be implemented to reduce energy consumption and to increase the sustainability of the building.

The measure sheets coverthe following areas:

- building envelope (external and internal walls, windows and transparent surfaces, roofs, basements, protection from sunlight, illumination using daylight);
- mechanical systems and plant (heating cooling and ventilation systems, domestic hot water systems, plumbing systems);
- electric systems and plant (generation distribution and utilization of electrical energy, lighting, building management systems);
- renewable energy sources (solar thermal systems, solar photovoltaic or PV systems, biomass);
- managing improvement actions and procedures.

The set of retrofit measures proposed in Chap. 13 is a guideline that the reader can use to promote concrete actions which, when applied, can improve the sustainability of the buildings concerned.

The reader, furthermore, can download from the Springer website (address to define) a complete set of 45 checklists to gather and organise information collected during the technical analysis of the acquired documentation and the field surveys.

References

1. Commission of the European communities (2006) Green Paper a European strategy for sustainable, competitive and secure energy Com 105 final, Brussels
2. Directive 2002/91/EC of the european parliament and of the council of 16 December 2002 on the energy performance of buildings
3. Directive 2010/31/EU of the european parliament and of the council of 19 May 2010 on the energy performance of buildings (recast)
4. Bauer M, Mösle P, Schwarz M (2010) Green building, guidebook for sustainable architecture, Springer, Munich

Part I
Methodologies

Chapter 2
Green Energy Audit: General Aspects

This chapter provides a general overview of the Green Energy Audit. Aims and objectives of the work are highlighted and particularly those aspects that make the process convenient, without ignoring the critical issues that may emerge from its incorrect application. The chapter explores the three operational levels of walk-through audit, standard audit and simulation audit, highlighting the differences between the different approaches. Organisational aspects are then examined, together with the human factors and contractual aspects, required to meet the expectations of the owner and/or user client. The differences between, but also the complementarily elements of energy audits and energy performance certification, are also highlighted. The chapter then considers some important aspects that relate the Green Energy Audit to quality assurance processes.

2.1 Building Energy and Environmental Enhancement Strategies

2.1.1 Defining the Operational Plan

It is possible, and almost unfailingly cost effective, to reduce the amount of resources (not limited to energy) that an existing building consumes. However, starting the process of change is not easy if the owner/user client has little technical knowledge, and little or no awareness about the building's energy quality and its actual chances of improvement.

Difficulties are not limited to the evaluation of possible strategies to improve energy performance. Defining an appropriate plan to implement such strategies is also necessary, tackling in a rational way not only all the technical, but also the economic, and sometimes the legal limitations that might apply.

The process is even more complicated when there is more than one decision maker, as in the case of a large block of flats or housing complex or a timeshare property, because any final decision will have to be agreed upon between all the parties involved. Before starting any project to enhance a building's energy and

G. Dall'O', *Green Energy Audit of Buildings*, Green Energy and Technology,
DOI: 10.1007/978-1-4471-5064-0_2, © Springer-Verlag London 2013

environmental quality, we need to conceive an operational plan that allows us for rational and coherent consideration of every aspect of the problem.

The diagram in Fig. 2.1 outlines a possible *operational plan*. Subsequent phases are identified down the middle, with actions detailed on the right and the different participants involved on the left.

Once the need for implementing action to improve energy and environmental performance is defined, we need to make an assessment of the building, from the energy point of view, in the current situation and we define the *baseline*.

The energy audit, carried out by the auditor, can make an overall assessment of the building in order to understand what are the causes of inefficiency, and to obtain useful elements for developing an energy retrofit plan. In the operational strategies for improving the energy performance of a building, the Energy Audit is a key element.

The next step is the selection of the *retrofit measures*. These may affect the building envelope (or its core), the systems, or the use of renewable sources. At

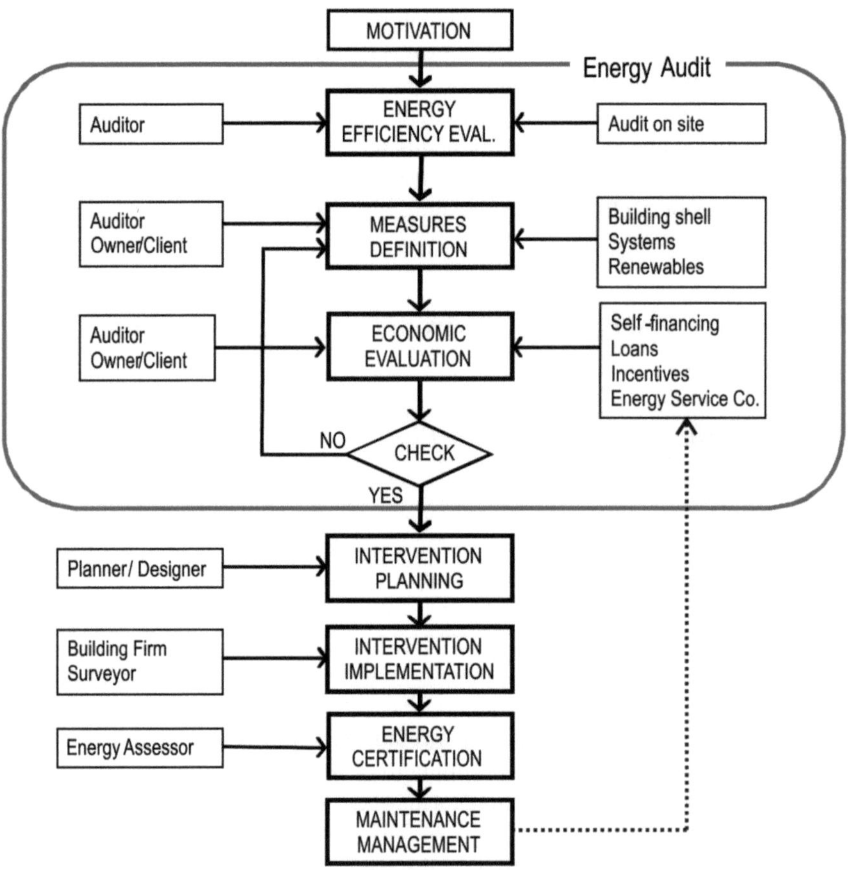

Fig. 2.1 Operational plan to design and implement an energy enhancement project

this point an auditor possesses all the technical and management information necessary in order to decide a strategy and to define the measures, however, a "comparing notes" with the owner/user/client, is appropriate.

Proposals must also be evaluated economically: at this stage of the process, the financing issues must be considered. If economic resources are not available, it is necessary to consider a bank loan. Third party financing by means of a energy service company (ESCo) is a good opportunity to evaluate. The availability of incentives for energy retrofit actions (e.g., low interest loans, tax reductions, etc.) play an important role, as they can render convenient retrofit work that, without incentives would otherwise not be.

If the economic plan is consistent, it is possible to proceed to the next stage which involves the executive design of the works, this may involve the auditor or a competent designer (e.g., architect, engineer). The construction phase involves the construction company and the project supervisor/construction and engineering manager. If the economic plan is not consistent, a modification of the choices is required.

For the EU Member States, Directive 2002/91 EC [1] makes energy certification of buildings mandatory. When the energy retrofit works have been completed, energy certification is convenient since the energy quality of the building has been enhanced.

Energy retrofit of buildings increases the energy efficiency and reduces emissions of greenhouse gases and pollutant gases, improving the sustainability. Aspects that are often overlooked, however, are the system management and maintenance issues: maintaining high performance over time is possible only if these aspects are considered.

2.1.2 Reasons that Lead to Energy and Environmental Enhancement

Reducing the consumption of resources of an existing building is often possible and convenient. However, initiating the process of implementing the changes is not simple if the owner/user customers have little expertise and little or no awareness of the energy status of their building.

The problem is not only to evaluate strategies for improving performance but also to define a correct path for their rational implementation, considering not only the technical constraints but also the economic and legal aspects. The decision to start any project stems from the motivation; the stronger the motivation, the easier it will be to realise.

The energy recovery process may be triggered by a number of factors, amongst which are the following:

- there are facilities that fail to provide acceptable environmental comfort;
- need or desire to reduce the consumption of energy and resources (e.g., water);

- the plant must be replaced (because it is malfunctioning, it is inadequate in terms of regulations or safety or because it no longer complies with emission limits);
- major renovations of the building are planned;
- energy certification has demonstrated a lack of energy efficiency.

2.1.2.1 Facilities that Fail to Provide Acceptable Indoor Environmental Comfort

A first reason can be determined by the knowledge that indoor air/environmental conditions are not satisfactory. Heating ventilation air conditioning (HVAC) systems which are not able to control the temperature, for example, can generate indoor conditions which are uncomfortable (e.g., too hot or too cold, too dry or too humid).

During the winter season, an internal air temperature higher than that considered optimal for comfort (e.g. 20 ÷ 21 °C), is a clear indicator of energy wastage. In continental climates, for existing buildings with external walls not properly thermally insulated, for each degree over the optimal temperature the energy consumption, and therefore the energy bill for winter heating, increases by 6 ÷ 8 % depending on the specific situation (energy efficiency of the building and climate conditions).

A lower temperature, on the other hand, makes the indoor climate uncomfortable and this is also a clear indicator that the system is inadequate.

Similar considerations could be made, of course, for summer air conditioning: in summer seasons, the highest energy consumption due to a lower internal air temperature determines an even greater energy consumption.

Sometimes HVAC systems are not able to control the indoor climatic conditions of a building, and in winter, for example, there are simultaneously zones which are too warm, with obvious wastage of energy, and others too cold, with a clear situation of discomfort. An energy audit, in this case, represents the best technical approach to establish whether the cause is the inefficiency of the building envelope in some parts of the building or on the inefficiency of the plant (unbalanced distribution systems, inefficient control systems, etc.).

2.1.2.2 High Consumption of Energy and Resources

It is not so easy, for a client who has no terms of reference, to determine whether energy consumption is high, but awareness of what affects the energy bill can be a powerful reason to investigate what might be the causes of any wastage.

Real estate managers or public managers, even if without specific skills in HVAC systems, can produce simple indicators, comparing energy costs with the air-conditioned area or the air-conditioned volume of the building.

2.1.2.3 Major Renovations of the Building are Planned

Retrofit measures on the building envelope (e.g., thermal insulation of external walls or roofs, window replacement, etc.), normally are the most expensive and economically less convenient if planned only with the goal of increasing energy performance. For these parts of the building, however, major renovation is periodically planned for maintenance purposes.

In these cases, the additional cost for improving the energy performance of the components is very low, hence cost-effective.

For example, if the simple renovation of the external plaster of a building facade is planned with a basic budget, the additional cost of additional insulation with external thermal insulation composite systems (ETIC) is economically acceptable, but the increasing of thermal performance of the facade is very high. Similar examples may include the windows or the roof.

The increasing of the thermal performance of the building, may furthermore suggest a review of the HVAC systems, since the thermal capacity could be significantly reduced.

We can say that when major renovation for the building are planned, an energy audit represents the best technical and economical approach to establish the correct way to plan the redevelopment that could involve not only the building envelope but also the related plant.

2.1.2.4 Energy Certification has Demonstrated a Lack of Energy Efficiency

The main scope of the energy certification of buildings is to inform the user, in a simple manner, about the energy performance of the building. A low performance indicator in the certificate, may suggest an energy audit in order to identify in greater detail the causes of inefficiency and also the best solutions to remove it.

2.2 Scope and Aims of the Green Energy Audit

2.2.1 *The Meaning of Auditing*

The term audit defines evaluation activity of an organisation, process, project or product. The audit activity is aimed at ascertaining the validity and reliability of the information collected and, at the same time, to configure an internal audit control system.

The purpose of auditing is therefore to use the information collected for achieving an improvement that can be defined in several ways, depending upon the application field: better performance, cost reduction, improved security or, more generally, an improvement of the overall quality.

The check and the evaluation, essential and fundamental steps of the whole activity, must then be transformed into concrete suggestions, leading to tangible improvements. The audit, therefore, must be considered to be a process.

To assess this improvement, the definition of objectives is essential. The achievement, or not, of these goals depends, of course, on several factors, beginning with economical ones: the improvement of the organisation, of the process, or of the quality of product, remain however key elements of auditing activities.

The figure in charge of performing audit activities, the auditor is usually a person from outside the system: the independence of the auditor is the added value of the whole process.

A useful reference for understanding the logic behind auditing activities in general is the ISO 19011-2002[1] [2] standard, which is applicable to all organisations needing to conduct internal or external audits of quality and/or environmental management systems or to manage an audit programme.

According to the above mentioned standard, audit activity is based on a number of principles, which refer to the auditor, and with which compliance is a prerequisite. This is to provide sufficient and relevant conclusions of the audit and auditors to ensure that different auditors, operating independently of each other, come to similar conclusions in similar circumstances:

- *Ethical conduct*: the foundation of professionalism

Trust, integrity, confidentiality and discretion are essential to the audit.

- *Impartial presentation*: the obligation to report faithfully and accurately.

The findings, conclusions and audit reports must reflect faithfully and accurately the audit activities.

- *Adequate professionalism*: the application of accuracy and discernment in the audit.

The auditor pays the appropriate level of attention to the importance of the task they perform and the confidence reposed in them by their audit clients and other stakeholders. It is essential that they have the necessary skills.

Additional principles refer to the audit process which is by definition independent and systematic:

- *Independence*: the basis for the impartiality of the audit and objectivity of its conclusions.

The auditors are independent of the activity being audited and are free from preconceptions and conflicts of interest. The auditors maintain a state of objectivity

---

[1] The Standard provides guidance on the principles of auditing, managing audit programmes, conducting quality management system audits and environmental management system audits, as well as guidance on the competence of quality and environmental management system auditors.

of thought during the audit process to ensure that the findings and conclusions are based only on evidence of the audit.

- *Evidence-based approach*: the rational method for reaching reliable and reproducible audit conclusions in a systematic audit process.

Audit evidence is verifiable. It is based on samples of the available information, since an audit is performed in a limited time and limited resources. The appropriate use of sampling is closely related to the level of confidence that can be placed on the conclusions of the audit.

The general principles set out in this paragraph may be applied to the activities of green energy audits or energy audits discussed below.

2.2.2 Energy Auditing

When the subject of audit are buildings, equipment connected to them or production infrastructure, and the purpose is to reduce the consumption of primary energy from fossil fuels, then the audit is referred to as an *energy audit*.

The definition provided in the Standard EN 16247-1:2012[2] [3], defines the energy audit as "a systematic procedure to obtain an adequate knowledge of the profiles of energy consumption of an existing building or group of buildings, an industrial and service private or public, in order to identify and quantify in terms of cost effectiveness of energy saving opportunities and the relationship of what is revealed".

This definition highlights the three elements that characterise an energy audit regardless the operational mode adopted:

- knowledge of the energy consumption profiles of the investigated system;
- identification of possible energy retrofit measures to reduce energy consumption;
- cost-effective evaluation of the retrofit measures;
- reporting activity (the return of the analytical work done).

Once these points are fixed, the energy audit can be articulated in different ways, with its complexity dependent upon the technical and economical effort that is required.

We must first make a distinction between *building audit* (considering, however, the building envelope and related plant) and *industrial audit*.

---

[2] This European standard specifies the requirements, common methodology and deliverables for energy audits. It applies to all forms of establishments, energy and use of energy, excluding individual private dwellings. This part covers the general requirements common to all energy audits. Specific energy audit requirements will complete the general requirements in separate parts dedicated to energy audits for buildings, industrial processes and transportation.

In the *building audit*, issues regarding thermal comfort, air quality, lighting comfort and acoustic comfort of the internal spaces, but also of the external spaces pertaining the building are mainly investigated. The energy audit normally concerns the building envelope but also the facilities such as HVAC systems and lighting systems using an integrated approach, and generally all the systems which are energy consumers e.g., lifts (elevators), safety and security, domestic hot water (DWH), etc.

Building energy audits concern not only residential buildings but also those with other uses such as commercial buildings, offices, schools, hospitals, etc.

In the *industrial audit* the energy used in the production cycle could be relevant, and indeed sometimes more significant, than that used for ensuring comfort and safety to the occupants. Industrial audits may require special skills in the specific production sectors of the industry investigated.

The *energy auditor interface* may change due to the complexity of the structure: for simple buildings normally the entity to report to is a person (e.g., the owner or the building manager), while for complex buildings the entity to report to may be a managing organisation with many people involved (e.g., facility management, building management, maintenance management).

The *object of the energy audit* may be different: with a global approach the auditor needs to consider all the facilities. In some cases, however, the requirements of the customer could relate to a specific service (e.g., only lighting system, only electrical facilities, only heating production, etc.).

Understanding how a building works from an energy perspective, is not simple because the building is a complex system, subject to variables not always predictable. The approach usually adopted in the *design phase* is to consider simplified boundary conditions (usage mode, standard conditions, standard specifications of the building notes), while the *actual operating conditions* are influenced by climatic variables usually different from the standard. Additionally management methods are not always related to single reference models. In other words, normally there is a difference between standard operating conditions and actual operating conditions and the energy audit approach must consider the second condition.

A simplified energy audit process can be structured into four steps

1. acquisition of documentation;
2. field surveys and monitoring;
3. definition of energy retrofit measures and
4. editing of the audit report.

2.2.2.1 Acquisition of Documentation

The basic information to make a proper energy audit address primarily the characteristics of the building (thermo-physical characteristics of the envelope, such as walls, windows, roofs, basements and characteristics of HVAC systems, DWH systems, lighting, electrical purposes, etc.).

Information on management procedures adopted (occupation schedules of the spaces, days of activation of the plant, operating temperatures, etc.) are very important in order to understand how the building is used by the occupants.

2.2.2.2 Field Surveys and Monitoring

Information to understand the energy behaviour of the building can be derived from project documents or, failing that, from field surveys and monitoring. So far as the management aspects are concerned, a monitoring campaign, albeit if limited in time of the main environmental parameters (e.g., air temperature, relative humidity, lighting, thermal and electric energy consumption, etc.) can provide important objective information on the actual behaviour of the users.

The analysis of energy bills through the actual consumption of energy, heat or electricity can complete the framework. The optimisation of contracts according to actual usage mode can help to reduce energy bills.

2.2.2.3 Definition of Energy Retrofit Measures Definition

The definition of energy retrofit measures, as previously mentioned, does not involve only technical aspects, namely the feasibility to realise the interventions in practice, but also financial issues, because measures not cost-effective will have little chance to be applied.

Normally, retrofit measures are not proposed as a list of single independent options, because the evaluation of the global effects, in term of energy saving or reduction in the use of resources, may be influenced by the synergies between the measures. If an auditor, for example, proposes as retrofit measures the wall insulation and the boiler replacement, it is clear that the evaluation of the independent effects of the two measures is different from the sum of the evaluation of the two measures taken together, since the wall insulation reduces the energy supplied by the new boiler.

Energy retrofit measures should be proposed to the owner/client as groups of consistent measures that we can define as *scenarios*.

2.2.2.4 Editing of the Audit Report

The success of an energy audit cannot be measured only by the quality of the study but in relation to the retrofit action that is actually implemented. The preparation of the report is therefore a strategic step.

The ability to communicate well to the owner/user client, the findings and proposals contained in the report is a clear indicator of professionalism and the key to the success of an energy auditing portfolio.

2.2.2.5 Defining of Operational Levels

A energy auditing activity well done could be expensive, dependent upon the skills of the auditors and the cost of instruments and the software used.

For maximum effectiveness at minimum investment, different operating levels of energy audit are used: from the cheaper but effective *walkthrough audit*, also known as one day audit, which involves less than a day for inspections and surveys, to the *simulation audit* which can require months of investigation. For a complete definition of the different operational levels, see Sect. 2.3.

2.2.3 Green Energy Audit

The word *green* can have many meanings depending upon the circumstances in which it is used. The original reference was to nature, because green is the colour of grass, plants and leaves. Over time, however, the meaning of this word has gone far beyond the original meaning, assuming a meaning that leads to a definition, even if symbolically, maintaining, or contribution to the maintenance of ecological balance.

The word green, then, identifies everything that contributes to the improvement of sustainability in its various meanings, not necessarily tied to nature.

Green building, for example, is used to identify buildings designed and constructed to minimise the environment impact. Similarly we define the *green economy*, as a new development model that contrasts the economic *black* model based on fuels fossil.[3]

2.2.3.1 Green Design

Green and *sustainable* are words often used as synonyms in order to emphasise an approach aimed at reducing the impact on the environment. Speaking of design processes, the American association of heating, refrigerating and air conditioning engineers (ASHRAE) GreenGuide [4] states that the difference between green and sustainable design is the degree to which the design helps to maintain ecological equilibrium. Indoor environmental quality (IEQ), for example, is an important issue of green design, but it has no impact on the ecological equilibrium of the environment.

---

[3] Green Economy is based on knowledge of ecological economies and green economies that face the problem of interdependence between the human economy and the natural ecosystem and taking into account the adverse effects of economic activity on climate change and global warming.

In our opinion, however, the boundary between green and sustainable design is very fleeting as all the issues related directly or indirectly to the use of energy are related to ecological balance.

Green and/or sustainable building design, but the same considerations apply to existing buildings under renovation, should have an approach, over the full life cycle, in the following areas:

- minimising natural resources consumption by mean of the valorisation of renewable energy and not energy resources;
- minimising atmosphere emissions, especially those related to the use of HVAC systems (greenhouse gases, pollutants emissions, acid rains, etc.);
- minimising solid and liquid waste discharge, during construction phase and operational phase;
- maximising of the quality of the indoor environmental conditions (thermal comfort, air quality, lighting comfort, acoustics/noise and visual aspects, etc.).

Green design is aimed at considering all the aspects in the areas listed above and, generally, to reduce the impact on the site's ecosystems.

The tools to assess the level of sustainability of a building designed with a green approach are the protocols for environmental certification such as leadership in energy and environmental design (LEED)®[4] or building research establishment environmental assessment method (BREEAM)®.[5]

These protocols, promoted on a voluntary basis, have been established as international standards for environmental certification of buildings.

2.2.3.2 From Energy Audit to Green Energy Audit

Energy audit has traditionally been aimed at targets of reduction of energy consumption with an approach very much based on economics, emphasising not so much the reduction in the environmental impact derived from the choices of greater efficiency, but the resultant reduction in operating costs. Energy audit, however, represents an important opportunity to contribute, through the measures proposed and possibly implemented, to reducing the overall environmental impact of the building or structure under investigation.

This consideration has given rise to the idea of giving a different interpretation of the energy audit. It emphasises those aspects which, in addition to still ensuring the achievement of the results of improved energy performance, lead to a reduced consumption of other resources that have little to do with energy but whose

---

[4] LEED®, Leadership in Energy and Environmental Design, is an environmental certification protocol developed by U.S. Green Building Council (USGBC).

[5] BREEAM®, BRE Environmental Assessment Method, BREEAM, promoted by building research establishment (BRE) is an environmental assessment method and rating system for buildings.

reduced usage generates benefits in terms of the overall sustainability of the building.

Green Energy Audit is thus not limited to providing tools and methods to reduce only energy consumption, but has a much more important goal:

to contribute to an overall improvement in the sustainability of the building.

This new approach to energy audit, consistent with the statements of the green design above discussed, has involved a series of choices that are summarised here:

- the definition of measures that lead to a reduction in the consumptions: *conservation of energy* becomes *conservation of resources.*
- criteria for the choice of actions to be taken/works necessary can be addressed from the outset with these indicators; the auditor then must have two objectives (or a mix of the two): to maximise energy performance and to maximise environmental quality;
- measures that use renewable energy are preferred (e.g., solar thermal, PV solar and biomass);
- when defining measures, the auditor should consider all natural solutions that can help control the climate and illumination in the building, such as green roofs, green facades, natural shading systems, passive solar and day-lighting systems;
- evaluation of sustainability targets according to the LEED® Standards.

2.2.4 The Green Energy Auditor

The term *energy auditor* (or *green energy auditor*) does not necessarily identify a person but, more generally, the *design team* that coordinates and manages the audit. In the event that several experts are involved in the same audit portfolio, the person who not only takes charge of the coordination but also takes on the responsibility of the entire work is appointed *lead-auditor.*

The auditor is primarily a professional figure and his minimum requirements and skills can be defined, in different countries, according to the local laws. In some countries, a membership a professional body/institution (such as those through which engineers and architects are recognised) is required.

An energy auditor operating with a green approach, should have acquired a solid experience in the following areas:

- green/sustainable design of buildings;
- design of energy systems (mechanical and electrical);
- energy management;
- energy accounting;
- international environmental protocols (LEED®, BREEAMS® or others).

For the auditor other skills should be considered:

- the ability to operate in the field;
- a knowledge of current security issues;
- competence in using survey and monitoring instruments;
- the ability to communicate and interact not only with the client but also with his staff;
- the ability to write the audit reports clearly and effectively;
- ensured continuing professional development (CPD), covering all updates in norms and regulations so that there can also be;
- the availability to allow for and handle continuous updating of the technical and legislative requirements and
- confidentiality in handling information.

In some countries, there are energy associations of energy with which the auditors may register. In the U.S., for example, the association of energy's engineers (AEE) have created as of 1981, the certified energy manager (CEM®) programme (www.aeecenter.org). Credentials belonging to this association are recognised both in the private and in the public sectors throughout the country. To join the Association, it is necessary to have a professional qualification and professional experience of a number of years ranging from 2 to 10. This depends upon the individuals' qualifications of membership and whether or not they are on the Register of Professional Engineers: they must also pass an examination. The inscription, moreover, must be renewed every 3 years.

A recent initiative is promoted by ASHRAE[6] called building energy assessment professional (BEAP), whose purpose is to enhance the professionalism of the technicians who work in the fields of certification and energy audits. In this case, the access to the list comes from passing an examination, while the necessary competence to qualify for remaining on the list of licensed professionals is checked every 3 years.

A new register for a green energy auditor, on a voluntary basis, was established in Italy from 2012 by SACERT®,[7] on the basis of the quality framework of the ISO/IEC 17024 [5] standard: the access to the register is given after passing of an examination and verification of maintenance of competence takes place every 4 years.

2.3 Definition of Operating Levels

An energy audit does not follow a codified or standardised pattern, but rather, it is a procedure that has been consolidated over time.

---

[6] ASHRAE—American Association of Heating, Refrigerating and Air Conditioning Engineers—www.ashrae.org.

[7] SACERT—Sistema per l'Accreditamento dei tecnici certificatori energetici, www.sacert.eu.

The approach must be effective and aimed at achieving the objectives, but at the same time, it must take into account an important factor for this type of activity: its cost. Therefore, the commitment of resources must be assessed on a case-by-case basis in a way that considers the return of investment (ROI) because the rule is that the investigation activities of an audit must be paid for by the benefits it will generate.

The experience gained from the 1970s allows us to group energy audit approaches into three categories that correspond to three different operational levels.

Table 2.1 shows features, instruments and procedures for green energy audit's three operational levels (walkthrough, standard, simulation).

2.3.1 Walkthrough Audit

The first category describes energy audits which can be realised in the short term. They are cost-effective because the auditor is a competent technician with considerable experience in the field. The terms used to define these types of investigation reflect the different dynamics of the procedures; a *walkthrough audit* gives the idea of just "passing through" during a field visit, whereas a *one-day audit* highlights that the time required is limited to approximately one day, and finally, a *preliminary audit* sets out that this approach is not final but is a first step towards subsequent energy audits at a higher operational level.

The walkthrough audit is used to describe this first operational category. The field survey is normally limited to one inspection: more inspections are possible but only where the building or its infrastructure is complex.

The prior planning of the survey should analyse the documents containing general information on the site. The inspection then assumes a double scope:

- a comparison of data with the client which serves also for requesting additional documentation;
- check directly features of building and facilities in order to identify areas of inefficiency for which it is possible to propose retrofit measures.

Once the data have been collected, the auditor analyses the situation: the result of the audit is a brief audit report, identifying plant and management inefficiencies, a first list of measures and suggestions on how to take the analysis into further detail.

The knowledge of the information relating to the consumption of both thermal and electrical energy during the last 3–5 years is essential since, for this type of approach, monitoring activities are not provided and the limited information acquired does not allow one to make a detailed calculation (e.g., calculation of heat losses or the electric loads) or a theoretical energy balance.

In the walkthrough audit, data regarding primary energy consumption are obtained directly from data of fuel or electricity consumption available in energy

Table 2.1 Features, instruments and procedures for three operational levels (walkthrough, standard, simulation) of green energy audit

Features, instruments and procedures	Green Energy Audit		
	Walkthrough	Standard	Simulation
Building size	Rough plan	Plans, cross-sections and detailed exterior elevation views	Plans, cross-sections and detailed exterior elevation views
Systems' features	Recommended (approx.)	Recommended (detailed)	Recommended (detailed)
Data on energy Required	consumption	Required	Required
Measurements	Size (approx.), air temperature, surface temperature, illuminance, electrical measurements	Size, air temperature, surface temperature, air speed, air flow, illuminance, power grid analysis, flue gas analysis, infrared audit, thermal conductivity	Size, air temperature, surface temperature, air speed, air flow, illuminance, power grid analysis, flue gas analysis, infrared audit, thermal conductivity
Monitoring	None	Data logger (recommended)	Data logger (recommended)
Forms	Basic check-list	Detailed check-list	Detailed check-list
Calculation tools	Nomograms, simple spreadsheets	Simplified calculation models, simple algorithms or simplified models	Dynamic simulation models (e.g., Energy Plus, ESP-r, etc.)
Expected results	Brief report pinpointing inefficiencies in the systems or their management; rough list of actions to be taken and works to be carried out; suggestions for further in-depth surveys	Detailed report including a description of the current situation (building and plants), pinpointing inefficiencies in the systems or their management; definition and description of actions to be taken and works to be carried out, economics	Detailed report including a description of the current situation (building and plants), pinpointing inefficiencies in the systems or their management; definition and description of actions to be taken and works to be carried out, economics
Estimated average time	A few days	A few weeks	A month or more
Cost	Low	Medium	High

bills. Starting from these data, it is possible to make energy or resources indicators (e.g., kWh/m^2 or kWh/m^3 per year) which are useful for a comparison with benchmark values that the auditor could directly derive from the literature, from reference sites or on the basis of his experience.

2.3.2 Standard Audit

The second category, the *standard audit*, defines a more challenging energy audit than the previous category. The term is most often used to mean that the audit standards are those of an energy audit and that no additional reports at this operational level are specified. The standard audit represents the best compromise between cost and effectiveness and, as we shall see below, it defines a comprehensive technical approach that requires a greater commitment of resources and more skills.

The standard audit is the most common type of energy audit. In this type of audit, much more information will be collected because the auditor needs all this information in order to make a theoretical model of the building. The comparison between the energy consumption of the theoretical model and the actual energy consumption derived from analysis of the energy bills, appropriately calibrated considering the actual operating conditions, allows one to build the *baseline.* This is a calibrated theoretical model of the building on which to check, by calculation, the effects of different retrofit measures.

The simulations do not necessarily have to cover the entire building. If, for example, we decide to propose the replacement of the windows, the estimate of the savings will cover the benefits that this particular action generates. If the project involved the replacement of incandescent light bulbs with energy saving light bulbs, then the evaluations to be carried out would focus on corresponding savings in electricity (with possibly a quick estimate of the energy savings for summer air conditioning, since the internal heat loads will be reduced).

The calculation models used for each of the measures are simple models: for thermal energy balances, for example, a steady-state simulation model is adequate.

In order to execute properly standard audit, a complete instrumentation tool is required (for insights on the selection and use of audit instrumentation, refer to Chap. 4). The monitoring activities may involve the evaluation of environmental conditions (e.g., air temperature, relative humidity, CO_2 concentration, etc.), or physical quantities that help to understand the operation of the equipment, from the energy standpoint, (e.g., absorption of electrical energy by a piece of equipment) or an entire subsystem. The measurement, usually limited in time (1 or 2 weeks) helps the auditor to understand the operation of the building-plant system over time.

In the standard audit, the auditor will make a report containing

- a description of the actual state (of the building and equipment);
- identification of structural and plant management inefficiencies;

- definition and description of action/works required and
- economic assessments.

The purpose of the audit report is to provide the information needed for the identification of the most convenient scenario of retrofit measures.

The green energy audit will also provide a preliminary assessment of the improvement of building sustainability on the basis of an international protocol (e.g., LEED®, BREEAM®, etc.).

2.3.3 *Simulation Audit*

The operational level of the *Simulation Audit* is more complex than that of the standard audit, because simulation audit provides dynamic simulation of a building-plant system in all its complexity.

In this case, a virtual model of the building is created, and based on this model, the effectiveness of the strategies adopted are verified. The simulation models used (e.g., Energy Plus, TRNSYS, ESP-r)[8] require high experiences and skills in building simulation, the process of model construction, furthermore, may require a lot of time.

Modelling of buildings with these tools, however, if well done, provides very precise information for predicting accurately the practical effects of the retrofit measures. Using the latest generation of simulation models it is possible to obtain a building simulation considering all the synergies and all flows of energy between different systems (e.g., it is possible to measure the effects of lighting and air conditioning in the summer or the effects of artificial lighting and shielding management).

For buildings with a complex envelope (e.g., extensive glass surfaces, sophisticated HVAC systems, extensive use of renewable energy sources), the building simulation approach is more reliable.

2.4 Organisational Aspects of the Auditing

Regardless of the established operating level, Green Energy Audit cannot be improvised, but must be carefully planned, taking into account both technical requirements and organisational aspects.

The check of the operating conditions must be made within the operational season: an audit aimed to verify plant or management inefficiencies of a HVAC

[8] For explanations and insights on simulation software refer to Sect. 2.7.4.

system in summer conditions needs a monitoring campaign during summer and energy audit, in this case, cannot be performed in wintertime. A complete audit of a building could theoretically even last a year, in order to properly consider summer and winter operation.

Many activities can also be performed independently of the season:

- the acquisition of technical and management documentation;
- the acquisition of primary energy consumption (energy bills) and
- necessary inspections for completing the information acquired through documentation.

Energy audit of electrical equipment, such as lighting or lifts (elevators), could normally be performed independently of the season, with the exception of that equipment related to HVAC systems (heat pumps, refrigeration equipment, auxiliaries, etc.). All the activities related to implementation of the information making use of simulation models, economic evaluations and the preparation of audit reports can be done in the office.

2.5 Contractual Aspects

2.5.1 The Relationship Between the Auditor and Client

In the professional relationship between client and auditor, a contract containing the following topics should defined:

- *audit objectives*;
- *the parts of the building that need to be investigated* (in the case of complex buildings or in the case of infrastructure consisting of multiple buildings);
- *infrastructure building or plant to be investigated* (for example the building envelope, the roof, the windows, or the electrical system, HVAC system, the lighting, etc.);
- *the technical documentation* that the customer is able to provide (e.g., floor plans, sections, plant designs, data on electrical energy or fuel consumption, etc.);
- *the execution time* of the Audit;
- *the expected results*, (what the auditor must provide to the customer) and
- *audit cost* and terms of payment.

The contract should also consider the organisational aspects, such as defining reference person for documentation supply or the persons responsible of the managing of the building or of the plants.

A correct definition of the contract issues that defines the obligations, commitments and responsibilities is essential to start a quality process in the implementation of the energy audit.

2.5.2 *The Implementation of the Retrofit Measures*

Green Energy Audit must be considered a good starting point to enable performance contracting with third-party financing (TPF). With this contracting formula, a third party, normally a ESCo, will assume the responsibility to perform the detailed design of the retrofit measures and their realisation within a management contract, for a period of time that is variable.

This contracting formula is important because it makes it possible for a client to accomplish the actions/works required without having to budget for them (the investment pays for itself in function of the energy savings obtained).

2.6 Energy Audit and Energy Certification, an Integrated Approach

Knowledge of the energy quality of a building is important to define a strategy of energy retrofit in order to reduce energy consumption, maintaining or possibly improving conditions of comfort. There are two tools available to achieve this goal, the energy audit and energy certification.

Both instruments seems similar, since they have common elements: a methodological approach on data acquisition regarding the characteristics of the building envelope and the plant, an energy balance, the definition of energy indicators and suggestions of retrofit actions in order to increase the energy performance.[9]

Significant differences, however, exist regarding the overall approach and hence also in the resources that are involved (technical skills, procedures, tools, etc.).

The main purpose of energy certification is to inform the user about the energy quality of the building through an energy indicator, normally expressed as a value of primary energy consumed divided by the net area or the volume of the building (kWh/m^2 or kWh/m^3 per year). The indicator thus obtained is then associated with a letter (e.g., A to G) within a rating scale that provides the user with a reference understandable similar to a benchmark (see Fig. 2.2).

Through the energy certification, the end-user can understand if the building energy performance of the building is average, if it is higher or is lower comparing with the rating scale. A low energy classification should suggest to the user the opportunity and the convenience of implementing energy retrofit actions.

The calculation of energy performance for certification purpose is based on standard conditions (e.g., internal winter temperature of 20 °C, operative conditions of 24 h a day, etc.). For this reason, the energy evaluation through a

[9] Although the goal of certification is to inform the energetic quality of the building, the EC-Directive 2002-91 provides that the certification is always accompanied by recommendations for improving building energy performance.

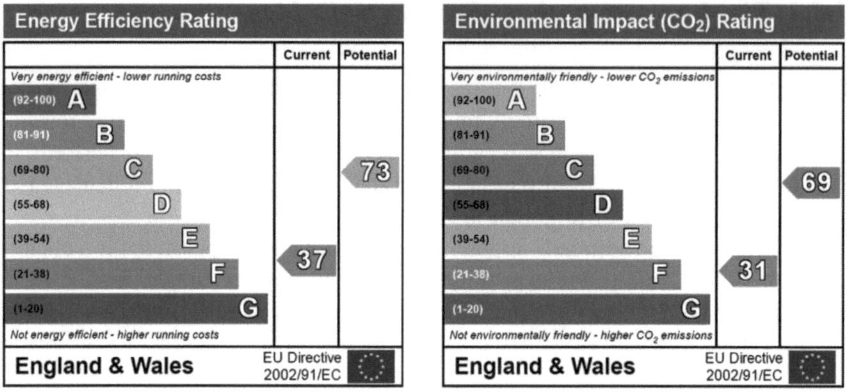

Fig. 2.2 Sample of energy performance certificate (*Source* SAP)

certification process can be very different from the real one and this difference becomes greater as a function of the difference between the real operating conditions and the standard conditions.

On the other hand, the purpose of certification is not to simulate a real condition but a good standard condition, in order to compare the performance of different buildings independently of how they are employed by the user.

The definition of standard boundary conditions becomes an indispensable choice if we are to ensure a high degree of reproducibility of results.[10]

The objectives of energy audit, one must remember, are quite different: starting from the survey on the building, with respect to methods of use under real conditions, it identifies inefficiencies, and therefore the causes of wastage, and then proposes solutions that will improve the overall performance.

Table 2.2 outlines the criteria for assessing the energy performance of buildings provided by the CEN European standard. In the case of existing buildings, evaluations can be of two types: standard assessment (asset rating) for the energy certification and that adapted only to users (tailored rating) aimed primarily at energy audit.

The difference between the two types of evaluation is in the input data relating to the mode of use and the climatic reference conditions: these are normalised in the first case and defined according to the purpose in the second one.

In all the countries of the EU energy certification is compulsory, in theory, since 2006, in accordance with Directive 2002/91 [1]. The energy audit, on the contrary, is rarely mandatory. Energy audit and certification, often considered one and the same instrument, are different but may be perfectly complementary, as shown in the diagram of Fig. 2.3.

[10] The concept of reproducibility is a key element in the certification process which is necessary to ensure that there is space for speculation on the technical definition of performance: ideally if more certifiers evaluate the performance of a same building, the result should be the same.

Table 2.2 Criteria for evaluation of energy performance of buildings (*Source* CEN)

Type of evaluation	Input data			Scope of evaluation
	Use	Climate	Building	
Design rating	Standard	Standard	Project	Design energy certification
Asset rating	Standard	Standard	Real	Energy certification
Tailored rating	Depending on the purpose		Real	Optimization, validation, energy audit and planning retrofit measures

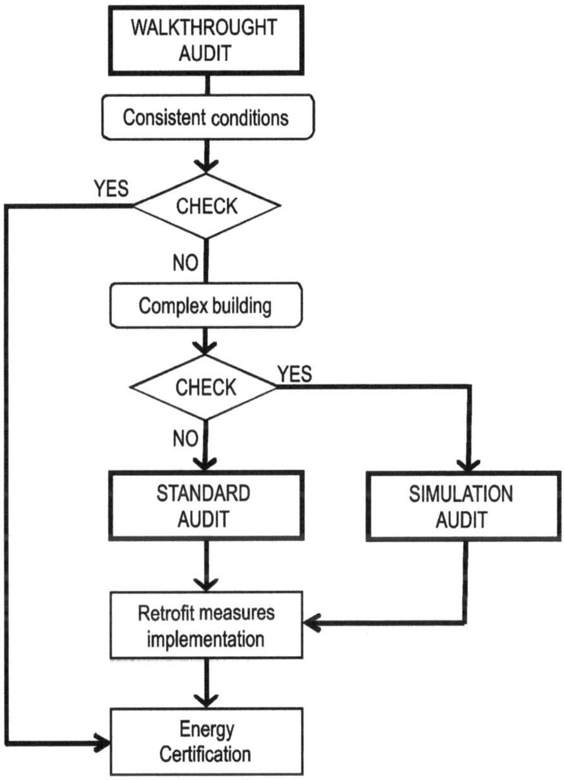

Fig. 2.3 Energy audits and energy certification, an integrated approach

Through a walkthrough audit, we assess whether the overall energy consumption is high comparing with buildings of the same type. The same survey allows a first assessment of the ways in which plants are managed.

From this initial examination, we can determine if the building's energy performance and management arrangements are acceptable or not.

In the first case, we can proceed directly to implement energy certification, whereas in the second case, depending upon the complexity of the building, we could suggest a standard audit or a simulation audit.

The next step is the implementation of the energy retrofit measures. When the works are finished, the energy performance of the building will be much higher and the energy certification will provide new performance indicators and new energy rating.

Energy audits and energy certification are two effective tools to promote the retrofit of existing buildings. The integration of the two procedures, according to the proposed scheme of Fig. 2.3, can lead to a benefit, the proper allocation of more resources necessary for the evaluation of performance.

2.7 Commissioning Authorities and the Green Energy Audit

Edited by Alessandro Speccher

There are two reasons why it is essential to explore the theme of the commissioning authority (CxA) within the Green Energy Audit process [6]:

- energy efficiency is the driver of mandatory requirements for new buildings and existing building stock;
- sustainability applied to building construction has become the new driver of innovation and economic recovery, and it is necessary to define multi-dimensional systems to measure sustainability.

Energy audit is the first stage of a Retrocommissioning process (Commissioning process applied to Existing Building); energy audit and RetroCx are mentioned within the LEED® EBOM[11] protocol; they are complementary parts of the same picture that allow a building be more energy efficient and environmentally friendly.

2.7.1 The Conjunction of the Two Methodologies

Taking on both energy efficiency and sustainability has become the winning combination for any country that aims to become an active partner in combating the ultimate world challenge: climate change.

The connections that exist between the methodologies introduced by the commissioning process and the energy audit process are many and articulated. The context in which these practices are applied cannot be separated from a parallel discussion that may lead to their synergy in what can be defined as a process of virtuous cultural growth, aimed at both energy efficiency and (primarily) sustainability.

The key conjunction points between the two methodologies (the Green Energy Audit and the commissioning process) can be summarised as follows:

[11] EBOM—Existing Buildings: Operations Maintenance.

- both are practices that apply in the international framework, defined by the requirements on energy efficiency and environmental sustainability applied to construction;
- LEED® certification requires the adoption of both a mandatory commissioning process for new buildings and a mandatory re-commissioning process for existing buildings: the re-commissioning process is based on considerations emerging from the energy audit;
- both practices are aimed at improving the energy and environmental performance of the building from the design process through to its management and maintenance.

2.7.2 Commissioning and Retrocommissioning

Here are some definitions that will facilitate comprehension of the next paragraphs [6]:

Commissioning (Cx): a systematic quality assurance process that spans the entire design and construction process, helping to ensure that the new building's performance meets owner expectations. Owner expectations are listed in the owner project requirement (OPR) document.
Retrocommissioning: a systematic method for investigating how and why an existing building's systems are operated and maintained and for identifying ways to improve overall building performance.
Re-commissioning: another type of commissioning that is applied when a building, which has already been the subject of commissioning, is subjected to another commissioning process. Ideally, a re-commissioning plan should be part of the original commissioning plan that is drawn up during the construction of the building.
CxA: an individual hired to lead a retro/commissioning process: the CxA is responsible for managing the process to ensure that the owner will obtain the required performance listed in the OPR document.
Commissioning Team: all persons involved in commissioning activities and that work together to complete the commissioning process. It ideally includes all the persons involved in the design, construction and management of the building, such as the client/user, the design team, the contractor, the construction supervision, the CxA, the operation and maintenance staff and many others.
Commissioning Plan: a document containing all the information required to re/commission the facility. The plan may include specific tasks, their descriptions and their schedules. Other information that may be helpful includes operational requirements for key systems, functional tests and documentation templates.
Commissioning Report: a document that provides an overview of the commissioning process.

2.7.3 Commissioning and Retrocommissioning: From New Buildings to Existing Buildings

The commissioning process was established long before the appearance of LEED® certification, and the benefits that it brings to a building's construction process are so important that it has been used since the first version of the protocol was introduced. It is part of the Energy and Atmosphere category, and it is a pre-requisite (i.e., it is a mandatory action).

ASHRAE Guidelines 0 and 1, which describe the Cx, refer to a continuum that starts with the definition of the client's requirements for the management and maintenance of the building. Because LEED® provides both a protocol dedicated to the design and construction phase (New Construction, School, Core and Shell, Commercial Interiors and others) and a specific protocol for maintenance and management (Existing Building Operations and Maintenance), the process of commissioning is mentioned in both. For new buildings, it is the commissioning process; when done to a building that has already been commissioned, it is re-commissioning, and when a commissioning process is activated on a building that was not commissioned during its construction, it is Retrocommissioning (Fig. 2.4).

Since the Commissioning Process aims to ensure that the client's requirements are satisfied, the fundamental documents for an effective and efficient Commissioning Process are the OPR and the Basis of Design.

The OPR contains the boundary conditions (i.e., what the client wants to develop and how) and the requirements in terms of performance, budget and certification.

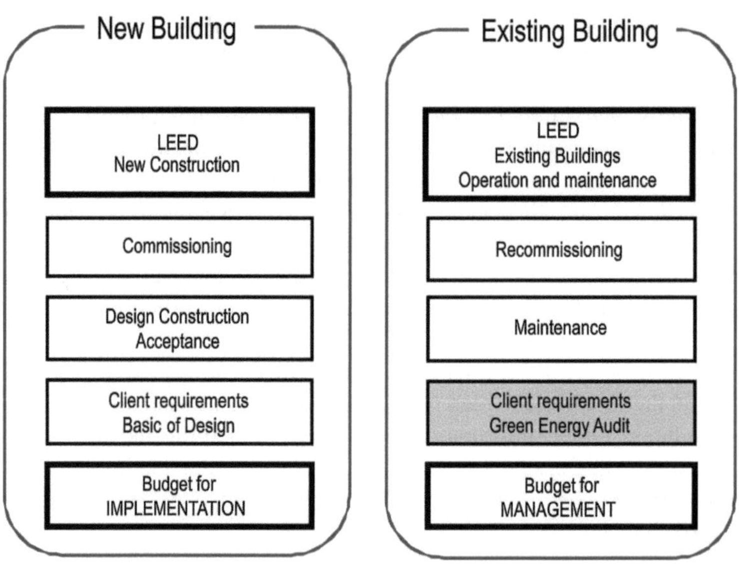

Fig. 2.4 Comparison between Commissioning and Re-commissioning process

The Basis of Design explains how a particular technical choice addresses the OPR. It also contains a description of the systems, the reasons for the choices made, the design criteria and the sequences of the control systems. This document is then delivered to the CxA, who examines it with the client.

2.7.4 The Re-commissioning Process for Existing Buildings

The Re-commissioning Process is the natural continuation of the commissioning process in terms of the maintenance and management goals set by the client. It starts from considerations emerging from an energy audit, and then it develops a plan to achieve the building performance objectives defined in the OPR.

As defined above, the commissioning process gives the client a detailed report on the design and construction phases, and Green Energy Audits are greatly

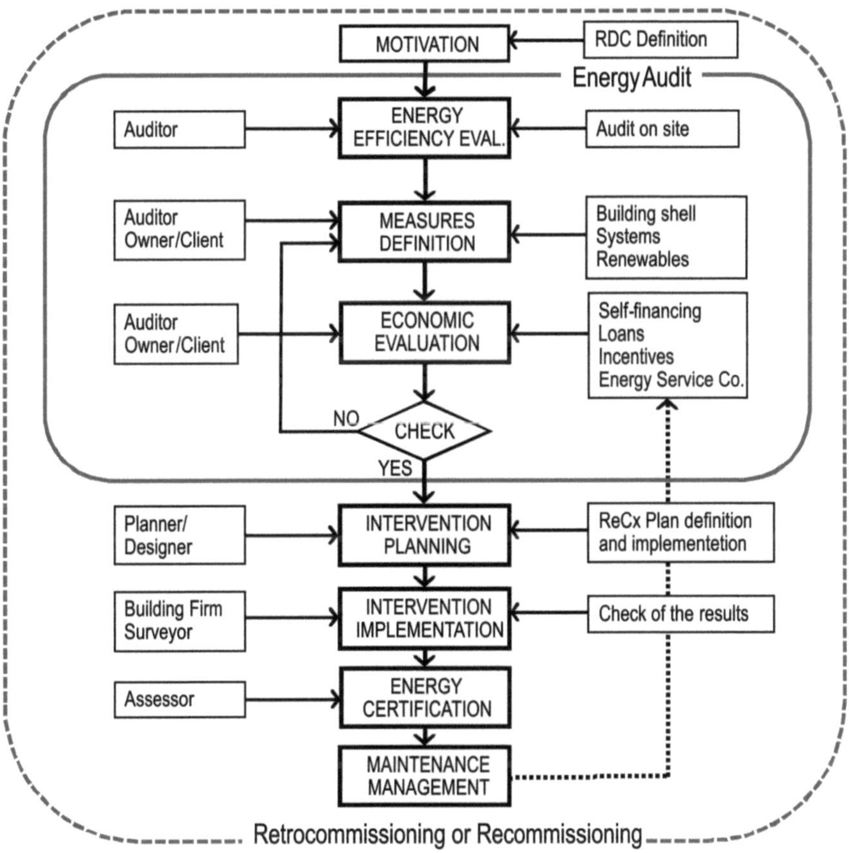

Fig. 2.5 Operational plan to design and implement an energy enhanced project

facilitated when the auditors have access to these documents before and during the analysis. There is a good chance of achieving very close synergy if the two processes are applied (with different timing) in the same building.

The planned activities in the re-commissioning process are structured in the following phases: the Planning Phase, the Investigation Phase (Audit), the Implementation Phase and the Hands-Off Phase. Further information is available about this subject.

Energy Audits and the energy Retrocommissioning process are very similar and often have large areas of overlap (see Fig. 2.5). Retrocommissioning analyses systems and building management so as to make them as efficient as possible. Energy audits examine the history of energy consumption and identify strategies to improve energy performance. The main difference between the approaches is the output produced. Commissioning generates a more efficient building, while energy audits generate a report that contains a series of recommendations to make the building more efficient owing to a series of strategies characterised by different returns on investment.

References

1. Directive 2002/91/EC of The European Parliament and of The Council of 16 December 2002 on the energy performance of buildings
2. ISO 19011 (2002), Guidelines for quality and/or environmental management system auditing
3. EN 16247-1 (2012), Energy audits—Part 1: General requirements
4. ASHRAE (2006) GreenGuide, the design, construction, and operation of sustainable buildings, ASHRAE. Elsevier, Atlanta
5. ISO/IEC 17024-2003—Conformity assessment—General requirements for bodies operating certification of persons
6. Speccher A (2011) Commissioning authority e Green Energy Audit. In Dall'O' G (ed. Ambiente), Green Energy Audit, (pp 47–56). Italy: Edizioni Ambiente degli edifici (pp 47–56). Italy

Chapter 3
Application of the Methodology

This chapter defines the methodological approach of the Green Energy Audit. The whole activity is divided into several phases, each of which is defined in a synthetic way within the various sections. As a function of the operating level that we want to apply, the elements contained in the phases may have greater or lesser detail. Some steps may finally involve some levels and not others. The chapter aims to provide the auditor with an overview, then in the subsequent chapters the necessary details for each step are covered fully.

3.1 General Criteria

The standardization of the auditing activities requires its own specific definition based on a series of elements:

- *the operational level of the audit* (walkthrough, standard or simulation);
- *the type of building or installation* (e.g., residential, commercial, industrial, etc.);
- *the size and complexity of building and/or facilities*;
- *the category of the system to be investigated* (e.g., building envelope, HVAC system, electrical, lighting, etc.);
- *the owner/client organization* (single person, management personnel, maintenance personnel, etc.).

The combination of the elements listed above greatly influences the strategy to be adopted, its complexity, its execution time, the commitment of resources and consequently the definition of the Activity Plan through which the audit is carried out.

Figure 3.1 illustrates a process flow for a Green Energy Audit. In the central part, the steps (phases) are arranged in sequence: on the left are indicated the details of each step whilst on the right are to be found the figures/actors with whom the auditor must ensure dialogue for the step in question. For simplicity, the flow is arranged in a sequential manner. In fact, exchanges of viewpoint and comparing notes with the parties involved, in particular with the customer, can generate recursive paths.

G. Dall'O', *Green Energy Audit of Buildings*, Green Energy and Technology,
DOI: 10.1007/978-1-4471-5064-0_3,

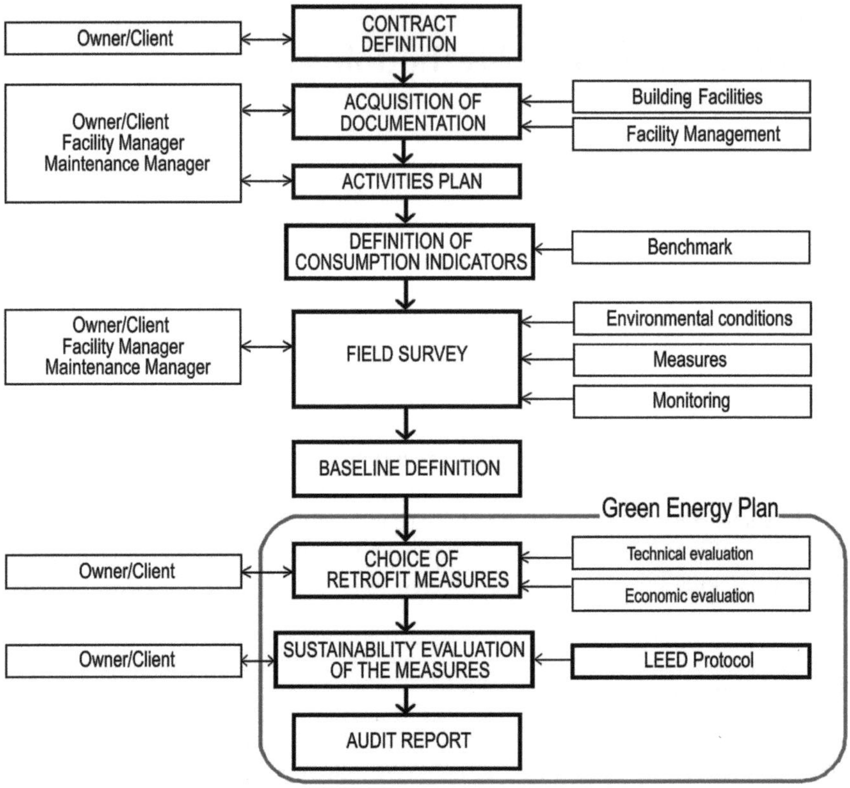

Fig. 3.1 Flow diagram of a green energy audit process

The methodological approach outlined in Fig. 3.1 [1] is a good basis for implementing a Quality System Audit: for the general requirements for a quality management system refer to [2].[1] As a means of defining the process, which is structured in stages, the following points are considered:

- *the purpose and scope of the process*: what must be done, the affected areas and those excluded;
- *liability*, in which it is indicated that the Organisation is responsible for implementing the contents of the document and achieving the goal;
- *actions and methods* to meet the requirements of the quality system that explain, step-by-step, what needs to be done;
- *documentation and references are given*, and the reference documents and forms associated with the use of such documents and data must be recorded;

[1] All requirements of Standard ISO 9001 are generic and intended to be applicable to all organizations, regardless of type, size and product provided: energy audit activities fit well with this standard.

- *records of which documents must be kept* and for how long are specified and generated.

The methodology proposed in this chapter divides the entire process into the following phases:

1. definition of the contract;
2. acquisition of documentation;
3. planning of activities;
4. definition of consumption and performance indicators;
5. field surveys;
6. verification of environmental conditions;
7. monitoring of climate parameters and energy consumption;
8. definition of baseline;
9. definition of Green Energy Plan.

For each of the above listed phases one must examine: the purpose and content, the actors involved, the tools, expected documentation and critical issues.

Amongst the tools of the Green Energy Audit procedure, there is a *checklist* of 45 items (downloadable from the Springer website) and 97 *retrofit measure sheets*, in Chap. 13, describing retrofit measures to increase the energy efficiency and sustainability of buildings. The measure sheets are divided into five areas:

- *building envelope* (roofs, basements, walls, windows, solar protection, day lighting, etc.);
- *mechanical systems* (heating, summer cooling, ventilation, hot water, water services, etc.);
- *electrical systems* (the generation, distribution and use of energy; lighting);
- *renewable energy* (solar thermal, solar photovoltaic, biomass, etc.);
- *management* (improvements to management, maintenance, energy accounting, etc.).

Each retrofit measure sheet is divided into three sections: a header, a descriptive part and a scoring section. The checklist is downloadable from the Springer website.

3.2 Definition of the Contract

The *terms of contract* between energy auditor and client is an important step of the process as the clarity of the relationships, defined from the beginning, is the best guarantee that the service (energy audit) fully meets expectations. The energy auditor shall undertake to establish and maintain documented procedures for contract review and coordination of related activities.

Before submitting any offer or accept a contract or order (reporting requirements), the offer, the contract or order must be reviewed to ensure that:

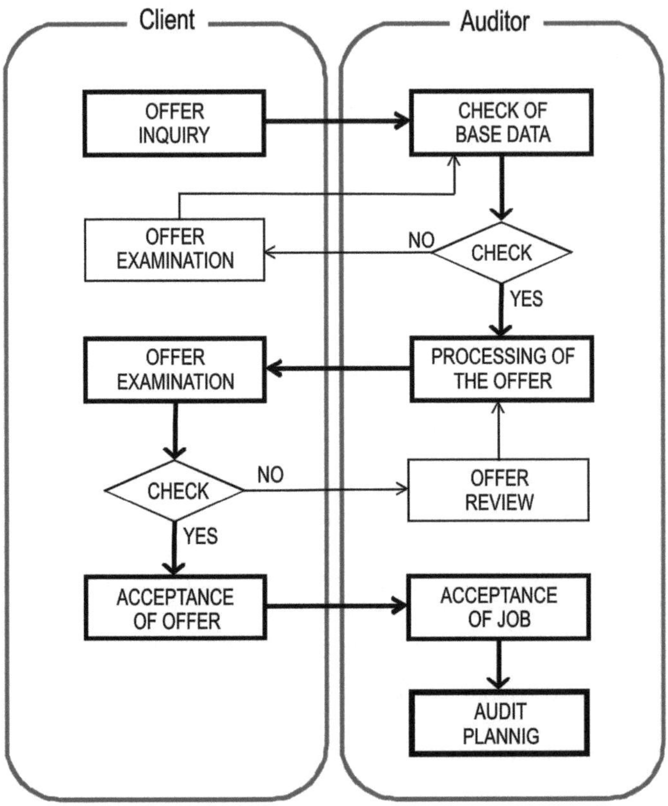

Fig. 3.2 Flow diagram of a procedure for the definition of the contract between auditor and client

- the requirements are adequately defined and documented;
- any differences between the requirements listed in the contract or order, and those reported in the offer are solved;
- the energy auditor, alone or with the support of his team, has the capacity and the skills to meet the requirements specified in the contract or in the order.

This step is aimed to understand, through meetings and through the examination of the documentation received, the characteristics of the work required and the deadline for its execution. The diagram of Fig. 3.2 shows an example of the activity relationship between auditor and client.

3.2.1 Actors Involved

This phase involves directly the energy auditor and the client. The energy auditor is qualified, as responsible, for all the signatures on the documents relating to contracts that provide services to undertake an energy audit.

3.2.2 Tools

Checklists can be used for the definition of the contract, the general data of the building and the list of documents.

3.2.3 Deliverable Documentation

In addition to the checklist, which must be preserved, entries made in this phase are the engagement letter, which formalizes the assignment of work, and possibly the contract. All processed products, including any verbal ones, must be recorded and maintained.

3.2.4 Critical Issues

The main critical items and issues at this stage arise when the contract has not been clearly defined and completed, and then emerge from the elements that can affect the relationship between the auditor and the client.

3.3 Acquisition of Documentation

The acquisition of the available documentation is an essential element for the success of the energy audit. A good basis of documents, in fact, allows one to considerably reduce the "on site" activities and, therefore, to reduce the time and costs (for this reason the availability of the documents should be verified at the time of definition of the contract).

It is important for the auditor to carry out inspections with a full picture of the situation that is going to be investigated: in fact by studying the documents provided the auditor is already well-informed and well-prepared

- to schedule major assets necessary to integrate the missing information;
- to identify critical areas in terms of energy and environment and then run the survey already with the first ideas about the measures to be proposed for the containment of resources;
- to have objective elements of discussion for the first meeting with the staff, when present, or with the designated representative of the client.

The documentation to request from the building developer may be of a technical nature (e.g., layout of the building, sections and elevations, HVAC schemes and drawings, technical characteristics of the plant components, technical reports, etc.)

or of a managerial/administrative type (records of fuel and electricity bills, data on water consumptions, operation times of facilities, air temperature levels, etc.).

3.3.1 Actors Involved

The documentation may be provided directly by the client or by the staff in charge of the operation and maintenance. For residential buildings, normally the documentation is provided by the building manager.

3.3.2 Tools

In order to facilitate this activity some checklists that auditors can use directly were drawn and reported in the appendix of the book. On the basis of the data acquired, the auditor can start filling in the checklist before the programmed inspection. This phase is reported in greater depth in this chapter.

3.3.3 Deliverable Documentation

The documents supplied by the client must be filed and stored in a folder, in order, separated according to type. The document representation concerns the completion of the checklist and a brief report which will provide relevant information for the final report.

In order to retain written evidence, for future reference, on the conduction of this activity, it may be useful to put on file the correspondence between the auditor and client or staff (e.g., e-mail, fax, etc.). It is important to demonstrate that the documents were actually requested.

3.3.4 Critical Issues

The documentation is not always available and, often, documentation management is not catalogued in order. Sometimes the documentation is available but is not supplied because of individual behavior inconsistent with company policy or associated with laziness.

A critical element of data concerns the energy consumption. For the electrical consumption (or consumption of heat from district heating) accounting is entirely transparent, since consumption is detected by the meter reading shown on the bills, and made available by a territorial body that provides electrical energy. Whereas in the case of heat consumption, if there is an active outsourcing contract of all inclusive heat management, the data on (heat or fuel) consumption must be obtained externally from the outsourcing company concerned.

External management organisations are not always available to provide data on energy consumption, since for some forms of contract a global service is provided.

3.4 Planning of Activities

The purpose of this phase is to precisely define, or to better plan activities, with particular attention to those involving a direct relationship between auditor and client. This phase defines the activities, programmes activities over time, checks that they have performed in line with the scheduling and records change-over time.

3.4.1 Actors Involved

The management of this activity is not limited to the initial planning stage, but develops in parallel with other phases, since it allows the energy auditor to conduct a continuing review.

The energy auditor is the figure in charge in this phase, but virtually all the actors working in staff with the client are involved, and in particular those responsible for the management and maintenance. The surveys, monitoring and eventually infrared audits, will be programmed to meet the needs of the auditor without interfering with normal activities of the client.

3.4.2 Tools

For the management of the activities plan, a checklist is drawn up containing the main elements for any phase involving interactions between the auditor and the client.

The checklist records both the scheduled dates and the rescheduled dates, hence permitting the auditor to control the entire audit process.

3.4.3 Deliverable Documentation

In addition to the programming document and verification activities, for this stage of the procedure, all relevant reports prepared on the occasion of various meetings are considered and processed.

In order to retain written evidence, for future reference, on the conduction of this activity it may be useful to put on file the correspondence between the auditor and client or staff (e.g., e-mail, fax, etc.). It is important to demonstrate that the documents were actually requested.

3.4.4 Critical Issues

During this phase there are no particular critical issues because it is a planning and control phase of all the activities of the energy audit.

3.5 Definition of Consumption and Performance Indicators

The energy performance of a building or facilities can be evaluated starting from the energy or other resources measured and extrapolated from the related bills. From these data, in fact, it is possible to verify the actual behaviour of a building and, on the basis of this, locate immediately the critical points and, therefore, the potential for improvement.

Energy consumption can be influenced by external factors such as climatic conditions, which may vary from year to year. An estimate, even if approximate, of the energy performance cannot be based on the value of a single season but must consider at least 3 years.

The possibility to build a monthly profile of energy consumption provides useful information for understanding the distribution logic of the shared utilities when many devices or parts of plant systems are connected to the same supply system. If a gas meter, for example, is in the supply line to both a heating system and a DHW system, the knowledge of monthly consumption allows one to compare energy consumption in summer and in winter. The energy consumption in summer, when the heating system is switched-off, is probably due to DHW.

In a similar way if an electricity meter simultaneously feeds a lighting plant and an air conditioning system for summer cooling, the highest consumption of electricity in the summer months, compared to those in winter, allows us to determine how much electrical energy can be allocated to the summer air conditioning.

The auditor should implement an accounting of energy by creating consumption profiles from the information available.

The total consumption of energy or other resources are not enough to understand whether a building or plant consumes a lot of or little. Hence the need for building up, just starting from the consumption, a set of the indicators that will then be compared with the benchmark.

The comparison, of course, is possible if the destinations of use or the type of buildings or plants are similar: a specific indicator for an office building, for example, cannot be compared with a benchmark for a residential building.

The availability of a reliable benchmark greatly facilitates the audit work as it permits rapid parameter assignment.

The consumption indicators needed to determine the baseline may subsequently become the benchmark to measure the performance improvements as a result of the retrofit.

3.5.1 Actors Involved

The definition of consumption indicators starting from the analysis of actual energy consumptions involves the client's administrative and technical organisational structure. The definition of the benchmark is a prerogative of the auditor who can refer to the relevant technical literature or to personal experience.

3.5.2 Tools

In the appendix of the book a useful set of checklists is available in order to gather and organize information about the consumption of (heat or electrical) energy or other resources and plant operating modes. This phase is also reported in greater depth in Chap. 4.

3.5.3 Deliverable Documentation

The documents elaborated include a report based upon which indicators are defined, the explanation of the choices and the comparison between indicators and benchmarks. The document produced, or a synthesis thereof, will be attached to the final report.

In order to retain written evidence, for future reference, on the conduction of this activity it may be useful to put on file the correspondence between the auditor and client or staff (e.g., e-mail, fax, etc.). It is important to demonstrate that the documents were actually requested.

3.5.4 Critical Issues

The retrieval of sufficiently disaggregated data on consumption of energy or resources is not easy.

However the main difficulties in this stage of the procedure are related to the search for reliable benchmarks. Some international websites provide interesting references (e.g., http://energybenchmarking.lbl.gov/).

3.6 Field Surveys

This phase is the most operative of the audit process: auditor takes direct vision of the building, installation or infrastructure in order:

- to integrate the technical and management information that could not be deduced from the documentation supplied by the client;
- to perform instrumental measurements to gather information that can provide direct evidence with which to evaluate the performance of components or building systems and plant;
- to undertake measurements to define the parameters of environmental conditions (comfort evaluation);
- to make a first selection of possible actions, verifying the applicability of the same.

The measurements can be more or less detailed, depending upon the operating level of the audit. In walkthrough audit, for example, the survey is completed in at most one day, which also includes meetings with the technical staff of the client, before and after the inspection.

In walkthrough audit the inspection should be organized in every detail. Based on information provided by the client, and processed in the office, the auditor must programme the "path" with the sequence of things that must actually be seen.

For a good success with this type of audit, a preparation phase preceding the survey should not be underrated or neglected.

For the standard and simulation audit this phase, depending on the complexity of the subject, could take several days. Table 3.1 shows a general outline of the information to be acquired in this phase.

3.6.1 Actors Involved

The field surveys must be arranged with the staff of the client in order to avoid the inconvenience that could otherwise be caused. Scheduling must obviously take account of two needs: on the one hand there is the necessity to make the best use of the survey, also considering environmental conditions, but on the other the need to limit interference with the normal activities.

In cases in which the customer outsources the management and maintenance services it is important that the company in charge will provide an effective technical support.

As safety is a critical element, the security manager, should be involved right from the start.

Table 3.2 shows the correlation between the variables which can be measured and the instruments to be used. A set of checklists is available in order to facilitate the survey work (a discussion of this topic is reported in Chap. 5).

Table 3.1 Overall scheme of the information to be acquired in the phase of survey

Type of information	Description
Characteristcs of the building	• Urban location of the building from which it is possible to highlight orientation of the different facades and the presence of other buildings that can generate shadows • Dimensions of the elements which constitute the opaque envelope (e.g., external walls, adjacent walls to premises at different temperatures or not heated, bases, covers, etc.) • Thermo-physical characteristics of the envelope elements (stratigraphies of structures) • Dimensions of transparent elements (e.g., windows, skylights, etc.) • Thermo-physical characteristics of the transparent components (glazing type, frames types, screens, overhangs, etc.)
Characteristics of the facilities	• Classification of facilities (e.g., HVAC, lighting, electrical, etc.) • Location of technological stations and general layout with the mapping of the main dorsal link • Identification of the energy meters (electricity and gas) and water meters within the site plan • Identification of the heating and cooling terminals related to the different thermal zones • Verification and updating of plant functional schemes and/or sketch of the schemes • Survey of the characteristics of the plant components
Management	• Contract type • Management mode

3.6.2 Tools

Many instruments are used in this phase for measuring the geometric or environmental characteristics of building and plants.

Choice of the right instruments, and of course the expertise in how they are employed, are critical factors since misinformation can lead to wrong conclusions and inconsistent proposals.

3.6.3 Deliverable Documentation

The acquired information should be classified using the checklist, useful for implementing the calculation procedures. Some of this information will be attached to the final retrofit report.

In order to retain written evidence, for future reference, on the conduction of this activity it may be useful to put on file the correspondence between the auditor and client or staff (e.g., e-mail, fax, etc.). It is important to demonstrate that the documents were actually requested.

Table 3.2 Correlation between the variables that can be measured and the instruments to be used

Variables to measure	Instruments
Characteristics of building	
Survey of façades	Digital camera
Dimensions (surfaces, volumes, wall thickness)	Tape-measure, digital meter, 3D laser scanner (optional)
Facades orientation	Compass
Wall surface temperature (spot measurements)	Infrared thermometer
Wall surface temperature (mapping)	Infrared camera
Walls conductance	Heat flow meter
Characteristics of HVAC systems	
Combustion efficiency	Combustion analyser
Wall surface temperature of equipment (spot measurements)	Infrared thermometer
Wall surface temperature of equipment (spot measurements)	Infrared camera
Temperatures of fluids	Thermometer
Air flow	Anemometer
Characteristics of electrical installations	
Efficiency of electrical supply network	Electrical power analyzer
Electrical absorption	Wattmeter
Electrical current	Ammeter

3.6.4 Critical Issues

The critical issues are considerable and mainly related to safety factors. The visits create interference because people outside access, for long or short periods, zones and spaces where the activities take place.

The auditor, when planning these activities, must notify the client before the inspections:

- what kind of data and what information need to be acquired;
- how and with which instruments these data should be acquired;
- any critical issues;
- the names of the persons responsible for, or in charge of specific activities at the time of the inspections must be identified with clearly visible badges.

The surveys also include access to equipment rooms normally not open to the public (e.g., power stations) in which the safety factor is essential. Moreover, they include plant surveys that, if made by inexperienced personnel, can cause damage to persons or property.

It is important that the instruments used for survey are working reliably and are properly calibrated. It is very useful during the survey to make a lot of digital images and short videos.

3.7 Verification of Indoor Environmental Conditions

The purpose of this phase is to gain valuable information on the ways in which plants are managed for comfort, how the air conditioning is used in summer and winter, and on the ways in which ventilation (HVAC) and lighting systems are managed. Although the target of the energy audit is to highlight the wastage of resources by proposing suitable actions, maintaining optimal levels of comfort should be considered a prerequisite.

The perception of the environmental comfort, deriving both from the thermo-hygrometric conditions and from the lighting, is mediated by the behaviour of the human body that has the ability to adapt itself to non-ideal situations.

The assessment of environmental comfort, which is also verification of the management arrangements, is an important step in energy audit procedure, since the information collected may be undertaken effective corrective actions.

The main reasons for energy waste, related to discomfort, that can be found are two:

- the plants are not able to control the environmental climatic parameters in an effective way (for example in winter air temperatures higher than those of reference for that type of activity);
- there is a phase displacement between occupied periods and periods in which the plants (HVAC, lighting) are activated but the spaces are not occupied.

A management and control system could not be very efficient owing to inefficient plant, and in this case specially targeted actions should be applied. Specially targeted actions are also applicable for wrong set-points on the indoor environmental parameters. In many cases we can greatly improve the situation, improving comfort and reducing fuel consumption, at no cost with measures of adjustment and setting.

3.7.1 Actors Involved

The verification of the indoor environment conditions must be arranged with the staff of the client in order to avoid the inconvenience that could otherwise be caused. Scheduling must obviously take account of two needs: on the one hand there is the necessity to make the best use of the survey, also considering environmental conditions, but on the other the need to limit interference with the normal activities.

In cases in which the customer outsources the management and maintenance services it is important that the company in charge will provide an effective technical support in this phase.

Post occupancy evaluation by means of checklists distributed to the end-users could be useful for the auditor to understand the comfort perception. In this case, the personnel manager or the office administration manager should be involved in planning the survey.

3.7.2 Tools

The assessment of indoor environmental conditions necessarily requires the use of instruments that enable the auditor to detect objectively the variables that define them. Table 3.3 shows the correlation between the indoor environmental variables which can be measured and the instruments to be used. To facilitate the survey activities, checklists are available in the appendix of this book (a discussion of this topic is reported in Chap. 5).

3.7.3 Deliverable Documentation

The acquired information should be classified using the checklist, useful for implementing the calculation procedures. Some of this information will be annexed to the final retrofit report.

In order to retain written evidence, for future reference, on the conduction of this activity it may be useful to put on file the correspondence between the auditor and client or staff (e.g., e-mail, fax, etc.). It is important to demonstrate that the documents were actually requested.

Table 3.3 Correlation between the environmental variables that can be measured and the instruments to be used

Variables to measure	Instruments
Thermal comfort/IAQ	
Air temperature	Thermometer
Wall surface temperature (punctual)	Infrared thermometer
Wall surface temperature (mapping)	Infrared camera
Air leakage	Blower Door, infrared camera
Air speed	Anemometer
Relative humidity	Psychrometer
IAQ (air renewal)	CO_2 meter
Global comfort	Microclimate control
Lighting comfort	
Illuminance	Luxmeter
Luminance	Luminance meter

3.7.4 *Critical Issues*

The critical aspects are practically identical to those already described for the step of the field surveys.

3.8 Monitoring of Climate Parameters and Energy Consumption

Almost all the environmental parameters may change over time. The environmental quantities (e.g., temperature, relative humidity, CO_2 concentration, etc.), in fact, express the dynamic behaviour of the system and provide, for example, information about:

how a plant is able to control the climate parameters;
how the building-plant system reacts to the variation of external parameters such as temperature or natural light.

The spot measurements of these parameters provide good information but it is useful only at the time that they occurred. Monitoring, on the contrary, returns the auditor an important framework of information: the change of the values over time that allows us to understand how things really work, what happens in the building or a facility and, therefore, allows us to discover problems and propose solutions.

Today's instrument market makes available monitoring systems of small dimensions, which are cheap but nonetheless very efficient. Technological innovation helps the auditor to acquire information essential for a complete and effective audit.

3.8.1 *Actors Involved*

The monitoring campaigns do not require the presence of auditing personnel on-site: the auditor places the devices, programmes them, activates them and recovers the data only at the end of the monitoring campaign. In some cases it is possible that the auditor carries out the inspections for download of data acquired over a certain period and begins to make the necessary calculations.

The monitoring campaigns usually do not last more than two or three weeks, except in specific cases where it is necessary to provide for longer periods.

The auditor will agree with the client or the technical staff the placement of mini data-logger, guaranteeing that the chosen location meets both the technical requirements and any limitations that the client may impose.

3.8.2 *Tools*

The monitoring of buildings or facilities necessarily requires the use of instruments (data-logger) that enable the auditor to detect objectively the variables that define them. To facilitate the survey activities, checklists are available in appendix of this book (a discussion of this topic is reported in Chap. 5).

3.8.3 *Deliverable Documentation*

The information collected are many and should be processed as quickly as possible, providing summaries that point out the elements useful for the choice of remedial actions.

The data files must be filed and stored in an orderly and methodical manner so as to guarantee ready access even at a later date. Once this phase has been completed, the auditor should prepare a short report which will then be used in the final audit report.

In order to retain written evidence, for future reference, on the conduction of this activity it may be useful to put on file the correspondence between the auditor and client or staff (e.g., e-mail, fax, etc.). It is important to demonstrate that the documents were actually requested.

3.8.4 *Critical Issues*

This phase presents no particular critical of the relationship between auditor and client. Some critical may arise for monitoring of plants, for example inside technological stations.

As the step of processing can be complex if too much data are collected, the auditor must carefully select the points to be monitored.

3.9 Definition of Baseline

The data collected through analysis of documentation provided by the client, integrated by field surveys, are the basis to understand the performance of buildings and facilities investigated.

With the term *baseline* we define the performance of the current reference situation with which we compare to identify and then evaluate the possible retrofit measures.

The different operational levels of the energy audit necessarily require different approaches.

For the *walkthrough audit*, generally speaking, a precise performance evaluation is not required. Being a preliminary audit, the baseline is defined on the basis of consumption indicators. The availability of benchmark values permits an assessment which is, for specific indicators, the difference between the real situation and the reference. Then one can understand, albeit approximately, the margins for possible improvements.

The definition of the baseline in the *standard audit* is achieved through a simplified analytical, partial or global calculation (steady-state models), or by using graphical tools such as nomograms. The baseline, in this case may relate the entire building, a part of it or even a single component (e.g., a wall, window or a length of pipe).

The availability of values of energy consumption can then refine the model that represents the building in the basic conditions by introducing correction factors.

With *simulation audit* the building-facility system is treated as a whole, creating a model using a dynamic simulation software that can simulate precisely the behaviour of the building, to assess its performance on (at least) an hourly basis.

The dynamic simulation models are complex to use both in implementation, since they require a considerable amount of data, and in their management. On the other hand the results obtained reflect the behaviour of the building in a manner which is very close to reality. Also in this case the comparison with the consumption allows one to "calibrate" the model.

3.9.1 Actors Involved

Since this is an activity which takes place in the office, the actors involved are internal to the staff of the auditor. If simulation audit is the operating level, the skills needed for the use of dynamic simulation software could be outsourced.

3.9.2 Tools

The definition of the baseline requires the use of instruments of various kinds as a function of the operating level adopted. As regards simulation software, see Sect. 7.4.

3.9.3 Deliverable Documentation

The results of calculations, partial or global, standard or with dynamics simulations have as output tables, charts and a technical report that can express concisely the results of the calculation. Template files of the models must be filed carefully to allow subsequent processing.

3.9.4 Critical Issues

Whether or not the auditor's skills are adequate, at this stage there are no critical issues.

3.10 Definition of the Green Energy Plan

The definition of the Green Energy Plan is one of the most critical stages of the process. In this phase the auditor, based on information obtained, including the results of the measurements, verification and monitoring, decides what are the measures to be proposed, assesses the costs and checks the benefits.

The possible retrofit measures are identified, although in a preliminary manner, from the earliest stages, when the auditor receives from the client technical and economical documentation.

Before initiating this phase a meeting with the client should be scheduled. The choices of what to do or not do, does in fact not only involve technical considerations but also economic, commercial and communication aspects. Once technical feasibility of the retrofit measures has been verified, it is necessary to make economic evaluations, in order to evaluate the convenience to implement them. An initial comparison can be based upon the definition of a list in which the measures are sorted by cost-effectiveness.

The analytical work, however, does not end here. Some measures, in reality, may be convenient but require high investments that are not compatible with the client's budget. Can the money still be found? Is it possible to obtain benefits or contributions? Is it possible to evaluate a third party financing by an ESCo?

These are the issues that the auditor should discuss with the customer during the meeting that precedes the development of the Green Energy Plan. This is a document that, rather than terminating a process, the audit, should be the document that contains the strategies, thus not so much a point of arrival but rather a starting point.

Retrofit measures on facilities in general are the most cost-effective. The retrofit measures on the building envelope require a longer payback time and sometimes are not convenient if considered only for energy saving. If the building however requires renovations, the margins of affordability can be high. This is because the cost to be considered in economic evaluations is the difference between the total cost of the remedial actions/works and the cost that would anyway be incurred for renovation.

One must remember that the energy audit that is proposed in this methodology is quite different from a conventional energy audit: one is talking about Green Energy Audit, a type of audit which considers therefore not only the energy aspects, but also all the aspects that can contribute to increase the sustainability of the building. For this reason the auditor should consider also the retrofit measures with a low or no economic advantage.

3.10.1 Actors Involved

Since this is an activity which takes place in the office, the actors involved are internal to the staff of the auditor. Once the possible scenarios have been defined, the auditor will present them to, and discuss them with the client and his staff. Another, final meeting will be done when the technical report will be formally presented.

3.10.2 Tools

In order to facilitate this phase, Chap. 13 shows 97 retrofit measure sheets that identify and describe retrofit measures for increasing energy and environmental performance of buildings.

The remedial actions envisaged collected within the following areas:

1. building envelope;
2. mechanical systems;
3. electrical systems;
4. renewable energy;
5. improvements in the management of system and building.

A detailed description of the structure of the measure sheets is shown in Sect. 8.1.2 of the book.

3.10.3 Deliverable Documentation

The documentation to deliver in this phase is the Green Energy Audit Report in its entirety. A sample scheme of the report is shown in Sect. 8.10. This document should contain the following information:

- analysis of the current situation;
- definition of the baseline;
- description of the proposed retrofit measures;
- economic evaluations of these retrofit measures;
- environmental assessments of these retrofit measures;
- definition of the management and maintenance plan.

3.10.4 Critical Issues

From the technical point of view, assuming that the auditor is competent and experienced, there are no critical elements in this phase. Critical issues arise with the external communication of the results obtained. Often a skilled auditor can have difficulty in writing a report in such a way as to be understandable even to those who do not have a technical background.

References

1. Dall'O' G, Speccher A, Bruni E (2012) The green energy audit, a new procedure for the sustainable auditing of existing buildings integrated with the LEED Protocols. Sustain Cities Soc 2:54–65
2. ISO 9001 (2000), Quality management systems—requirements

Chapter 4
Acquisition of Basic Information

The basic information allows the auditor to take a look of the actual situation of the building and facilities and can be gathered from the technical documentation and administrative documentation that the client provides. This chapter has the objective of providing criteria and methods to effectively manage, in an orderly and efficient manner, this phase which, although preliminary, is nonetheless strategic. In order to support this phase, a set of checklists, discussed in the chapter, are made available as downloadable files for this book.

4.1 General Data

The general data include the information needed to define the operating framework for building and facilities covered by the audit.

The first part of *checklist CL1* (General Data) contains information regarding

- building location;
- its intended use,
- year of construction;
- year in which any major renovation works on building or facilities were made.

The building or the building complex, is then divided into *thermal zones*: this term stands for independent parts of the building, characterised by the different usage made, HVAC or electrical facilities, with different criteria of usage or with independent indoor environmental control systems and management.

For each thermal zone one must indicate (Table 4.1) the gross heated and/or cooled volume, the gross heated and/or cooled floor area and the provision of facilities which will be indicated using the following abbreviations:

- HS Heating System
- DHW Domestic Hot Water System
- AC Summer Air Conditioning System

G. Dall'O', *Green Energy Audit of Buildings*, Green Energy and Technology,
DOI: 10.1007/978-1-4471-5064-0_4, © Springer-Verlag London 2013

Table 4.1 Geometric characteristics and facility equipment of the building complex

Thermal Zone No.	Description	Gross Volume (m^3)	Net floor Area (m^2)	Facility equipment

- ST Solar Thermal System
- PV Solar Photovoltaic System
- CHP Combined Heat and Power (Cogeneration system).

4.2 Technical and Operating Documentation

Checklist CL 1.2 (List of documents) contains a list of typical documents and by using this list, the auditor can check the availability of the useful documentation. The compilation of this data should be made even before the quotation process for the audit because its cost can vary depending on availability of documentation.

Table 4.2 Checklist of technical documents for building features

Documents	Description
Territorial framework	[ ] Plan with territorial framework of the building from which is possible to define the guidelines and the context surrounding area (eg buildings that could cause shadow, vegetation, etc..)
Design drawings	[ ] Plans (scale ________)
	[ ] Sections (scale ________)
	[ ] Elevations (scale ________)
Building envelope	[ ] Characteristics of opaque envelope
	[ ] Characteristics of transparent envelope
Other documents (specify)	[ ]
	[ ]
	[ ]
	[ ]

Table 4.3 Checklist of technical documents for building facilities features

Documents Description	
HVAC System	[ ] Functional diagrams
	[ ] Project on plan
	[ ] Technical report
	[ ] Safety report
	[ ]
	[ ]
Electrical systems	[ ] Functional diagrams
	[ ] Project on plan
	[ ] Technical report
	[ ] Safety report
Documentation for other facilities (specify)	[ ]
	[ ]
	[ ]
	[ ]

The table is divided into three parts: building, facilities and management. So far as the building is concerned, the checklist of the documents to be requested is shown in Table 4.2

As regard building facilities features, the checklist of the documents to be requested is shown in Table 4.3.

Documentation required, that should be attached to the checklist, is useful in making the energetics calculation.

The technical documentation concerning energy management is very important because it reflects the way in which plants are managed and provides a "photograph" of the actual consumption of energy.

It is useful to recover and include amongst this documentation:

- contracts for the supply of all utilities;
- any management contracts;
- energy bills for the last few (at least 3, better if 5) years of operation.

Among the information to be acquired is important to identify a person who can see or at the time of the inspections but also during all stages of execution of the audit.

4.3 Energy Usage for Electrical Appliances

The electrical consumption can be deduced directly from the information contained in the energy bills. If the objective is to estimate a value of the average

Table 4.4 Annual electric energy accounting

Historical record of consumption

	RY - 2	RY -1	Ref. Year. ___	Average value
Contract (meter) n° 1				
Contract (meter) n° 2				
Contract (meter) n° 3				
Contract (meter) n° 4				

Average values

Contract (meter) n° 1		kWh_e
Contract (meter) n° 2		kWh_e
Contract (meter) n° 3		kWh_e
Contract (meter) n° 4		kWh_e
Total annual consumption		kWh_e

electricity energy accounting, a table provided in the *Checklist CL 2.0* (Energy consumption), which is shown in Table 4.4, could be used.

Corresponding to each contract/metre, one must enter the annual consumption of electricity from the last year, called the reference year, and going back in time at least a further 2 years. The last column will then be compiled with consumption values averaged over the years considered, the sum of these will provide the value of the average total energy consumption.

The availability of monthly values of energy consumption allows for more precise analyses from which one can extract useful information to highlight abnormal situations, and therefore margins for management improved and reduction of energy costs.

Checklist CL 2.1 (Energy Accounting—electricity) can be used as an example to build-up a monthly balance sheet of the electricity consumption. The electricity bills usually contain all the information necessary to compile Table 4.5, which is constructed in the following manner:

Column 1 *Days of billing period*: (days of the month during which service is used).

Column 2 *Electric usage*: energy consumption in kWh electricity, which can be detected directly from the bill (check which months refers the bill).

Column 3 *Electric demand*: expressed kW, should not be confused with the available power (the power used represents the maximum value of the power taken during the course of a certain time interval).

Column 4 *Electric cost*: total cost in € (or local currency).

Column 5 *Electric unit cost*: average value of the cost of electric energy expressed in € / kWh: the value obtained by dividing column 4 to column 2

Table 4.5 Energy accounting form (electricity)

Area/Zone ______________________ Net Area(m^2) ____________ [9]
Electric meterNo. ______________________ Utility ______________________
Reference year ————

		Electricity					Indicators	
Month	Days billed (invoiced) [1]	Electric Usage kWh [2]	Electric Demand kW [3]	Electric Cost [4]	Electric Unit Cost /kWh [5]=[4]/[2]	Load Factor [6]=[2]/([3]*[1]*24)	Specific Cost /m² [7]=[4]/[9]	Specific Consumption kWh/m² [8]=[2]/[9]
Jan								
Feb								
Mar								
Apr								
May								
Jun								
Jul								
Aug								
Sep								
Oct								
Nov								
Dec								
Annual								

Column 6 *Load factor*: obtainable from the relationship between electric usage in kWh (column 2) and the product between the electric demand (column 3), the number of days (column 1) and 24 (the meaning of the load factor will be discussed later).

Column 7 *Specific cost*: economic indicator, expressed in €/m^2, which expresses the monthly cost for electricity referred to a unit of usable area covered by the electricity meter.

Column 8 *Specific consumption*: usage indicator, expressed in kWh/m^2, which expresses the monthly consumption of electricity referred to a unit of usable area covered by the electricity meter.

The average monthly cost per kWh in column 5 of Table 3.4 can vary from month to month depending upon the manner in which the fee is built-up and the items of which it is composed (e.g., commitment power, transportation, excise, measurement, etc.).

Non-residential users, usually have a monthly bill, while residential ones may have a different bill, for example on a quarterly basis. In this case it is still possible to draw-up an energy account, similar to that in table similar to 4.5 but with some simplifications.

A preliminary evaluation of energy efficiency of an electrical installation can be made by analysing the *load factor* which represents the relationship between the consumption of electrical energy in a certain period and the electric demand for the same period.

An ideal load factor for a user that operates in the 24 h is 1 while decreases to 0, 5 to an end user who operates in a period of 12 h. A low load factor indicates that certain users are experiencing peak consumption periods in limited circumstances and that these peaks could be avoided through a more rational distribution of electrical loads.

It may be useful to establish in advance what is the maximum permissible load factor and then monitor installations so that the value of the actual load factor can be maintained always below that maximum (for example, by installing the power limiters or implementing a management system which is able to disable, in certain periods, those utilities that are not strictly necessary.

Monthly energy consumption derived from Table 4.5 can be used to graphically display the trend of monthly loads. In this way is possible, for example, to make a distinction between the loads that are the basis and those that occur at some times.

Figure 4.1 shows the curve that defines the total electrical load. It can be observed that summer energy consumption increases until it reaches a peak in July. Given that the graph represents the energy balance of a user equipped with a summer air conditioning system, it is conceivable that the increased consumption is really due to this facility [1].

Basing oneself upon this hypothesis, it is possible, starting from the aggregated data (a single counter) to define a base load and split a load to be associated with a particular type of user (in this case summer air conditioning).

The utility companies (electricity providers) generally charge a penalty for low power factors, such information can also be acquired by studying the electricity bill.

The *power factor* ($\cos\varphi$) is the cosine of phase shift angle φ between current and voltage in an electrical AC (alternative current) system.

In a purely resistive system, the phase shift is zero, so $\cos\varphi = 1$. In a true inductive type system, that is with non-zero resistive component (for example, an electric motor or a power supply for fluorescent lamps), the phase angle is between 0 and $\pi/2$ (phase shift behind).

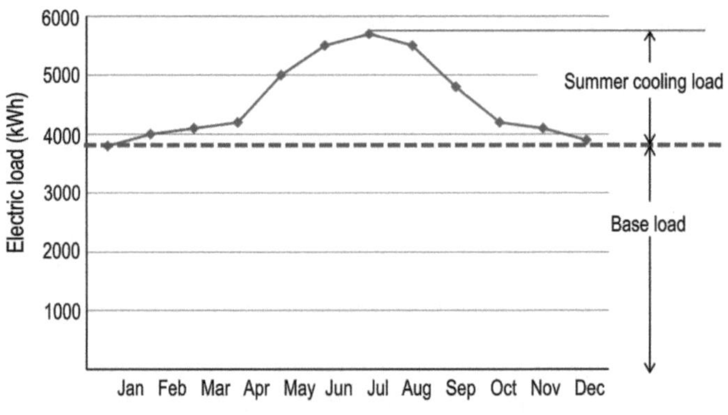

Fig. 4.1 Example of energy balance of annual *base load* and seasonal load (*summer cooling*) [1]

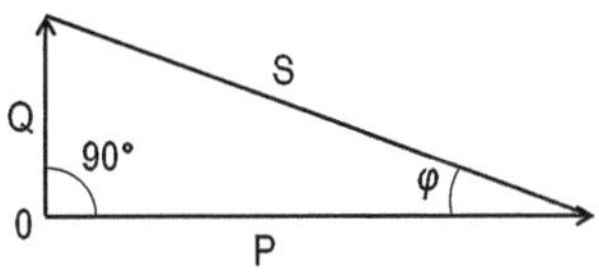

Fig. 4.2 Graphical definition of power factor or cosφ (S apparent power, P active power)

In a system with capacitive component, the phase shift is between 0 and $-\pi/2$ (phase shift in advance). In both cases, the value of cosφ power factor has a value lower than 1, with a minimum of (the theoretical value of) zero.

Cosφ is defined *power factor* as it is equivalent to the ratio between the active power (P) and the apparent power (S) as shown in the diagram of Fig. 4.2.

A power factor = 1 means that the apparent power corresponds to the active power and reactive power is zero. Since the reactive power is always undesirable, a value of power factor becomes progressively more undesirable as it deviates from 1. Since the inductive and capacitive phase shifts occur in opposite directions, appropriately combination of the two components in a circuit can be made in order that their effects are mutually cancelled, bringing cosφ close to unity.

On electric motors and other loads used in electronics which are almost exclusively inductive, the value of the power factor produced by the equipment item is always indicated, so that the designer can calculate the value of a possible capacitor to be inserted to perform the so-called *power factor correction* [2].

4.4 Energy Usage for Thermal Appliances

Even for thermal loads, data on consumption can be retrieved starting from the supply bills. In this case, however, we need to differentiate between the user's connected network and those parts which are not connected, and those which are equipped with a storage system.

The supply of natural gas (CII_4) or the connection to a district heating and/or cooling system are equipped, respectively, with a *flow meter* or a *heat meter* and through these devices one can make an instantaneous reading of consumption, in analogy with electrical supply. For other fuels (for example diesel fuel), this is not possible owing to the nature of the supply line to fill the storage tank and the seasonal (or with a greater frequency depending on the storage capacity) occurrence of filling, which is therefore independent of the timing of the user's consumption of the fuel.

Accounting of the thermal energy output is possible if heat meters are installed.[1]

Each fuel is characterised by a *calorific value* which expresses the maximum amount of heat that can be extracted from the complete combustion of 1 kg of fuel

[1] Heat meters are devices that integrate, into the value of the continuous flow of fluid, the difference between outlet and return temperature

(or 1 m^3 in case of gas fuels) at 0 °C and 1 atmosphere of pressure. For all fuels, there are two calorific values: the *lower calorific value* (LCV) and the *gross calorific value* (GCV).

The *gross calorific value* is the amount of heat available from the complete combustion, at constant pressure and the density of the fuel, when the products of combustion are returned to the initial temperature of the fuel and the combustion air. Gross calorific value can be obtained approximately by calculation, based on elemental analysis of the fuel, or directly through the use of special calorimetric instruments.

Typically, under normal combustion conditions, products of a combustion process are released at a higher temperature than that of the reference fuel. Thus, a part of the heat theoretically available is dispersed owing to the heating of fumes and, above all, in the vaporisation of the water produced by the combustion. For each degree of increase in fume temperature, the process takes approximately 1 kJ/kg of fumes, furthermore for each kg of water-vapour in the flue gas the process takes approximately 2500 kJ owing to latent heat of vaporisation at 100 °C.

The *lower calorific value* is the gross calorific value decreased by the heat of condensation of water vapour during the combustion process. This is the value to which one usually refers when generally discussing the calorific value of fuel and the efficiency of a heat engine or a heating device.

In *condensing boilers* it is possible to recover part of the latent heat of water vapour. In this way, from a unit of fuel it is possible to obtain a quantity of heat greater than that obtainable when referring to lower calorific value (thus with 100 % nominal efficiency), even if a part of the heat theoretically available (gross calorific value) continues to be dispersed with the fumes.[2]

For determining the lower calorific value, by elemental analysis, one must first determine the gross calorific value and then subtract this from 2500 kJ per kg of water vapour contained in the fumes. The water vapour in the flue gas will be due to the combustion of hydrogen and the humidity initially present in the fuel.

The calorific value of commercial fuels is very variable: it depends on the origin of the material and the subsequent treatment. The values shown in Table 4.6 are therefore only indicative.

The preparation of the annual thermal energy accounting records, according to the scheme shown in Table 4.7, and inserted in the *Checklist CL 2.0*, is relatively simple.

For each type of fuel one needs to know the value of annual consumption for the reference year and for the previous 2 years, at least. The average value thus obtained, for each type of fuel (or energy, in the case of district heating) must then be multiplied by a conversion factor (in this case the lower calorific value expressed in kWh rather than in MJ).

[2] One says that condensing boilers have an efficiency greater than 100 % simply because the performance evaluation of boilers is based upon the lower calorific value LCV

Table 4.6 Lower calorific value and density of some fuels

Type of fuel	Units	Lower calorific value (LCV)		Density
		MJ	KWh	kg/dm^3
LPG (liquefied petroleum gas)	kg	46,05	12,79	0,565
Gasoline	kg	43,96	12,21	0,670
Diesel oil	kg	42,71	11,86	0,835
Thin fuel oil LSC[1]	kg	41,03	11,40	0,923
Dense fuel oil LSC[1]	kg	40,19	11,16	0,950
Natural gas (methane)	Sm^3	34,54	9,59	0,720
Fossil coal	kg	30,98	8,61	–
Kerosene	kg	43,12	11,98	0,791

[1] *LSC* low sulphur content

Once the values of consumption homogenised to the same unit of measurement have been derived, it is possible to obtain the value of the sum of the total annual energy consumption in kWh.

Even for thermal loads, as already observed for those electrical, the ability to break down consumption into the different months of the year allows one to make more accurate assessments.

The difficulty lies not in the calculation, but in the availability of disaggregated consumption data. *Checklist CL 2.2* (Energy Accounting—Natural gas) contains an example of table that can be used to construct a monthly energy accounting scheme for thermal energy consumption using natural gas supplied by the utility companies. The gas bills usually carry all the information necessary to compile Table 4.8 which can cover every area served by a meter in the following way:

Table 4.7 Annual thermal energy accounting

Historical consumption

Typology	RY - 2	RY -1	Ref. Year ___	Average value
Natural gas (methane)				
Diesel oil				
LPG (Liquefied Petroleum Gas)				
Wood				
Heat				

Average values

Natural gas (methane)		m^3 x 9,59 =		kWh_t
Diesel oil		l x 11,86 =		kWh_t
LPG (Liquefied Petroleum Gas)		l x 12,79 =		kWh_t
Wood		kg x 2,91 =		kWh_t
Heat		MJ x 0,37 =		kWh_t
		Total annual consumption		kWh_t

Table 4.8 Energy accounting form (natural gas)

Area/Zone ____________________

Net Area (m^2) ____________ [9] Fuel type: **Natural gas (CH_4)**

Meter No. ________ Location ________________ Reference Year __________

Utility ______________________________ LCV= kWh/Sm^3 ____________ [10]

		Fuel usage				Indicators		
Month	Days billed (invoiced) [1]	Usage quantity Sm^3 [2]	Usage energy kWh [3]=[2]/[10]	Energy cost € [4]	Unit cost €/Sm^3 [5]=[4]/[2]	Specific cost €/m^2 [6]=[4]/[9]	Specific usage 1 Sm^3/m^2 [7]=[2]/[9]	Specific usage 2 kWh/m^2 [8]=[3]/[9]
Jan								
Feb								
Mar								
Apr								
May								
Jun								
Jul								
Aug								
Sep								
Oct								
Nov								
Dec								
Annual								

Column 1 *Days of billing period*: (days of the month during which service is used)

Column 2 *Usage quantity*: energy consumption expressed in Standard m^3 of natural gas, which can be detected directly from the bill (check which months refers the bill)

Column 3 *Usage energy*: expressed in kWh, calculated dividing usage quantity of column 1 by the lower combustion value [10]

Column 4 *Energy cost*: total cost in € (or local currency)

Column 5 *Unit cost*: average value of the cost of natural gas expressed in €/Sm^3: the value is obtained by dividing column 4 by column 2

Column 6 *Specific cost*: economic indicator expressing the average value of the monthly cost of natural gas referred to a unit to usable area, expressed in €/m^2, the value is obtained by dividing column 4 by the zone area [9]

Column 7 *Specific usage 1*: consumption indicator expressing the monthly consumption of natural gas, in Sm^3/m^2, referred to a unit of usable area covered by the gas meter, the value is obtained by dividing column 2 by the zone area [9]

Column 8 *Specific usage*: consumption indicator, expressed in kWh/m^2, which expresses the monthly consumption of natural gas to a unit of usable area covered by the gas meter, the value is obtained by dividing column 3 by the zone area [9].

The average monthly cost indicated in column 5 of Table 4.8 may vary depending upon the manner in which the charge is made and items that compose it.

The non-residential users usually have a monthly bill, while residential ones may have a different bill, for example on a quarterly basis. In this case, it is still possible to build a table similar to 4.8 but with some simplifications.

Using *Checklist CL 2.3* (Energy Accounting—other fuels), similar to *CL 2.2*, it is possible to make monthly energy accounting records for other fuels or energy sources (e.g., district heating or cooling).

4.5 Parameterisation of Performance Through Benchmarking

4.5.1 The Scope of Benchmarks

The parameterisation of the energy performance of a building through the benchmarks is a good starting point for the energy audit with the purpose of optimisation of management, and definition of improved efficiency targets, on the basis of which to plan actions.

The objective of benchmarks is usually to be able to compare the energy consumption (or consumption of resources) of a building with those of another set of similar buildings. The comparison can be based on the overall consumption i.e., electrical or thermal, but also on partial energy consumption, for example those which refer only to the lighting, the summer air conditioning or to the domestic hot water.

The benchmarks also allow easy and effective monitoring of the progress of specific actions or strategies and identification over time those buildings, or parts of the building or plants (defined as services) that require additional efforts to reduce energy consumption or other resources.

Generating benchmarks is relatively simple, more complex, however, is their proper use. There are different methods and approaches to generate and use benchmarks but there is nevertheless a common basis on which they can be compared.

LBL Lawrence Berkeley National Laboratory is promoting interesting research and development of energy benchmarks for residential, commercial and industrial (http://energybenchmarking.lbl.gov/) applications, creating a database accessible from the web that is constantly fed and updated by real cases. The limitation of this database is that it relates primarily to North American building types and users.

The need to promote these important tools, but also to promote their proper use, is the basis of the European project *European Green Building Programme* (www.eu-greenbuilding.org) which has issued guidelines on benchmarking [3]. The objective of this work is also to create the "state of the art" for this issue in Europe.

4.5.2 Factors that may Affect Benchmarks

The results of an analysis conducted using benchmarks can help to identify buildings with low performance characteristics; the correct interpretation of data, can, however, be a real change in the approach.

There are many factors that can affect energy use in buildings, such as the occupation, installed equipment, climatic conditions and design features. The difficulty in the use of benchmarks is precisely this: to carefully analyse a number of parameters before making the comparison.

Buildings can be compared if objective basis for comparison really exist. In a residential building, for example, the energy consumption for heating and cooling depends very much, for the same climatic zone, on how the facilities are handled. It is clear, for example, that if during the winter the user spends several weeks in another location consequently switching off the plant, the energy consumption and the indicator that is constructed on the basis thereof, are not reliable. Neglecting to consider these important aspects, in such cases, may generate indications of high energy performance that are, in reality, quite false.

It is also necessary to pay attention to how the energy is being used. The electrical energy, for example, is certainly used for lighting, for the supply of electrical equipment but also for HVAC plant. If, in a building space, heating is supplied by an electric heat pump, the benchmarks may become unacceptable: it is not always true, in fact, that energy heat demands are always provided using fossil fuels.

So far as energy uses are concerned, the construction of the benchmark ideally should take account of four factors

- indoor environmental conditions (temperature, relative humidity, lighting, etc.);
- hourly and daily profiles of occupation;
- characteristics of plant systems;
- how the facilities are used.

The higher the level of detail in the construction of the benchmarks, then the greater is their reliability.

4.5.3 The Selection of Parameters

The comparison between buildings based on the annual energy cost (e.g., €/m^2, €/m^3) can be useful if the energy cost is comparable between the different buildings. Unfortunately, this is almost impossible because utility companies (e.g., energy or fuel providers) may be different and the form of the supply contracts may also differ (e.g., the complexity with which the supplier determines the cost of electricity which, even for the same building, is variable from month to month).

Assessment of energy consumption based on energy units, for example kWh/m^2 or MJ/m^2 is instead more accurate. One must again make a distinction between electricity and energy from fossil fuels: if the two quantities of energy, which of course generate two different indicators, are managed separately, there is no problem.

If the aim is to build a unique indicator of energy consumption (and then sum the electrical energy with the thermal energy) it is necessary to convert electrical energy into electrical primary energy to take into account of the fact that to produce 1 kWh of electric energy, more primary energy from fossil fuels must be consumed.

An energy benchmark normally is built-up by dividing an amount of energy by a geometric size. The geometric size of the most common reference is the floor area: when comparing different indicators, however, based on the same units, we can still make mistakes even above 20 % if we do not specify the meaning of the geometric size reference. In the case of areas, for example, one might consider:

- the *gross area* that includes the thickness of the outer and inner walls;
- the *net area* that is the gross area minus the thickness of the outer and inner walls;
- the *useful area* (for example excluding the connection areas as corridors, stairs, technical spaces, etc.);
- the *air conditioned area* (in some buildings, in fact, the air conditioning does not apply to all the spaces of the building)

Those listed above are just examples of how the same term, in the chosen case the floor area, may have different meanings, and if it is not properly specified, can generate inconsistent benchmarks.

Table 4.9 shows some examples of indicators for assessing the energy performance.

The floor area or volume are not always the best criteria of benchmark reference for buildings. The energy intensity can be conveniently assessed by comparing the amount of energy consumed to other parameters.

Some examples are listed below:

- the number of nights or the number of beds (in the case of a hotel);
- the number of pupils (in the case of a school);
- the number of employees (in the case of a building used for productive activities);
- the number of products (for industrial audits).

The use of multiple indicators is preferable because the information provided, being the most complete, permits assessments not necessarily only related to energy performance but also to the ways in which the spaces available are used. For example, two school buildings compared may have resulted in similar levels of energy performance when using an indicator expressed in kWh/m^2 year, but

Table 4.9 Examples of indicators for assessing the energy performance

Benchmark	Units	Comments
Energy usage/gross volume	KWh/m^3 year	Can be referred to the consumption of thermal energy, the consumption of electrical energy or to the global consumption. In this case, it is appropriate that the value of the electric energy consumption, before being added to that of thermal energy, is transformed into primary energy. This benchmark fits well for buildings characterised by high internal free height of the room, warehousing or manufacturing spaces (offices, commercial, industrial, etc.)
Energy usage/useful area	KWh/m^2 year	The reference to the value of the surface, and not to the volume, makes this indicator particularly useful for residential buildings
Energy usage/user	KWh/user year	Can be referred to the consumption of thermal energy, the consumption of electrical energy or to the global consumption. This benchmark could be interesting for administrative management if the availability of running costs depends upon the number of user-occupants (for example, schools, hospitals, etc.). It can be used also to make comparisons between buildings with the same intended use in order to highlight situations of energy wastage and start promoting energy saving policies
Energy usage/useful area/Degree Days	KWh/(m^2 DD year)	This benchmark is useful for comparing consumption data measured in different seasons and/or in different locations

different energy performance, when considering the energy used related to the number of pupils.

The comparison of energy consumption needs to consider the external weather conditions. Energy consumption, indeed, can vary over the years depending upon the change in the environmental parameters. One of the methods used to standardise the information relating to seasonal or annual energy consumption for air conditioning is the parameterisation by mean of Degrees Day.

The calculation is quite simple: once the value of energy consumption of the reference period is acquired, it is divided by actual Degree Day of that period, which may be provided by weather stations, and is multiplied by the value of conventional Day Degrees.

The values of Day Degrees for most locations (both in winter and in summer) may be obtained from the website www.degreedays.net. that contains an application online that allows to generate Degree Days as a function of the internal temperature reference. A discussion on the meaning of Degree Days and on application of this method is discussed in Chap. 6.

4.5.4 International Benchmark Tools

The Lawrence Berkeley National Laboratory (LBNL), founded by several U.S. Government agencies, including the Department of Energy and the Environmental Protection Agency, have been engaged for several years in research and development of energy benchmarks. The research results are made accessible through the web portal, Energy Benchmarking for Buildings and Industries http://energybenchmarking.lbl.gov/ mail address.

An interesting application online for non-residential buildings, is *Energy IQ*, a web platform that allows not only to compare the data of energy consumption of a building with other buildings belonging to the same category, but also to guide the user to identify measures for energy upgrading.

For the comparison of a building with other buildings, using the Energy IQ platform is possible select filters, for example:

- the type of user (e.g., offices, data centers, restaurants, schools, etc.);
- the construction period (age) of the building;
- the type of appliance;
- the location;
- the gross floor area, etc.

The energy evaluation can be made:

- on the basis of gross energy consumption (electricity, heat, total);
- based on end-use (i.e., use electrical, thermal use or total): in this case it is necessary to have the actual consumption broken down for each user in question;
- on the basis of peak electricity.

More detailed information can also be specified about the construction features of the building or plant characteristics. Once all the values have been inserted, the application allows to compare the building with the buildings belonging to the categories selected by the filters: in this way, one can determine how to place the indicator of energy consumption. The graphical representation can be modified by means of some display options.

Energy IQ is a platform in which updates are being continually implemented through the insertion of new buildings: the greater their number, the more representative become the benchmarks that are derived from it. However, a valid use of this platform is limited to buildings located in the United States and particularly in California.

Figure 4.3 shows an example of graphical representation of the results of a benchmark made with this tool: the source energy intensity of the audited building is compared with the source energy intensity of a sample of buildings of the same category. The comparison is an important starting point for understanding the energy quality level of the audited building [4].

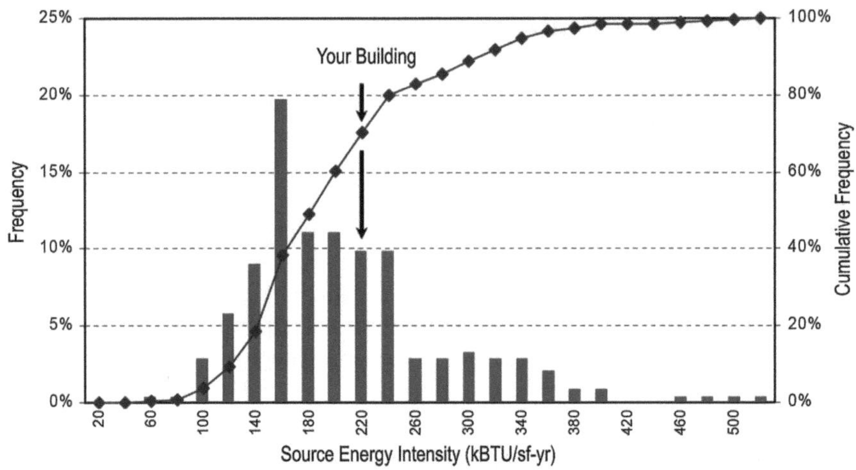

Fig. 4.3 Graphical representation of the results of a benchmark with Energy IQ Tool [4]

Again in the State of California, a tool for benchmarking, simpler than that described above, has been developed: the Cal Arch 2.1 (California Building Energy Reference Tool) (http://poet.lbl.gov/cal-arch/compare.html.).

Another interesting benchmarking system is promoted by Energy Star. The benchmarking starter kit is available directly on the official web site (www.energystar.gov). Using the kit online it is possible to:

- track energy and water consumption;
- identify under-performing buildings;
- set priorities;
- monitor progress;
- verify improvements;
- receive EPA (Environmental Protection Agency) recognition.

Once the necessary data have been collected directly by the user (or the auditor), benchmarking is quick and simple.

References

1. Thumann A, Younger WJ (2007) Handbook of energy audit, 7th edn. Fairmont Press, Lilburn, p 19
2. CRES (2000) Energy audit guide, part B: system retrofits for energy efficiency, project financed by the European commission. Athens, p 8
3. Benke G (2006) Austrian energy agency, The European green building programme—benckmarking. www.eu-greenbuilding.org
4. Mattew PA (2006) Action-oriented benchmarking, using CEUS data to identify and prioritize efficiency opportunities in California commercial buildings, Lawrence Berkeley National Laboratory, Berkeley (presentation published on the web)

Chapter 5
Survey Instrumentation and Methods

The field survey permits the completion of the phase of acquisition of the information concerning the characteristics of the building and facilities, but at the same time is an important opportunity to acquire information on the management procedures. The proper use of the survey instrumentation is fundamental at this stage. In the first part of the chapter, survey instruments required for the Green Energy Audit are selected and described with their characteristics and methods of use. In the second part, methodologies that the auditor should apply to carry out the audit properly are described and discussed. The last section covers the technical and application monitoring, which is strategic for audits.

5.1 Instrumentation for Measuring Thermal Comfort

The interactions between a person and their surroundings (environment) are due to fairly complex mechanisms of mass and heat transfer.

The subjectivity of the reactions to a given external thermal stress, can however be different from one person to another. The purpose of HVAC systems is to ensure (indoor) environmental conditions that satisfy the greatest number of people: according to *Fanger*[1]*'s theory*, the best value of the comfort, the indicator is *predicted mean vote* (*PMV*), is obtained when the *predicted percentage of dissatisfied* (*PPD*) index is equal to 10 %, that is when at least 90 % of persons are satisfied.

In the Fanger's theory, the basis of 7730 ISO Standard [1] the environmental parameters that affect interactions between a person and their indoor environment, and thus influencing the PPD indicator, are:

- the air temperature;
- the mean radiant temperature of walls;
- the air speed;
- the relative humidity.

---

[1] Povl Ole Fanger (1934–2006), University professor at Syracure University (USA), was the maximum expert in the field of the health effects of indoor environments.

G. Dall'O', *Green Energy Audit of Buildings*, Green Energy and Technology,
DOI: 10.1007/978-1-4471-5064-0_5, © Springer-Verlag London 2013

Other two parameters, according to Fanger's theory, affect the PPD indicator: *Met* that is the index which takes into account physical activity and therefore the metabolism, and *Clo* that takes into account the thermal resistance of clothing. (The theory of thermal comfort will be briefly discussed in Chap. 12).

A parameter that affects not so much the thermo-hygrometric comfort but rather the internal air quality (IAQ) (see Chap. 12) is the percentage of carbon dioxide (CO_2) detected in the air.

All these parameters can be measured with instruments either instantaneously (spot measurement) or continuously (monitoring).

5.1.1 Thermometers

The measurement of air temperature within a space is not enough to determine whether thermal comfort conditions are guaranteed, because the temperature, as perceived by the human body, is affected also by other parameters, such as the mean radiant temperature of the walls. The *operating temperature*, that is the temperature that a person is able to perceive, is in fact calculated, using a simplified but acceptable approach, from the average value between the air temperature and the mean radiant temperature of the walls.

A value of the internal air temperature of 20 °C in winter may not be sufficient for thermal comfort if, for example, the value of the radiant mean temperature of the walls is 16 or 17 °C (in this case, in fact, the mathematical average would be equal respectively to 18 and 18.5 °C).

The air temperature, however, remains one of the most important parameters to assess whether indoor environmental conditions are acceptable, or even if they occur in a situation of energy wastage (such as an inefficient regulating system which is not maintaining the set-point values) and thermometers are the instruments to use for measuring this parameter. Portable electronic thermometers report in a clear way, on their display, the value of the temperature (Fig. 5.1).

Before choosing a thermometer one needs to consider a number of parameters including the measurement scale (for the environmental survey and plant survey it

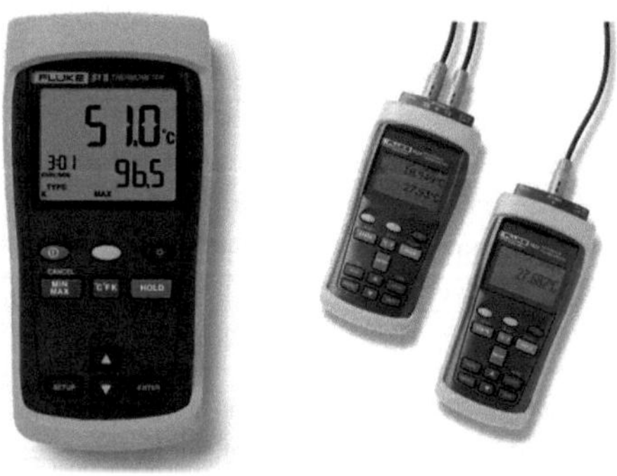

Fig. 5.1 Example of portable digital thermometers (courtesy of Fluke Co.)

is sufficient to have a range between −10 and 120 °C), the measurement accuracy and the durability. The accuracy of the measurement mainly depends on the sensor that is employed. Almost all electronic thermometers are fitted with interchangeable sensors, which allows one to perform different measurements (e.g., air sensors, sensors for liquids, contact sensors etc.).

There are many types of thermometer sensors, the most common being the following:

- *Thermocouples*: the temperature measurement is based on the thermoelectric effect. The thermocouples consist of two conductors welded together and made of different metals or different metal alloys. The basic values of the thermoelectric voltages and maximum manufacturing tolerances of thermocouples are defined by the IEC 584 Standard. The most common thermocouple is the NiCr-Ni (type K designation).
- *Resistance thermometers (Pt100)*: the platinum resistance thermometer has the property of changing its resistance as a function of the temperature. The thermo resistance is powered by a constant current whose voltage drop changes proportionally as a function of the temperature measured. The basic values and tolerances for these resistance thermometers are defined by the IEC 751 Standards.
- *Thermistor (NTC)*: temperature measurement with the use of thermistors is also based on a variation of the resistance as a function of the temperature of the sensor element. Unlike resistance thermometers, thermistors have a negative temperature coefficient (i.e., the resistance decreases with increasing temperature). In this case there are no standardised characteristics.

Table 5.1 shows measuring ranges, class and maximum tolerances of typical thermometer sensors.

Table 5.1 Characteristics of typical thermometer sensors (courtesy of Testo AG)

Sensor type	Measuring range	Class	Maximum tolerance	
			Fixed value	Related to the temperature (t)
Thermocouples	−40… +1,200 °C	2	±2,5 °C	±0.0075 × (t)
Type K (NiCr-Ni)	−40… +1,000 °C	1	±1,5 °C	±0.004 × (t)
Type T	−40… +350 °C	1	±0,5 °C	±0.001 × (t)
Type J	−40… +750 °C	1	±1,5 °C	±0.004 × (t)
Pt100	−100… +200 °C	B	±(0.3 + 0.005 × (*t*))	
	−200… +600 °C	A	±(0.15 + 0.002 × (t))	
NTC (Standard)	−50… -25.1 °C	–	±0.4 °C	
	−25… +74.9 °C		±0.2 °C	
	−75… +150 °C		±0.5 % of the measured value	
NTC (High temperatures)	−30… -20,1 °C	–	±1 °C	
	−20… 0 °C		±0.6 °C	
	+0,1… +75 °C		±0.5 °C	
	+75,1… +275 °C		±0,5 % of the measured value	

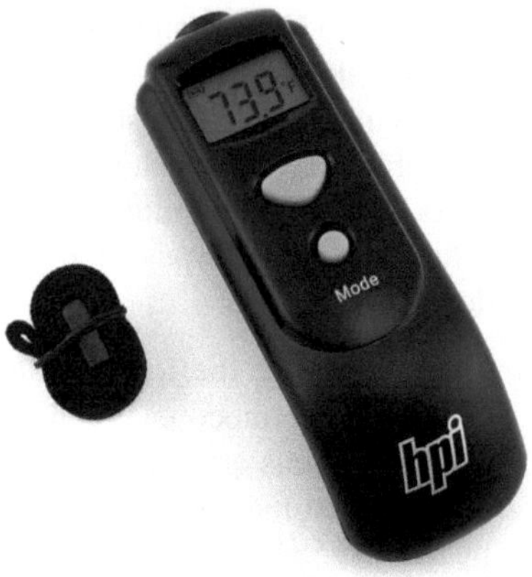

Fig. 5.2 Example of a pocket infrared thermometer (courtesy of Tempgun Co.)

The sensors for measuring the air temperature are made in such a way as to reduce as much as possible the thermal inertia of the sensor element and thus the time of measurement which, nonetheless, is never instantaneous.

The ambient temperature must be measured with a thermometer protected from solar radiation. The mean radiant temperature could be performed using a special spherical sensor of diameter 150 mm, called a *globe thermometer*.

Some thermometers are equipped with multiple inputs, therefore the same device can be connected to several sensors, in such a way that the auditor is able to simultaneously detect multiple values of temperature (Fig. 5.1 on the right).

Some thermometers return a single measured value, while others can be programmed and are therefore able to acquire a series of temperature values which are stored and, through an interface, can be transferred to a Personal Computer for further processing.

5.1.2 Infrared Thermometers

The surface temperature can be detected instantly through an infrared thermometer that can safely measure hot, hazardous, or hard-to-reach materials without touching, contaminating, or damaging the material's surface. Temperature sensitivity ranges available are typically from −40 to 700 °C.

The instrument does not detect directly the value of the temperature but, through an optical sensor sensitive to infrared, the thermal energy emitted as radiation from the body.

The analytical calculation to pass from the infrared energy emitted by a surface to its temperature requires the knowledge of some environmental parameters but mainly the emissivity ε of the surface. The accuracy of the measurement, therefore, is not very high but more than enough for a fast measurement.

In the simplest models (Fig. 5.2) the emissivity has a default value that cannot be changed: in this case the measurement is approximate. In models with higher

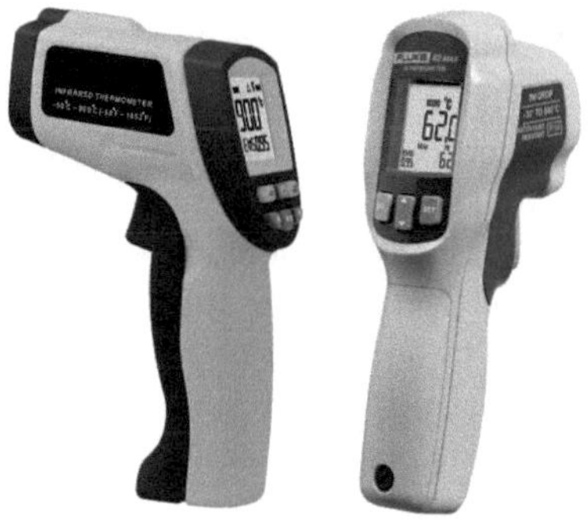

Fig. 5.3 Example of portable infrared thermometers (courtesy of Fluke Co.)

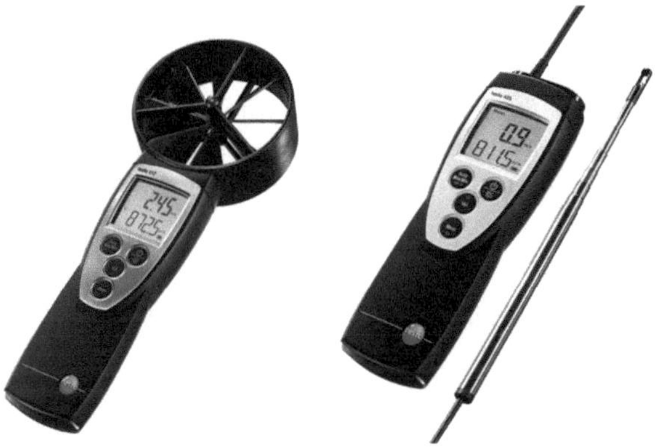

Fig. 5.4 Example of portable anemometers: on the left vane type on the right hot-wire type (courtesy of Testo AG)

performance (Fig. 5.3) is possible to set manually the values of emissivity and the air temperature, the measurement in this case is more accurate.

If the instrument is equipped with a laser pointer, we can pinpoint the measurement area. Optics are stated in a ratio indicating the distance where a spot size will be 1 m in diameter: a ratio of 8:1, for example, means that from a distance of 8 m, the sensor will be reading an area 1 m in diameter. The quality of the instrument also depends on the cone of measurement, which should be as narrow as possible.

Using these instruments, which are very economical, the auditor could make a preliminary measurement of the surface temperature of the walls, identifying for example the coldest points in correspondence to which a thermal bridge may be present, or the hottest points where there may be leakage from the distribution system.

These instruments, obviously, does not substitute the infrared camera but provide interesting information that may lead to a more in-depth diagnosis (e.g., infrared audit). With good approximation one can perform the mapping of surface temperatures in order to estimate an indicative value of the mean radiant temperature of the walls.

5.1.3 Anemometers

The air speed measurement can be performed with an anemometer. Figure 5.4 shows two models of anemometer: on the left a vane type, on the right a hot-wire type.

A vane type anemometer uses a small fan that is turned by air flowing over the vanes. The speed of the fan is measured by a rev counter and converted to a air speed by an electronic chip. Using this instrument the volumetric flow rate may be calculated if the cross-sectional area is known: it is so useful to measure the air flow rate output from diffusion grilles in HVAC systems.

Hot-wire anemometers operate on the basis of another principle: they use a very fine wire (diameter of the order of several micrometers) electrically heated up to some temperature above the ambient. Air flowing past the wire has a cooling effect on the wire. As the electrical resistance of most metals is dependent upon the temperature of the metal (normally tungsten) a relationship can be obtained between the resistance of the wire and the flow speed. Having a sensor of small dimensions, these instruments can be used more easily for the measurement of the air speed inside a duct (only a small hole on the side of the duct is necessary).

5.1.4 Hygrometers and Psychrometers

The moisture content in the environment can be measured with an instrument called a hygrometer. In these instruments the measurement of the relative humidity is not direct, but based on the measure of other environmental parameters such as temperature or pressure: from these measured quantities, through a calibration and calculation process, the measurement of humidity can be derived.

A type of instrument widely used in the HVAC sector for the control of indoor environmental conditions is the *psychrometer*. The instrument consists of two thermometers, one of which is dry and the other is kept wet with distilled water on a sock or wick. The first thermometer measures the dry-bulb temperature whilst the second one measures the wet-bulb temperature. Relative humidity can be established by locating the intersection of the wet- and dry-bulb temperatures on a psychrometric chart or using a table normally supplied with the instrument.

In the *sling psychrometer* the two thermometers are mounted together with a handle attached on a chain (Fig. 5.5). One thermometer is ordinary, while the other has a cloth wick over its bulb and is called a wet-bulb thermometer. When a reading is to be taken, the wick is first dipped in water and then the instrument is whirled around in the air by the operator. During the whirling, the water evaporates from the wick, cooling the wet-bulb thermometer. Then the temperatures of both thermometers are read and from that reading, and using a table of these differences for each degree of temperature, the auditor can easily determine the relative humidity.

The sling psychrometer, sometimes used for field measurements, is gradually being replaced by more convenient instruments with electronic sensors. One of

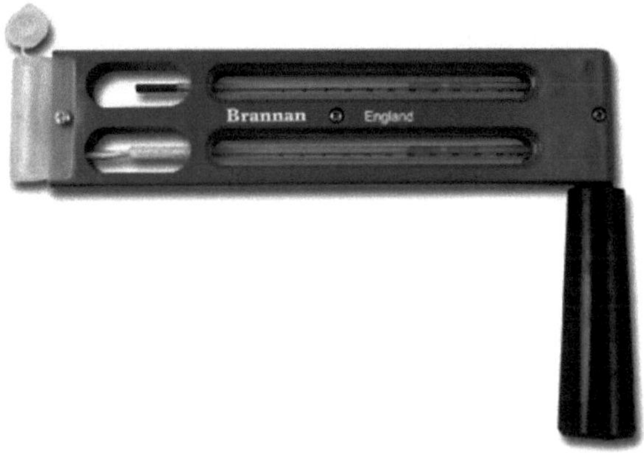

Fig. 5.5 Example of sling psychrometer for measuring relative humidity (courtesy of Brannan Co. www.brannan.co.uk)

these is the *chilled mirror hygrometer* that measures the temperature of a mirror at the point when moisture (dew) begins to condense on it, thus giving a measurement of the dew-point temperature.[2]

The temperature of the mirror is controlled by electronic feedback to maintain a dynamic equilibrium between evaporation and condensation on the mirror.

An certain accuracy is attainable with these devices that do however require frequent cleaning, a skilled operator and periodic calibration to attain these levels of accuracy.

The electronic hygrometers directly return to the auditor, besides value of relative humidity, also the temperature. For this reason they are also called thermo-hygrometers (Fig. 5.6). These instruments are also typically used to determine the dew-point air temperature.

5.1.5 Microclimate Analysers

Each of the instruments described above normally detect one or two microclimate parameters. For a complete assessment of the indoor environmental conditions it would be necessary the simultaneous detection of all the parameters required to calculate the comfort indicators PMV and PPD.

Microclimate analysers are multifunction instruments equipped with several inputs in order to manage signals that come from different external sensors. These instruments have also the capability to make calculations and can be programmed to record many measurements on the basis of a schedule defined by the auditor.

Figure 5.7 shows a typical portable microclimate analyser. The instrument is equipped with three digital inputs from sensors: each sensor is provided with an

[2] The dew-point temperature is the temperature below which the water vapor in a volume of humid air, at a constant barometric pressure will condense into liquid water.

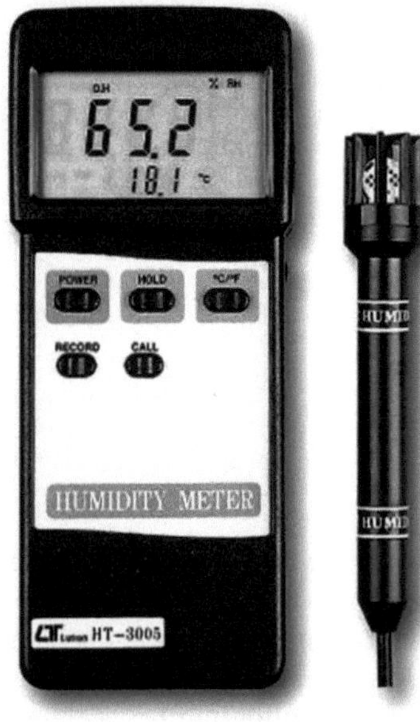

Fig. 5.6 Example of electronic portable hygrometer for measuring relative humidity (courtesy of Lutron Corporation)

electronic circuit that communicates with the instrument, which has calibration data of the sensor stored in its solid state memory. The instrument can detect simultaneously the following parameters:

- globe thermometer temperature;
- wet-bulb temperature with natural ventilation;
- air temperature;
- relative humidity;
- air speed.

On the basis of the measurements detected the instrument can calculate and display the following indexes:

- mean radiant temperature;
- PMV;
- PPD.

5.1.6 Carbon Dioxide (CO_2)

The carbon dioxide (CO_2) in the air is generated by the people present themselves, its measurement, preferably detected continuously with a monitoring system, can provide important information about the quality of ventilation of confined spaces in relation to the mode of usage (Fig. 5.8).

```
========================
  ISO 7730  PMV Index
========================
Model HD32.3 WBGT - PMV
Firm.Ver.=01.00
Firm.Date=2008/12/05
SN=12345678
ID=0000000000000000
------------------------
Probe ch.1 description
Type: Hot wire
Data cal.:2008/10/15
Serial N.:08109460
------------------------
Probe ch.2 description
Type: Pt100 Tg 50
Data cal.:2008/10/01
Serial N.:08109452
------------------------
Probe ch.3 description
Type: RH
Data cal.:2008/10/15
Serial N.:08109464
========================
Date=2008/11/21 15:00:00
Va              0.00 m/s
Tg              22.0 °C
Ta              22.0 °C
RH              39.1  %
MET             1.20
CLO             1.00
PMV             0.10
PPD             5.10  %
========================
```

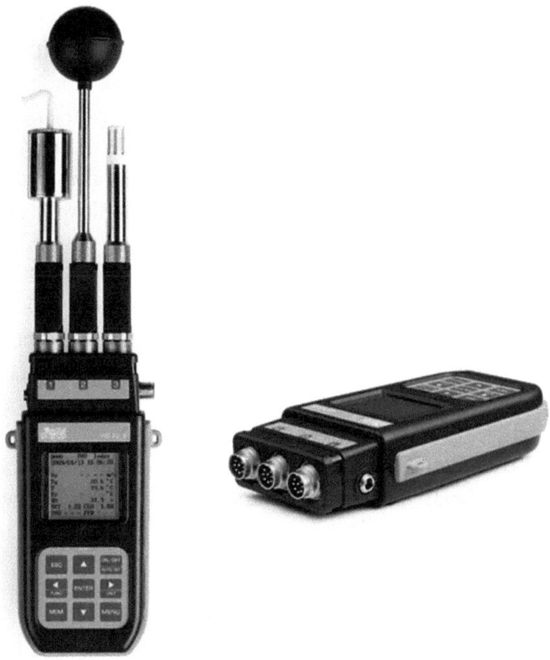

Fig. 5.7 Example of microclimate analyser (courtesy of Delta Ohm—www.deltaohm.com)

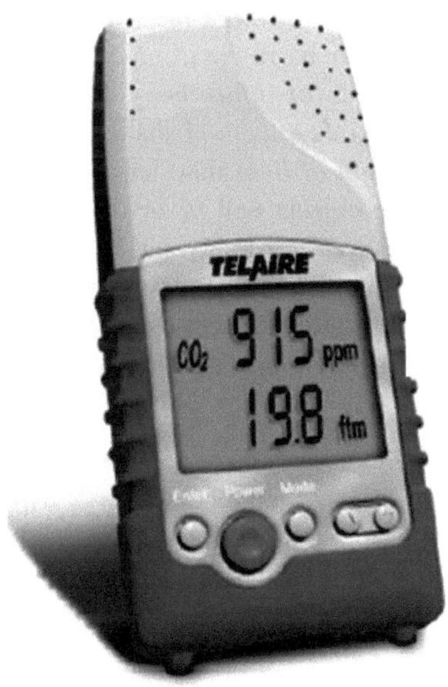

Fig. 5.8 Carbon Dioxide meter (courtesy of Telaire Co.)

CO_2 concentrations of the outside air are normally between 250 and 350 parts per million (ppm). The presence of people within the confined spaces can be detected indirectly through the variation of this parameter.

A monitoring of CO_2 concentration gives useful information for understanding whether ventilation is working properly. With a CO_2 detector that can monitor the indoor environment conditions, the information could be transmitted to the control system that could adjust the air ventilation flow to meet the real needs, so avoiding unnecessary wastages of energy.

According to ASHRAE Standard 62.1 [2], CO_2 concentration levels greater than 1,000 ppm may indicate that ventilation is poor.

If within a room space, despite the presence of people, CO_2 concentration levels are maintained at values similar to those of the outside air, probably ventilation is excessive, and this has a negative impact on energy consumption and management costs.

The detection of carbon dioxide concentration, can be made instantaneously or continuously with an instrument such as that shown in Fig. 5.7: for continuous measurement the instrument need to be connected to a mini data-logger.

5.2 Instrumentation for Measuring Lighting Comfort

Lighting comfort of a space depends on several factors:

- *amount of light* that expresses the possibility to perform a given visual task;
- *spatial variation of lighting level* that expresses the hierarchy of lighting conditions in an environment;
- *degree of diffusion of light* that refers to the directionality of the light, the possible effects of shade achievable in a space and the strength of the shadows;
- *origin of light* that defines the direction of the shadows and then which part of the objects will be in shadow;
- *light colour* that refers to the colour of light produced by the lighting used and its ability to change the perception of objects in the visual field;
- *variability of light conditions*: that expresses how and where the light conditions are stable or variable over time in the space.

A proper analysis of the quality of the light generated by a lighting system within a certain indoor environment is quite complex.

As far as the related energy aspects are concerned, in a first evaluation it may be appropriate to make a quantitative analysis by making measurements of the entity of light which invests a certain surface.

Luminance is a measure of how much luminous flux is spread over a given area. One can think of luminous flux (measured in lumens) as a measure of the total "amount" of visible light present, and the luminance as a measure of the intensity of illumination on a surface. A given amount of light will illuminate a surface more dimly if it is spread over a larger area, so luminance is inversely proportional to area.

Fig. 5.9 Lux-meter for measuring illuminance (courtesy of Testo AG)

The instrument to be used for this purpose is the lux-meter that can measure the illuminance level expressed in *lux* (Fig. 5.9).

A lux-meter works by using a selenium photocell to capture luminous energy which is converted into the energy of an electric current, which is then measured by a pointer-type microammeter with scales calibrated in luxes (lx). Measuring this current allows the device to calculate the lux value of the light captured.

Different scales correspond to different ranges of the illumination being measured; scale changes are made by a switch that changes the resistance in the electric circuit. Higher illuminations may be measured by using a light-diffusing attachment on the photocell, which attenuates the incident radiation by a certain factor which is constant over a wide range of wavelengths.

The measure of the level of illumination can provide interesting information to the auditor not only about the luminous comfort but also about the efficiency of the lighting system.

The measurements will be taken at significant points of the illumination of the room, for example on the work surface in the case of an office building or a school.

Measurements must be taken at several points, considering both the effect of natural lighting, both of those particular situations in which shadows could be generated.

Portable data-loggers equipped with lux sensors are also available to record variation in light levels over time.

5.3 Instrumentation for Measuring Building Losses

The measure of the physical parameters of the building is done in order to verify its thermal performance. Knowledge of the thermo-physical characteristics of the envelope is necessary to build the model based upon which, through the use of the software, the energy performance of the base situation can be estimated, but is also useful in assessing the possible improvement action to be taken.

If the geometric characteristics of the building are obvious, then there are hardly ever difficulties in determining the thermo-physical characteristics of the envelope building structures. The purposes of this survey are:

- evaluation of the performance of the opaque envelope, then conductance of walls, roofs and basements and define a hypothetical stratigraphy of the same;
- detection of possible pathologies, and degradations in the walls, basements and roofs (e.g., plaster detachment, detachment of external insulation, fissures, moisture and moulds etc.);
- detection of thermal bridges;
- identification of the technical characteristics, and energy performance of windows;
- measurement of the air tightness of the building envelope.

The survey of the geometrical characteristics of the building, but also of plant components, requires simple instruments for example take measurements, such as the *meter*, the *measuring tape* and the *digital meter*.

The pictures acquired during the inspection are very important, not only because they allow us to document in an accurate way details that may be overlooked, but because they can be used for subsequent digital processing. The photograph of a facade, also complex, for example, processed with specific software "rectification" can be turned into a precise, to-scale survey of the facade.

Many of the functions required for the survey activities, for example photography, video or voice recording, are normally available on a standard Smartphone.

This paragraph does not deal with thermographic diagnosis since a discussion of this interesting topic will be presented in Chap. 6.

5.3.1 Endoscope

The endoscope is an instrument that allows one to "see" objects or situations that would otherwise be unobservable. It is constituted by a flexible cable, or sensor, made with optical fibres. These have both the function of carrying the light to the end of the cable in which is placed a lens, through which the image is transmitted to a display through which the technician handling the instrument can observe the object, or to a different, remote display through which the auditor can observe the object.

In its simplest shape, an endoscope has at its end, on the observer side, a lens: for the auditor the effect is somewhat similar to that of looking into a telescope. Other endoscopes have a small monitor, others are equipped with a PC interface that allows not only the viewing of images in a larger size but also to save them as image files, or short films, a standard for the most efficient appliances.

The section and the length of the sensor may vary depending on the model and the application: the diameter vary from a minimum of 4–12 mm or more. Of course, the greater is the diameter, the better is the image quality. Also with small diameters, minimally invasive, the instruments can give a good images. The sensor length can vary from one meter to more meters.

The applications of the endoscopic diagnosis, conceptually quite similar to that used in the medical field, are many:

- achievement of images of points that otherwise could not be reached without removing the equipment;
- cavity inspections;
- duct inspections;
- inspections of zones inside false ceilings or plenums.

So far as the performance of the building envelope is concerned, the endoscope allows investigation of the stratigraphy of a wall. Through a small hole, of 4–5 mm width, one can look inside the brickwork, to understand what is its real stratigraphy, if there is insulation or not, if there are air spaces.

It is not always easy and possible to make a hole in a wall, even when from the inside, but in all cases the author believes that, given its low cost and great flexibility, this small instrument should be part of the base instrumentation of an energy auditor.

Figure 5.10 shows two types of endoscopes: the one on the left is without display and must be directly connected to PC via a USB port. That on the right is an independent instrument and therefore has a display that allows viewing of images.

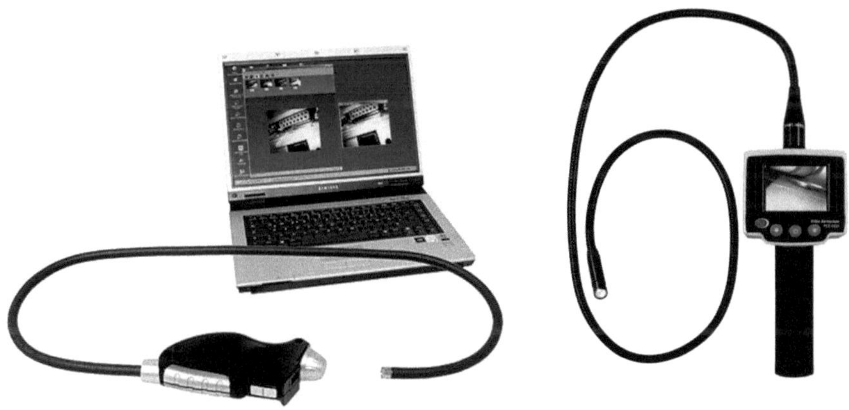

Fig. 5.10 Examples of endoscopes used for the energy audit (courtesy of PCE Instruments Co.)

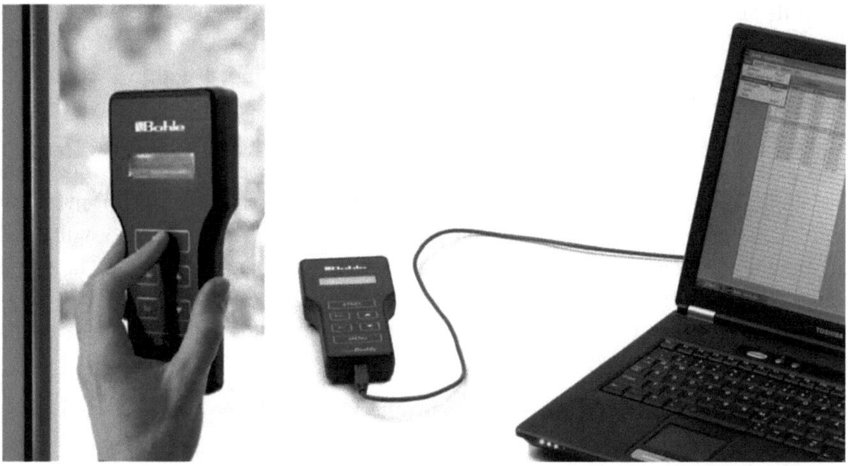

Fig. 5.11 Example of thickness gauge for glass (courtesy of *Glass Buddy—Bohle)*

5.3.2 Thickness Gauges for Panes of Glass

When a single glazed section of a window is detected, there are no particular problems, but if the subject of the check is a double or triple glazing unit (i.e., with double or triple panes) it is not so easy to identify the thickness of the panes of glass and the thickness of the air interspaces. Additionally it is difficult, to check the presence of surface treatment of the glass (e.g., low-emissivity coating, reflective coating etc.).

Instruments able to detect all of this information, of course without damaging the glass, have recently become available. They are called thickness gauges for glass, although this term does not do justice to their true potential. With an instrument such as that illustrated in Fig. 5.11, it is in fact possible to check:

- the thickness of the panes of glass;
- the thickness of the air interspaces between the panes of glass;
- the presence of a surface coating treatment;
- number, position and thickness of films.

The device also comes with software that allows to easily manage the data acquired through an interface. The software displays a list of all data, including date and time of measurement and the total thickness.

5.3.3 Heat Flux Meter

Energy audits of existing buildings often raise the problem of determining the value of transmittance or U-value of opaque envelopes of which the thermo-physical characteristics are not known. Direct analysis through holes is the method that guarantees the best results since it permits the evaluation of the correct stratigraphy of the various materials present. If this cannot be done and if one needs to know an accurate value for the transmittance, the heat flow meter is the more reliable instrument.

The complexity of the test method depends upon many factors:

- the heat flow within a wall is not unidirectional but has different components in the three directions of space;
- generally speaking, the element considered, for example a vertical wall or an horizontal structure is not isolated, but has a three-dimensional shape which may even be complex;
- the boundary conditions are not stationary but vary continuously, in particular temperature variations, humidity and air speed have an influence on the dishomogeneity;
- the element is not dry and isolated, but there may be abnormal phenomena such as evaporation or condensation (due either to alternating day night or to sudden weather changes such as rain), or abnormal conditions of irradiation;
- since the transient phenomena occur, the detection of the U value is an average and therefore requires an appropriate period of observation (measurement cannot be instantaneous).

If the object of the measurement is a building component, the following rules should be observed:

- the detection must be carried out on a portion of the wall as "average" as possible that is equidistant from other constructional elements (such as doors, windows, corners etc.) and representative of the entire wall, taken as a whole. A preliminary thermographic inspection is useful to identify the best position for the measurement;
- detection must be done in order to avoid thermal bridges and correspondence with their zones of influence;
- detection should be continued for a reasonable amount of time (minimum 72 h) and in conditions of the presence of heat flow (the difference in temperature between the inner and outer surfaces should never fall below 10–15 °C);
- the wall should not be subject to direct sunlight, rain or snow during the reporting period (for north latitude locations, north wall should be preferred, together with the placement of protective screens).

The instrument consists essentially of a data-logger and thermocouples (plates that detect the heat flow through an opaque building component) (Fig. 5.12). For the measurement of heat flow, there are different types of plates of different sizes, materials and sensitivity, depending upon the application required. The coupled plates (one internal and one external) need to be fixed on the surface layer of the wall whose the transmittance is to be measured and then left in place for some days.

The detection time varies depending upon the stability of the external climatic conditions: the smaller the changes in temperature, the higher is the reliability of

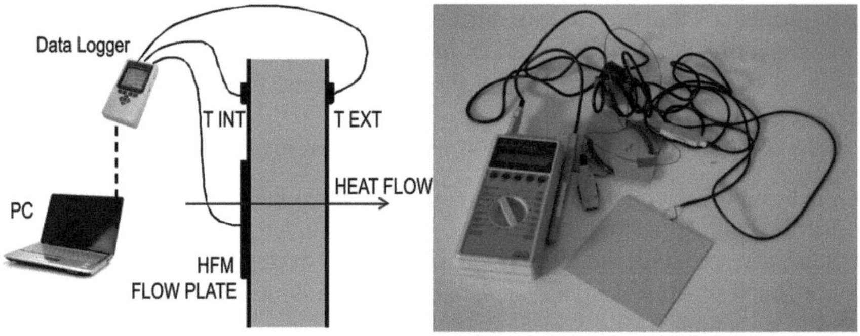

Fig. 5.12 Installation scheme of a heat flow meter and example of a flow meter

the result. Obviously, the optimal measurements are obtained during the winter period, when the temperature difference inside/outside is greater and hence the heat flow is greater.

The quality of the software supplied with the heat flow meter is an important element in obtaining values as close as possible to reality. To obtain a correct measurement of the transmittance it is necessary to apply the sensors in a portion of the surface which is representative of the wall. The instrument must be used by experienced operators who are able to understand building structures.

The results obtained with the use of a heat flow meter are generally good, and the data given in the bibliography place the value of the errors between 1 and 15 %, with an average value of about 8 %. The measurement error is greater, the lower is the thermal resistance of the wall concerned (in Fig. 5.12 the plate is indicated with HFM).

The Fig. 5.13 shows the display of the processing of the information acquired by a heat flow meter from which it can be observed that in the first phase fluctuations of data relating to the conductance are considerable (an indicator that the measurements are unreliable) whereas after a certain period of time they stabilise.

The conductance value determined by the method of heat flow meters is not that calculated in the design phase from the known thickness and thermal conductivity of the materials, but rather that value really expressed by the structure in place, in a given *context and in a given period.*

The heat flux meter provides reliable values on the thermo-physical characteristics of the building structures and can then be used for audit purposes. However, owing to the constraints that arise in the application phase, needed to obtain reliable measurements, this instrument is not always used in a audit process.

5.3.4 Blower Door Test

The air tightness of the envelopes of buildings is very important because the vacuum generated between the environment inside and outside can generate

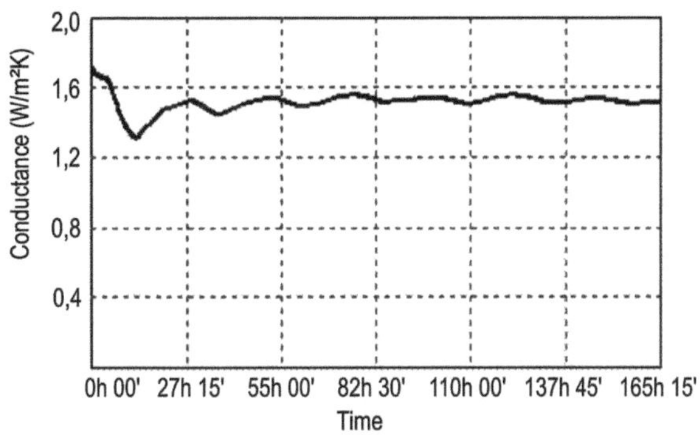

Fig. 5.13 Processing of information acquired by a heat flow meter

uncontrolled air flows. Air infiltration usually occurs as a result of unsealed windows and in particular at joints between wall and window frame, through the external roller-blind fitment-box or through joints between the actual walls.

When the infiltration is evident, chemical smoke or infrared survey may help to highlight the problems. For a complete verification of the seal of the housing, however, the blower door test can be used, a test that mechanically, using a fan, places the internal environment under pressure.

It seems useful to discuss this opportunity, not just to persuade the auditor to adopt systematically this instrumentation, but also with the aim of informing him about this test, that is becoming widespread. If necessary, the auditor may request this service in outsourcing.

The blower door test is used by some protocols, such as Passive-house, to complete the testing of high performance buildings, and avoid any unwanted infiltration.

The blower door test assesses the degree of air tightness of the building through the measurement of the flow of air exchanged due to infiltration, mechanically generating a pressure difference. If the test is performed during the construction phase of the building's potential, weaknesses can be solved readily and more effectively.

The complete test is performed in the following steps:

1. In the first step a pressure difference between the inside and the outside constant of 50 Pa is created and maintained; during this stage the entire surface of the building is inspected in order to seek out non-hermetic points that may cause the greatest thermal losses for infiltration;
2. In the second step a decreasing under-pressure is generated, starting from values of approximately 70 Pa and proceed in steps of 5 Pa to reach a final value of 25 Pa. For each step, volumes of air which are lost through the points of permeability are recorded and the index amount of air penetrated in one hour (n50) is calculated;
3. In the third phase overpressure is created and the sequence is repeated as per the previous phase. This is to evaluate also the dispersal due to the different seals. The final result of the test represents the number of air changes of internal air through the envelope, the sum of which defines the total loss.

5.4 Instrumentation for Measuring the Performance of Mechanical Systems

During the field survey it is often necessary to perform instrumental investigations on HVAC systems, DHV systems and mechanical plant in general, in order to verify whether they are functioning properly and to acquire any missing information necessary for the preparation of a proper energy balance.

Some of the instrumentation which has been described in previous sections can also be used to detect values, take measurements and acquire information concerning the plants, namely (but not exclusively):

- the *thermometer* which allows one to detect, using the appropriate sensors, the temperatures of the fluids in question (for example, the temperature of the air supplied from a grille or a diffuser or the temperature of the water exiting from a tap);
- the *infrared thermometer* which permits detection of the surface temperature and, for example, verification of the adequacy of the thermal insulation of a duct, a pipe or a storage system;
- the *endoscope* that can be used to effect the inspection of a part of the plant that would otherwise be unachievable;
- the *anemometer*, with which we can measure the air velocity through a duct or the air velocity at the discharge from a grille or a diffuser (in these cases knowing the cross-sectional area it is possible to derive the flow rate).

Amongst the instruments that the auditor can use for plant inspections one can also consider the infrared camera: the portable compact models are very convenient for the inspection of plant parts or components (the topic elating infrared audit will be discussed in Chap. 6).

In this section other tools which can be used to detect, directly or indirectly, the energy performance of a system or affect its operational features will be discussed. An energy auditor does not necessarily has to acquire all the equipment described here, since for special investigations it is more convenient to refer to external specialists (under sub-contract).

5.4.1 Combustion Analysers

The analysis of the combustion is made to verify whether or not combustion in a boiler occurs properly. In order to maximise the combustion efficiency it is important to know the composition of the flue gas: only by obtaining a good fuel–air ratio is possible to assure the maximum performance and save energy.

The combustion testing is made using *combustion analysers*, instruments capable of determining the concentrations of the combustion products in the exhaust gas. The constituent compounds usually considered are: carbon dioxide

(CO_2), carbon monoxide (CO), oxides of nitrogen (NOx), not however required for the performance test. Oxygen (O_2) is also tested for in order to ensure adequate excess air levels.

This test also requires the evaluation of other parameters, such as the temperature of the fumes and the temperature of the environment (surrounding air).

The knowledge of smoke temperature, the type of fuel and CO_2 or CO concentration allows one to derive, using a chart, the combustion performance. A combustion is usually performing well if the CO_2 and O_2 concentrations are low and there is little or no trace of CO.

Modern combustion analysers are entirely electronic: electrochemical cells are used for the measurement of the concentration of various gases in the flue gas, and there are sensors for temperature and pressure (Fig. 5.14).

To perform the test it is simply necessary to insert the sensor of the analyser into the flue and start the measurement, by means of a pump: the sample will be automatically engaged and analysed.

The instrument directly processes the data acquired by providing the results of the analysis: Stack losses, excess air, CO_2 and NO_X.

From the values so obtained it is determined whether the combustion takes place correctly and if the combustion efficiency of the boiler is in compliance with design values, otherwise it will proceed with the calibration of the burner and the cleaning of the hearth and the heat exchange surfaces of the boiler.

The electrochemical cells need to be replaced after a certain number of uses and temperature sensors and pressure recalibrated as required by the manufacturer.

It is important to remember that the combustion analyser occurs as the combustion takes place and therefore does not allow to determine the overall performance of boilers, intended as the ratio between the energy delivered to the heat transfer fluid and energy to hearth. In order to estimate the overall performance of a boiler it is necessary to know also the losses in the chimney and the losses through the shell of the boiler.

5.4.2 Flow Meters

The flow meters used to determine the mass flow rate or the volume flow rate of a fluid in a pipe or in a duct, typically do so by measuring its speed. Portable instruments utilise measurement techniques (e.g., ultrasonic) that allow measurement from the outside of the pipe, so no interference with the functionality of the system (Fig. 5.15).

The measurement of the velocity of the fluid must take account of a series of items of information that the auditor can provide to the instrument before the measurement:

- type of fluid;
- temperature;

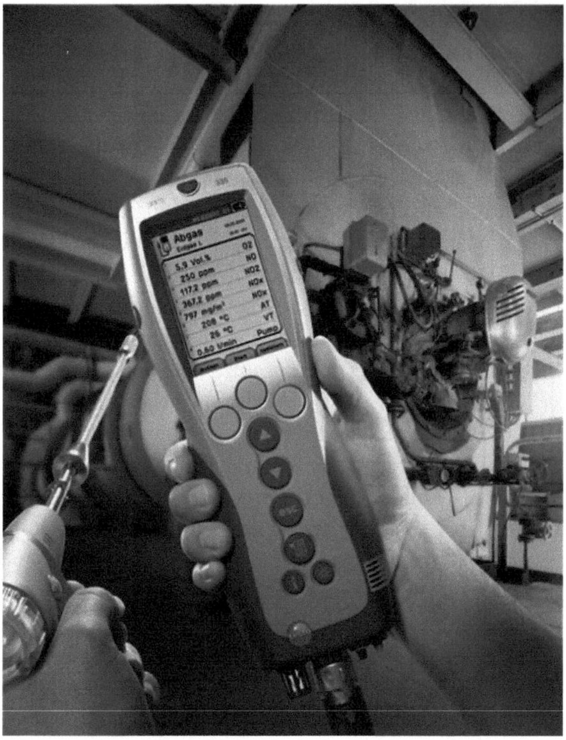

Fig. 5.14 Combustion analyser (courtesy of TESTO AG)

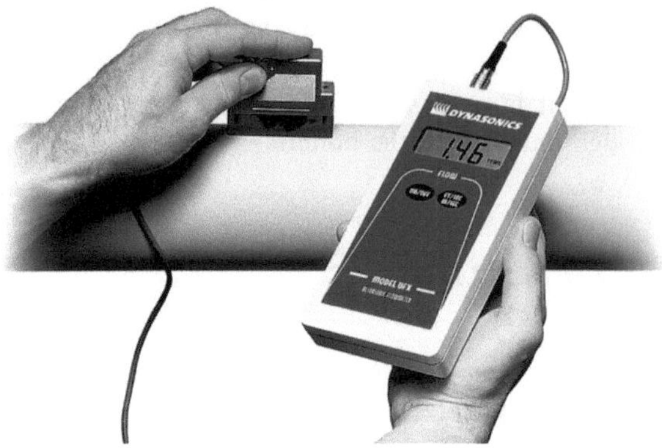

Fig. 5.15 Ultrasonic flow meter (courtesy of Dynasonic Co.).4.3 Chemical smoke

- material and thickness of the pipe;
- diameter or section of the pipe.

Once the value of the cross-sectional area of the pipe is known, on the basis of the speed detected by the instrument, it is possible to evaluate the fluid flow.

Chemical smoke generators can be used in building audits to highlight the presence of air infiltration and verify, in the HVAC systems, the effective air distribution in order to ensure uniformity of temperature and efficient air renewal. Chemical smokes can be used to identify which are the supply and which are the return air grilles, and to check distribution losses from air ducts.

There are to be found on the market portable chemical smoke generators, the size of a shoebox, which can produce up to 200 m^3 of smoke per minute. To operate, these devices require electrical power and charging with a specific liquid, usually based on glycol.

Other models, for specific users can be powered by gas, use oil-based fluids, carbon dioxide (CO_2) or nitrogen (N_2).

The smoke is dry, non-corrosive, non-toxic and leaves no residue, exposure to it however should be limited because it can irritate, albeit temporarily, the respiratory system.

5.5 Instrumentation for Measuring Performance of Electrical Systems

In electrical systems it is possible to detect the consumption of *active energy* directly from the meter that, in the case in which the contract provides for this service, is also able to measure the *reactive energy* which will be discussed later.

The meter, however, returns the value of total consumption, whereas for energy audit purpose is necessary to understand how energy is used for different uses (e.g., lighting, internal loads, HVAC etc.) or in different areas of the building.

The instruments presented here have the purpose of informing the auditor about the maximum power delivered through the electricity cables which supply selected users.

The survey on the electrical systems requires experience and expertise: if it is made by inexperienced personnel it could cause serious damage both to persons and either to the plant or to equipment served.

If the energy auditor does not have experience and expertise in the electricity sector, or is not authorised to act personally, for this type of analysis the best choice is to involve an experienced colleague, ideally already on the technical staff which supports the management and maintenance of the building.

The instruments presented here allow one to perform measurements which are instantaneous or in any case over a limited period of time that requires the presence of the operator. Even for the electrical measurements it is possible to monitor, with apparatus that, whilst functioning on the same principle, is equipped with a special internal memory. Measuring systems based on data-loggers can be equipped with sensors for electrical measurements: this item will be discussed in Sect. 5.9.2.

5.5.1 Electric Energy Meters for Individual Equipment

These meters for electric energy are simple to use and require no special skills. For these reasons they can be used by end-users who want to gain greater knowledge about their energy consumption and energy management.

Their installation is simple, just inserting them into the electrical socket and then connecting to them the electric equipment. The display, with which it is possible to interact by means of buttons, shows the following values:

- instantaneous voltage (Volts);
- current intensity (Amperes);
- instantaneous power absorbed by the equipment connected (W);
- power factor ($\cos \phi$) (not for all models).

The devices provide the value of electrical energy consumed by the equipment, expressed in Wh or kWh, starting from the moment in which, by acting on the buttons, the accounting record was launched.

If the cost per kWh of the electricity consumed is entered, the display provides the user directly the economic value of the energy consumed. The market offers a wide range of devices of this type, some even at low prices. The more sophisticated models, such as that shown in Fig. 5.16, permit the recording of the values on a memory card that can then be removed for the transfer of data on to a standard PC.

5.5.2 Electrical Network Analysers

The apparatus illustrated in the above paragraph may be used to measure and do the accounting of electric energy absorbed by a single device. A complete audit of electric systems requires more sophisticated and professional instrumentation, in order to measure the performance of sections of plants, for example lighting or HVAC, up to the performance of the whole electric system.

The intensity of electric current can be measured using an instrument which is called *ammeter*. Once the voltage is known, the electricity power can be calculated by multiplying the two values.

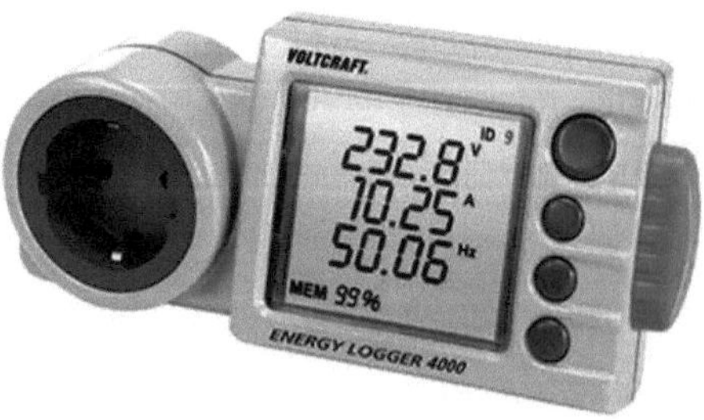

Fig. 5.16 Electric energy meter for individual equipment, performed to record information acquired (courtesy of Voltcraft Co.)

The measurement made in this way, however, allows the calculation of what is called *apparent power* which does not correspond to the power transported (to the user devices). To achieve this latter value, which is what is needed for energy accounting purposes, we must multiply the value of apparent power for the power factor $(\cos \phi)$.

The value of the voltage itself, on the other hand, is subject to variations and, therefore, the use ammeter is not suitable for applications related to the energy audit. The correct measurement of the electrical power delivered by an electrical cable, in W or kW, thus depends on the following variables:

- instantaneous voltage (Volts);
- intensity of current (Amperes);
- power factor $(\cos \phi)$.

Voltage and intensity of current can be measured with an instrument called the *multimeter*. The most suitable instruments to perform these type of measurements, which is what an energy auditor needs, are devices that can detect all of these three variables simultaneously.

These devices exist and are called *electrical network analysers*. Modern versions, such as that shown in Fig. 5.17, are lightweight, portable and permit the detection of many other variables.

Using these tools, however, requires specialist skills in electrical engineering. The measurement, however, is not straightforward, since it is necessary to isolate a single conductor, for bi-phase current, or all three conductors, for three simultaneous measurements, for three-phase current supply.

Before performing the measurements the right cable must be found. It is therefore necessary, before starting the survey, to have to hand the plant layout and, more generally, a knowledge of the geometry of the electrical system.

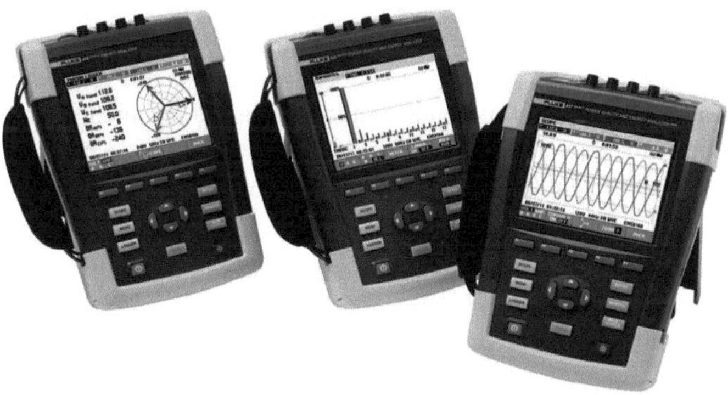

Fig. 5.17 Electrical network analyser (Courtesy of Fluke Co.)

5.6 The Building Envelope Audit

The field survey must be coordinated and programmed in advance (see checklist *CL 1.3—Plan activities*). The time dedicated to its correct organisation phase is largely compensated for by the rapidity with which the surveys can then be carried out. Before starting the survey is important to handle the following issues:

- identify a contact person (building manager, maintenance manager, facility manager etc.) with whom to schedule the date and time of the inspection;
- ensure that the contact person is authorised to have access to all of the most significant locations and areas, particularly to those containing technical installations;
- if the survey covers buildings for collective use or buildings which are strategic or with restricted areas (e.g., schools, hospitals, barracks, airports, government etc.), specific authorisation issued by the competent authorities must be obtained in advance;
- the compilation of the cards must be completed in a short time: in fact, only in this way is the possibility of some oversight minimised. It is nonetheless useful to keep in touch with the reference contact person: a missing item of information may then simply be requested by telephone, thus avoiding further inspections;
- if the survey is performed by auditors with differing skills (e.g., an expert in HVAC systems and an expert in building design and construction) it is appropriate that the lapse of time between the two inspections be minimised. As far as possible it is convenient to concentrate the requests into a single day, in order to create only a minimum of disruption to the client.

The building survey could be a rather complex task. The following aspects should taken into account:

- if the shape of the building is particularly articulated, the audit of the envelope may take a long time;
- the presence of surrounding buildings and other structures should be checked as these could generate shadows;
- concerning the thermo-physical characteristics of the dispersant structures of the envelope (stratigraphy of the walls) it is almost never possible to perform destructive tests: a good photographic documentation could assist the auditor in seeking out an answer on these characteristics in the technical literature;
- the building envelope audit is mostly done outside and the weather (rain, cold, wind etc.) could make the task more difficult by increasing the time required.

In order to facilitate the audit work, a set of checklists to be employed directly in the field have been prepared. Their use can reduce dramatically the survey time and help to make more efficient data collection.

For the building envelope the following checklists are presented:

CL 1.1—General data;
CL 4.0—Geometrical characteristics of the building;

CL 4.1—Photographic survey;
CL 4.2—Envelope-opaque walls;
CL 4.3—Envelope-doors and windows;
CL 4.4—Envelope-roofs and basements.

All checklists should be compiled prior to the survey, in as far as possible, using the information (technical documentation) that the client has provided to the auditor. This step, besides reducing the survey time, helps to organise the survey properly.

5.6.1 Geometrical Characteristics of the Building

For each building, or each part of a building, a planimetric scheme should be defined which indicates the different thermal zones or the zones with different intended use.

The facades of the envelope, as well as the roofs and basements, should be numbered sequentially. In the layout the North must be specified so that it is possible to determine the orientation of each facade. In the audit form is also useful to note the position of the central heating plant and the main technical equipment rooms.

A general map of the building complex, obtained from photogrammetric maps, if possible, or from Google Earth® or similar websites, could facilitate the auditor in the survey activity.

5.6.2 Photographic Survey

The photographic survey is an important step and must be carried out properly in order to avoid the need to return subsequently to the site for missing data or information. Using a digital camera it is possible to make small videos both from the outside and from the inside of the building. Videos and photos will then be filed in an orderly way.

Wherever the auditor considers it necessary to use the software to facilitate straightening of images (Fig. 5.18), photos of each facade will be made, limiting the possible distortion effects.

The straightening of photographic images finds its preferred application on relief of objects that are related by a single flat surface, such as building facades or structures that are developed in a plane. The elements of which are located in the chosen plane will be identified and returned with the greatest possible precision, whereas those that are not in the plane will be progressively less accurate the further they are distant from that plane. This will be affected by an error which is the sum of two terms, and a one-dimensional position (parallax error).

Fig. 5.18 Straightening process of a digital still picture (Courtesy of Cadlandia.com)

The accuracy achievable in the measurements performed on the plane that is chosen for the straightening can vary within very wide limits, depending upon several parameters:

- geometry of the shooting (the camera's distance from the object, camera angle);
- quality of the scanned image (in the case of a digital camera: lens quality, sensor and, if it is a video camera, lens quality, video, photographic printing and scanner), use or not of the tripod for shooting;
- support measures (number, arrangement and precision).

Operating properly, even if only using a standard digital camera, with this methodology one reaches an accuracy around 0.1 %. Even in a more expeditious rendering a precision of around 1–2 % can be expected.

5.6.3 Building Envelope, Opaque Walls

The availability of floor plans and elevations of the building facades greatly facilitates the survey phase of the opaque walls. However in some cases drawings supplied by the customer appear in accurate or have not been updated. The auditor's responsibility is to verify the correctness of the information (if the quality of the drawings is good then a few checks are enough to determine whether the graphic documentation is reliable).

A typical set of measuring instruments comprises double meter, strip meter standard disk, 20 m tape-measure, 50 m tape-measure and digital meter (Fig. 5.19).

To facilitate the survey of geometric and typological characteristics of the envelope, checklist CL 4.2 should be completed for each zone.

Table 5.2 shows the table for collecting the data of building characteristics that are:

- progressive number of the facade;
- orientation of the facade (N, S, E, W etc.);
- code of facade;
- average thickness of the wall;
- base size of the facade;
- height of the facade;
- gross area (including windows);
- code of boundary;
- type of windows.

The coding of the façade may be referred to a base abacus, for example standardised walls types e.g., ASHRAE [3], or to an abacus prepared previously by the auditor on the basis of the documentation supplied by the client.

The code of boundary refers to a code list on which is indicated the environment on the external side of the wall (for example not heated or not cooled spaces,

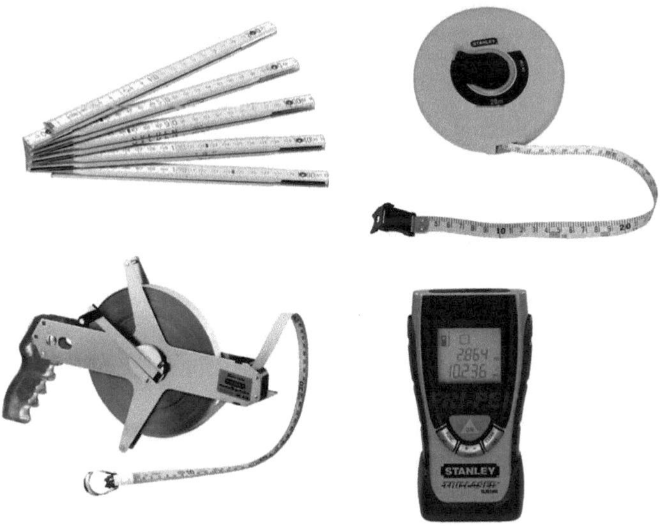

Fig. 5.19 Typical set of measuring instruments (courtesy of Stanley Co.)

Table 5.2 Table for collecting data of building envelope characteristics

Zone No. _______

No.	Orientation	Wall Code	Thickness (m)	B (m)	H (m)	Area (m^2)	Boundary Code	Type of Windows (*)						
								1	2	3	4	5	6	7

spaces maintained to a different air temperature, unheated or uncooled stairwells, unheated or uncooled attics, cellars etc.).

Table 5.2 makes provision for up to a maximum of 7 different types of windows for each facade, each of these will then refer to an abacus schedule contained in checklist 4.3. Each of the boxes should be shown the number of windows of a given type referring to each of the 7 categories.

The table is structured with the goal of minimising the activities on-site, since for each facade the dimensions of gross area will be entered whereas the actual calculations that return the net values of the facade areas should be performed, by subtraction, back in the office.

The characteristics of the U values of the walls of existing buildings can be obtained in several ways:

- on the basis of the technical documentation supplied by the client;
- conducting an invasive survey (for example by making a hole and using a endoscope);
- using a heat flow meter;
- referring to databases contained in technical literature.

5.6.4 Building Envelope, Doors and Windows

Checklist CL 4.3 shows the schedule of individual types of windows (and doors) and, for each of them, all the information, as seen in the diagram of table 5.3. The schematic defines a drawing of the window and synthetic samples to identify the number of wings and glass panes.

The schedule additionally collects information regarding the characteristics of the external roller-blind fitment-box, the presence or absence of internal or external shading systems, the air tightness and the degree of maintenance.

The compilation of the abacus of the windows, greatly facilitates any replacement, repair or renovation since all the information necessary to request a detailed quotation has already been provided therein.

5.6.5 Building Envelope, Roofs and Basements

Even for the acquisition of data relative to shell and base a data scheme (CL 4.4) has been drawn-up which can be inserted in both the geometrical data thereof and the thermo-physical characteristics by reference to the codes, analogously to the system discussed for walls and windows.

5.7 Audit of Mechanical Facilities

The checklist of mechanical facilities is sufficiently self-explanatory and their use does not involve particular difficulties. However the following aspects should be emphasised:

- compilation of the checklist can be difficult if the equipment examined has data-plates with incomplete information or, as often happens, their data-plates are totally lacking. For such equipment the auditor can still collect all the information (usually the brand name and the model no.) consulting technical catalogues or consulting the manufacturer or the contractors who installed the unit;
- some important information, especially that concerning parts of plant that are not visible, can be obtained by interviewing the plant operator or by contacting the facility manager.

Audit of mechanical facilities is carried out partly inside the technical plant rooms and partly inside the living spaces of the building. The logical approach for the compilation of the checklist is as follows:

1. understand how the system works;
2. identify the circuits that are fed by the boilers or the chillers, then associate them to the various areas or zones of the building;
3. detect thermal substations and circuits which they supply;
4. identify the logic used by the control system;
5. identify the criteria by which the distribution of thermal vectors is effected (passages of the pipes or ducts, etc.);
6. detect the type of terminals used in the different air conditioned areas of the building;
7. detect the technical specifications of major plant components.

For the survey of mechanical facilities the following checklists have been provided:

Table 5.3 Table for collecting data of windows and door characteristics

Zone No. ________

Window Code ________

Scheme			
		Glass type	
		Window type	
		External shading	
		Internal shading	
		Air tightness	☐ good ☐ average ☐ poor
		Maintenance	☐ good ☐ average ☐ poor
		h. shutter box (m)	
Base (m)	Height (m)	h. sub window (m)	
Area (m^2)		NOTES	

- CL 5.0—General data on the heating system;
- CL 5.1—Schematic layout of the HVAC system;
- CL 5.2—Boilers;
- CL 5.3—Heat pumps;
- CL 5.4—Heat exchangers;
- CL 5.5—DHW production;
- CL 5.6—Thermal zone heating systems;
- CL 5.7—Chillers;
- CL 5.8—Air handling units;
- CL 5.13—Thermal solar system;
- CL 7.0—Instrument survey of the heating systems.

The sketch of the functional diagram can be very useful to understand the logic of production and distribution of thermal vectors to different users. To reduce the inspection time, nameplate data of major equipment items (e.g., boilers, pumps, chillers etc.) can be reported directly on the diagram and then later inserted in the checklist.

5.8 Audit of Electrical Facilities

The electricity consumption is in some cases, such as that of commercial buildings or offices, a significant proportion of total energy consumption. If the aim is to reduce the amount of primary energy, in accounting for the electricity we have to apply the coefficients of production (electric energy efficiency) i.e., those coefficients that take into account the fact that to produce 1 kWh of electricity 2.5 times the equivalent energy from fossil fuels is consumed.

The electricity user-items are also those that offer great margins for easy-to-realise retrofit that allow one to achieve some interesting benefits in terms of investment needs. These are just some of the reasons to carefully consider the electrical side of the audit.

The audit of the electrical systems will consider the following aspects:

- verification of lighting systems;
- verification of the load factor and energy demand;
- verification of the conditions of electric motors;
- evaluation of the electrical loads.

Audit of electrical facilities is carried out partly inside the technical plant and partly inside the living spaces of the building. The logical approach for the compilation of the checklist is as follows:

1. understand how the system works;
2. identify the circuits that feed the various users;
3. identify on the site plan the location of main electrical substation and any subsidiary substations;
4. define by detection the specifications of major plant components.

For the survey of electrical facilities the following checklists have been provided:

CL 5.9—Electrical-General;
CL 5.10—Electrical-lighting;
CL 5.11—Electrical-appliances;
CL 5.12—Electrical-motors;
CL 5.14—Solar PV system;
CL 5.15—CHP (Combined heat and power).

The auditor must also detect the presence of user-equipment (e.g., computers, printers, monitors, photocopiers etc.) for which it is possible to prepare an adjustment or replacement schedule.

In those cases where the lighting systems are not very efficient, the survey should have a more detailed approach (for example through the identification of all the lighting equipment and associated lamps) in order to provide information useful for taking retrofit action.

An important area in which it is possible to improve energy efficiency is that of the electric motors which can be replaced with more efficient models, obtaining a considerable improvement in performance.

5.9 Monitoring of Indoor Environment and Facilities

The Energy Audit is very useful to see what happens over time. Some parameters such as air temperature and relative humidity can be monitored for a certain period using small, portable and inexpensive devices called mini data-loggers.

These devices have typically the following characteristics:

- small dimensions (on average they can fit in the palm of a hand);
- integration into a single unit of the sensor (or sensors), of the transducer and of the data acquisition system;
- integrated clock;
- battery;
- programmable periods of data acquisition;
- interface to transfer data to an external personal computer.

Mini data-loggers are sold with a software thanks to which one can make a schedule, acquire the stored information and perform calculations. The frequency of the measurements may vary, dependent upon the specific needs, from one second to several hours, hence their flexibility is high.

Some systems are equipped with concentrators and interfaces that allow transfer of the information acquired remotely through the normal mobile phone network (GPRS, UMTS).

The market offers a fairly wide range of data-loggers with different performance levels and a diversity of costs. The cheapest ones, used in the food industry, are limited in performance: they can usually only detect one measurement (temperature), and it is not possible to replace the battery.

The costs of the most professional data-logger, however, is not high: between € 100 and € 200, dependant upon the performance.

The most recent data-logger models also have the ability to "communicate" with each other using wireless, so it is possible to create real acquisition networks which are space-saving, economical and reliable.

5.9.1 Monitoring of Environmental Conditions

Environmental parameters that can be monitored using mini data-loggers are:

- air temperature;
- relative humidity;
- illuminance;
- CO_2 (parts per million).

Some devices detect a single parameter while others detect several variables simultaneously. Others are equipped with inputs that can be connected to external sensors, thus ensuring the systems great flexibility of application.

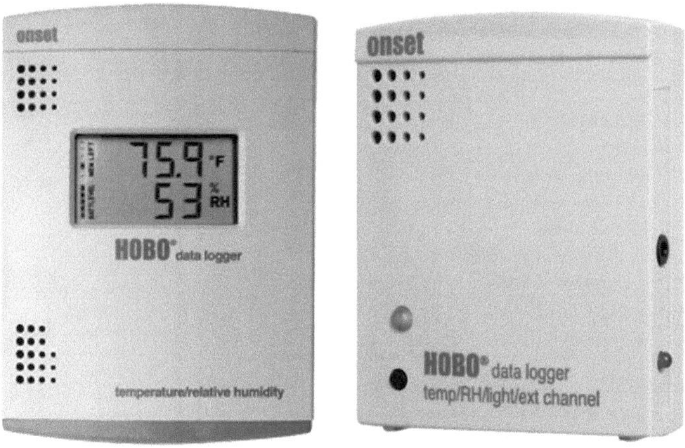

Fig. 5.20 Data-logger for monitoring of environmental conditions (Photo courtesy of Onset)

Figure 5.20 shows two models of data-logger. The one on the left, with a display, simultaneously acquires data on the temperature and relative humidity. The right hand unit is not equipped with a display but acquires data on temperature, relative humidity and luminance.

It is also designed for connection to external sensors and could then, for example, be connected to a CO_2 detector (see Fig. 5.21) thus providing a more complete assessment of the parameters of comfort and environmental quality.

The acquisition software allows the auditor to programme the entire monitoring phase. Once the data-logger is connected to the PC can be defined the start time and end of the monitoring campaign and the acquisition intervals.

The software can read the charge of the battery and, therefore, allows verification of the maximum duration of the acquisition period which is of course longer if the scans are less frequent. Once the logger is activated it becomes an entirely independent unit.

Fig. 5.21 Connection scheme between a CO_2 sensor and a data-logger for monitoring (Photo courtesy of Onset)

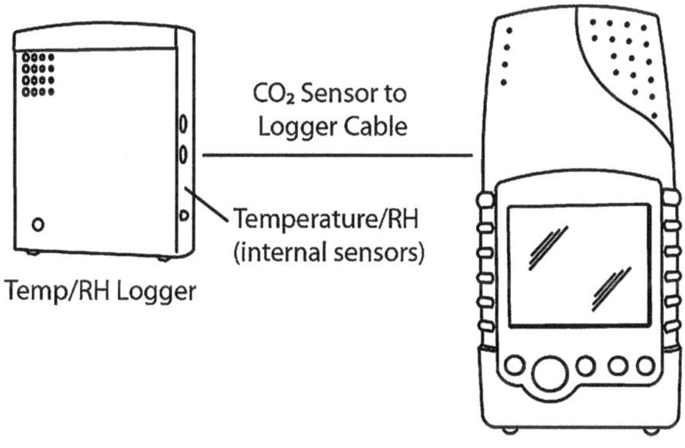

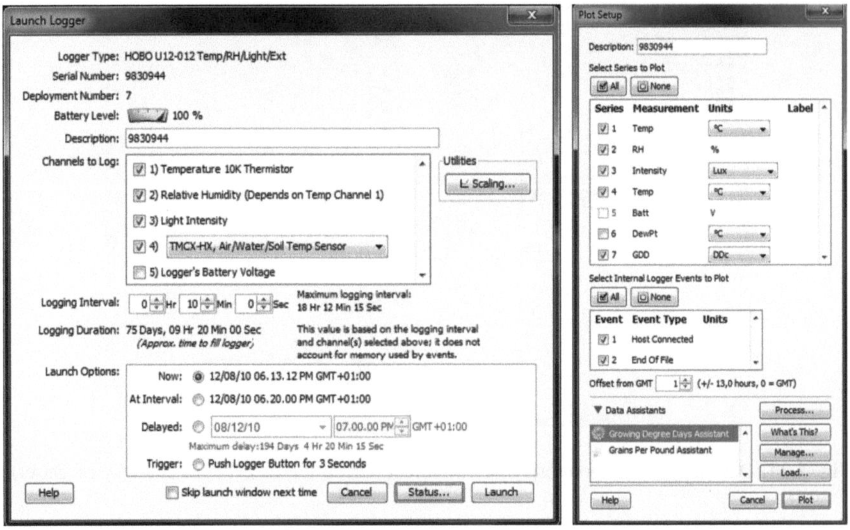

Fig. 5.22 Graphic interface of a software for data acquisition (software interface of Onset)

The software is also used for data downloading. The data can be saved in standard formats as plain text files, to then be processed by external software or can be displayed via the graphical interface of the same acquisition programme (Fig. 5.22).

Monitoring allows detection of phenomena that with instantaneous measurements could not emerge. Figure 5.23, for example, shows the monitoring of two indoor environmental parameters: relative humidity (upper curve) and dry bulb temperature (lower curve), for a period of three days.

The variation of the internal temperature is controlled by a heating system (the terminals in this case are radiators) and during the night the temperature is lowered. The relative humidity is influenced partially by the variation of temperature (at the same conditions, thus without the variation of the specific humidity decreasing the air temperature, the relative humidity increases) but in part by the amount water vapor emitted by the persons.

5.9.2 Monitoring of Electrical Facilities

Electrical quantities can also be monitored using portable data-loggers, which however must be connected to specific sensors. Figure 5.24 shows two sensors, the sensor on the left detects the voltage (volts) whilst the sensor on the right detects current intensity (amperes). The sensors must be connected to a transducer that, processing the information, calculates the electricity in the cable which is being monitored (Fig. 5.25). The transducer is connected to the data-logger.

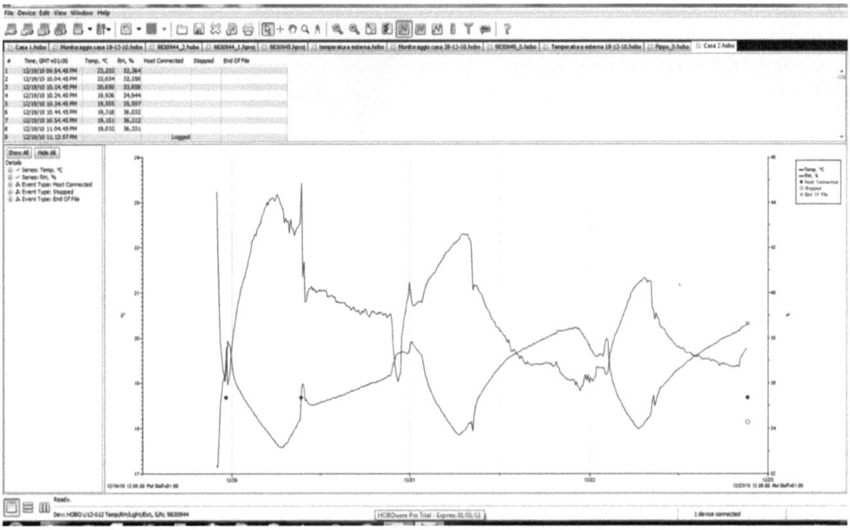

Fig. 5.23 Monitoring of environmental parameters, dry bulb temperature and relative humidity, for a period of 3 days (software interface of Onset)

Fig. 5.24 Sensors for monitoring of single-phase electric user-equipment (Photo courtesy of Onset)

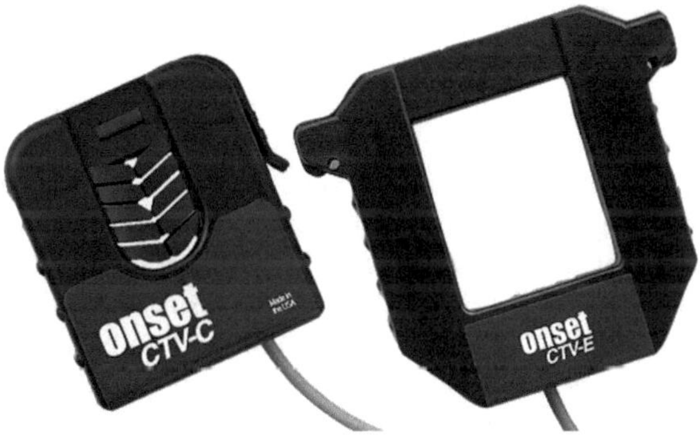

Fig. 5.25 Transducer for the detection of the electricity (Photo courtesy of Onset)

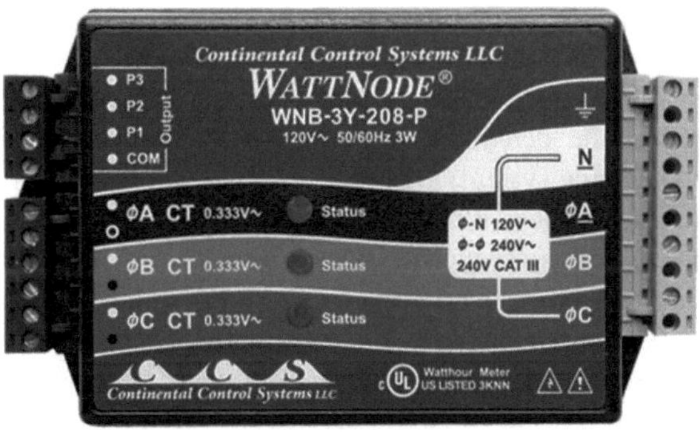

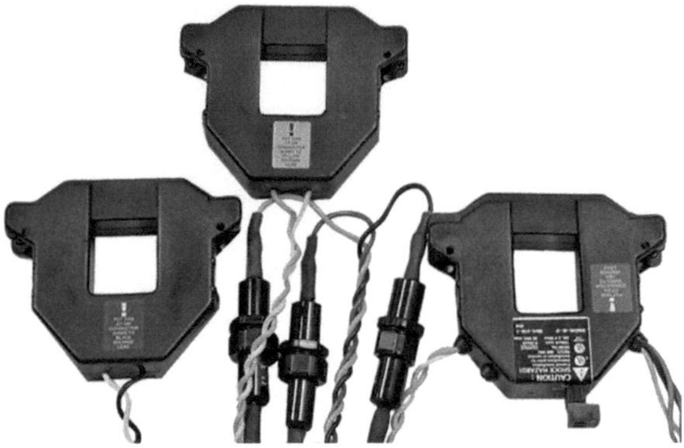

Fig. 5.26 Sensors for monitoring of electric users three-phase (Photo courtesy of Onset)

If the current is single phase then it is possible to monitor only one cable, whereas in the case in which the current is three-phase all three cables should be monitored. The complete kit is shown in Fig. 5.26. Belonging to these monitoring systems there are also sensors of pulses that are used to verify when an electrical device is operating or not.

5.10 Energy Signature

The energy signature of the building is a method of assessment in which the heat consumption is correlated with climatic variables. It describes the real behaviour of the building by means of a graphical representation of energy consumption (heating, cooling, hot water etc.) or of power consumption as a function of an external parameter, typically the temperature. It is used to visualise the actual consumption of a building, to compare the calculated values with the actual consumption or to verify the validity of an energy retrofit measure.

The method proposed in Annex B of EN 15603 [4] assumes that the internal temperature is constant and that the outdoor temperature is the parameter that most affects the energy behaviour of the building. The method is therefore not suitable for buildings characterized by heat gains, internal variables or by contributions from passive solar systems.

To apply this method it is firstly necessary to collect data on energy consumption and outside air temperature at regular intervals over a certain period. When the detection of data is performed automatically (e.g., by means of a monitoring system) intervals of one hour are sufficient, while in the case where the data are collected manually it is appropriate to define a longer time, for example a week, so as to reduce the possibility of errors. In this case, the external temperature data can be recovered from those available from a weather station.

The parameter used for the graphical representation of energy consumption is the so-called *average power*, obtained by dividing the thermal equivalent of the energy consumption expressed in kWh by the duration of the period in hours.

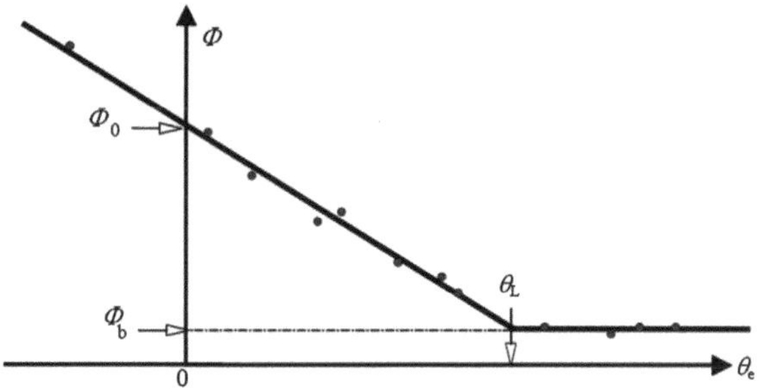

Fig. 5.27 Example of energy signature: the graph constructed on the basis of actual consumption and outside temperature of a heating system (*Source* EN 15603)

The values of the average power of the plant and the average outside temperature during the period are then plotted on a graph.

The diagram shown in Fig. 5.27 was constructed by applying this procedure to assess the energy performance of a building in the heating season: each of the points represented refers to a measurement campaign within the same season.

The diagonal solid line represents the linear regression of these points during the period in which the plant actually provides heat to the building in question. The portion of the horizontal line, instead, represents the *base power* of the plant, that is the minimum power required independent of the outside temperature.

If the system is dedicated exclusively to space heating, the base power corresponds to the system losses during the period when it is not active. If instead the system is also used for DHW, the base power represents the power required for the production of domestic hot water.

In Fig. 5.27 the symbols have the following meaning:

Φ is the average power in the interval between two measurements;
Φ_0 is the power corresponding to $\theta e = 0$ °C;
Φ_b is the base power;
θ_L is the outside temperature limit for the ignition of the heating system;
θ_e is the average outside temperature between two successive detections.

The line drawn during the heating season (or conditioning) is characterised by a power Φ_0 at 0 °C and followed by a gradient H:

$$\Phi = \Phi_0 - H \cdot \theta_e \tag{5.1}$$

where:

Φ is the average power;
θ_e is the average outside temperatures
H is obtained from the relationship $(\Phi_0 - \Phi_b)/\theta_L$

The slope H represents the reactivity of the building compared the outside temperature. Equation (5.1) can be compared to the overall energy balance evaluated in a simplified manner:

$$\Phi = H' \cdot (\theta_i - \theta_e) + \Phi_a - \eta(A_e \cdot I_s) \quad (5.2)$$

where:

H' is the overall heat transmission coefficient of the building;
θ_i is the average internal temperature;
Φ_a includes system losses and energy uses not related to climate. As a first approximation this power is not dependent on external temperature and therefore can be evaluated by the average power during the intermediate seasons without heating;
η is the utilisation factor of solar gains;
A_e is the area subject to solar radiation;
I_s is the solar flux

Comparing Eq. (5.1) with Eq. (5.2) we get that $H' = H_e$:

$$\Phi_0 = H \cdot \theta_i + \Phi_a - \eta(A_e \cdot I_s) \quad (5.3)$$

The energy used during the heating season can be evaluated from the equation:

$$Q_h = (\Phi_0 - H\theta_e)t \quad (5.4)$$

where:

θ_e is the average outside temperature;
t is the duration of the heating season

With the methodology described above, an estimate of the energy requirements can be obtained without waiting for the entire heating season. In any case, to have accurate values of H and Φ_0 and then to be able to best characterise the curve of consumption, it is necessary to detect a sufficiently extensive range of external temperatures θ_e.

Regardless of the analytical method used, the evaluation approach to energy consumption based on the energy signature is able to provide some important information. The energy signature, however, gives no indication on the optimal consumption of a building, it is merely a tool for qualitative analysis.

Those that follow are some considerations

Envelope heat losses. The slope of the straight line obtained is a function of the average U value of the building envelope. In the method proposed by the standard EN 15603, the parameter H obtained corresponds for example to the heat transmission coefficient of the building. Thus the data obtained are comparable to the legal limits.

Base power. As mentioned, the portion of horizontal line represents the base power of the system. If the system is dedicated exclusively to space heating, the base power represents the system losses during the period when it is not active. In this case, one has an immediate verification of the losses that one expects to eliminate, maintaining, however, the quality of the service. If the system is also used for the DHW, the base power represents the power required for the production of sanitary hot water. In this case, the energy signature allows very easy separation of domestic hot water consumption resulting from those due to heating, and allows an improvement in the estimation of the effects of energy efficiency measures that one imagines to accomplish.

Average design power. The value of the average design power at the outside temperature is easily obtainable starting from the right of the interpolation that has been identified. This figure is comparable to the thermal power needed to operate the plant under the design conditions, and then allows assessment of whether the plant was oversized.

For example, if the thermal power plant is higher than the average power in the design conditions by more than 25 %, the system is always below its maximum potential. Most old boilers, when running at reduced power, operate with a lower efficiency. In this case, therefore, it will be appropriate to evaluate the replacement of the old boiler with a new one of smaller size.

5.11 Verification and Calibration of Instruments

An important part of the activities of the Energy Audit is based on the use of instrumentation: the information collected, with spot measurements or monitoring, helps the auditor in defining remedial action strategies.

The measuring instruments must be regularly verified and possibly calibrated. The timing depends on many factors, including the instrument type and timing of use. Usually in fact more than one instrument is used and more tests are needed.

Guidelines on how and when these checks should be carried out are normally provided by the companies which manufacture and sell the instruments.

Hereunder is simply reported what Standard ISO 9111 [5] provides under the section "Control equipment for testing, measuring and testing":

"The supplier shall establish and maintain documented procedures to control, calibrate and maintain equipment for testing, measuring and testing (including test software) used in order to demonstrate that the product conforms to specified requirements. The equipment for testing, measuring and testing must be used in order to ensure that their measurement uncertainty is known and is compatible with the requirements of measurement requests.

In the case in which a test software or comparative references, as in the test equipment, are used for activity testing or testing, they must be checked, to ensure its suitability to verify the acceptability of the product, before its use in production, installation or service and must be checked again at prescribed intervals. The provider must determine the extent and frequency of such checks and shall maintain associated records to give evidence. In the event that the availability of technical data relating to test equipment for measuring and testing is a specified requirement, such data must be made available when requested by the customer or his representative, to verify the adequacy of the equipment from the point of functionally".

Verification and calibration of instruments used for the energy audit and all related activities must be performed at an accredited calibration centre.

References

1. ISO International Standard 7730 (2005) Ergonomics of thermal environment—analytical determination and interpretation of thermal comfort using calculation of the PMV and PPD indices and local thermal comfort criteria
2. ASHRAE Standard 62.1-2004 (2004) Ventilation for acceptable indoor air quality (ANSI Approved)
3. ASHRAE Handbook, fundamentals (2009), Chapters: 25—thermal and water vapour transmission data, 28—residential cooling and heating load calculation, 29—non residential cooling and heating load calculation procedures, Atlanta
4. EN 15603 (2008) Energy performance of buildings—overall energy use and definition of energy ratings
5. ISO International Standard 9001 (2008) Quality management systems—requirements

Chapter 6
Infrared Audit

The infrared audit is a technique of investigation that allows one to obtain, in an indirect way, a thermal mapping of a surface. Applications of thermography in the energy fields are very widespread, since through such an investigation, which is non-invasive, and a correct interpretation of infrared images, it is possible to highlight inefficiencies on buildings and facilities. This chapter describes the application of infrared audits in support of energy audits. The potential of this technique, but also its critical aspects due essentially to errors during the shooting and mistakes in thermogram interpretation will be discussed.

6.1 Introduction

The ability to visualise the heat emitted by a surface or a device is one of the major technological innovations in the field of non-destructive surveys.

The thermographic survey techniques, through the use of infrared cameras, are not recent. What is relatively new, however, is the diffused use of these instruments thanks to the reduction in costs, nowadays affordable, combined with ever better performance characteristics of the equipment.

Owning a reliable, high performance infrared camera is certainly a good starting point for the execution of an infrared audit, however, it is not sufficient to guarantee the quality of the result. There are really two essential elements for achieving a good infrared audit: certainly a high performance instrument but also considerable technical skill of the infrared auditor.

Regarding the first point, the market offers a wide range of IR cameras that can be used for many purposes. Lightweight, portable and low cost infrared cameras can be used, for example, to detect failures or malfunctioning which manifest themselves as an increase in temperature.

Luca Sarto and Angelo Martucci contributed this chapter.

G. Dall'O', *Green Energy Audit of Buildings*, Green Energy and Technology,
DOI: 10.1007/978-1-4471-5064-0_6, © Springer-Verlag London 2013

If the objective of the audit is to investigate, for example, the correctness of installation of a ETICS on a building's facade, detailed infrared images are essential to highlight possible structural pathologies. In this case we need to use an high performance infrared camera able to manage high resolution images.

Even the skills required to use the two instruments are different. In the first case the instrument is used as an infrared thermometer. In the second case the auditor needs the skills to carry out the IR-shooting in the best possible way, considering all the boundary conditions: moreover he should understand correctly the information contained in the infrared image, avoiding any interpretation error.

The correct choice of instrument is obviously very important: also if the operating principles are more or less the same, each infrared camera has its own typical application and, of course, different costs.

Once the two aspects, quality of the instrument and skills of the auditor are satisfied, infrared audit is an important and strategic tool to support energy audit. For this very reason a chapter is dedicated to this issue with the aim of stimulating the interest in using infrared audit within the energy audit activities, without underestimating the implications of professional usage.

6.2 Essential Thermal Radiation Principles

All bodies with a temperature above Absolute Zero (−273.15 °C) emit electromagnetic waves, or radiation, as a function of their temperature: however, energy is transported. This aspect allows the professional auditor to measure contactlessly, the temperature of an object simply making use of the thermal radiation emitted.

As mention previously, the intensity of the emitted energy varies with temperature and radiation wavelength. In addition to emitting radiation, an object reacts to incident radiation from its surroundings by absorbing and reflecting a portion of it, or allowing some of it to pass through. The Total Radiation Law can be written as follows (Kirchhoff Law):

$$G = \alpha G + \rho G + \tau G \quad [W] \tag{6.1}$$

which can be simplified to:

$$1 = \alpha + \rho + \tau \tag{6.2}$$

The coefficients α, ρ and τ describe the object's incident energy absorption α, reflection (ρ), and transmission (τ) (Fig. 6.1). Each coefficient can have a value between zero and one, depending upon the physical characteristics of the material under consideration. For example, if $\rho = 0$, $\tau = 0$, and $\alpha = 1$, then there is no reflection or transmission, and 100 % of incident radiation is absorbed. This is the case of the so-called *perfect blackbody*, an ideal object that constitute a criterion for

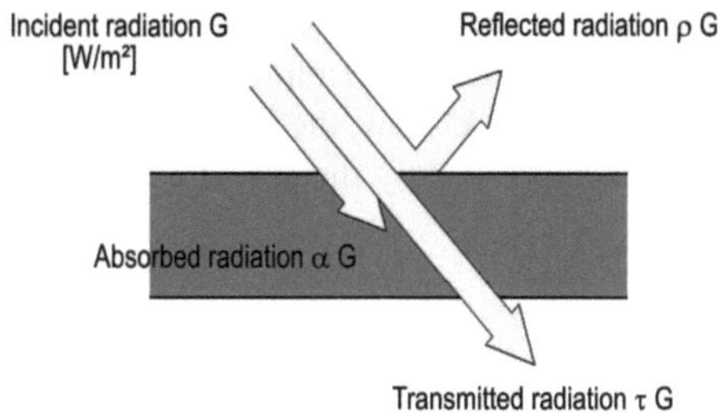

Fig. 6.1 Absorption, reflection and transmission of incident radiation

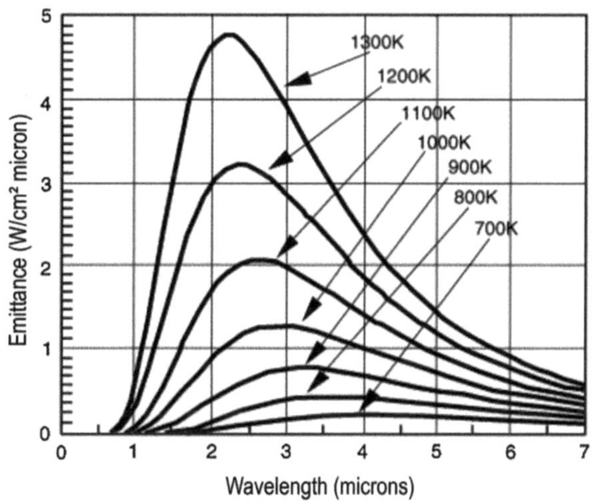

Fig. 6.2 Blackbody spectral radiant emittance

comparison in order to classify the characteristics of existing materials. In the real world there are no objects that are perfect absorbers, reflectors or transmitters. Therefore, the concept of a perfect blackbody is very important for infrared auditing, because it is the functional link between IR radiation and an object's temperature. This concept is reassumed by the Kirchhoff's Law, presented above [1].

The radiative properties of a body are denoted by the symbol ε, the emittance or emissivity of the body. Kirchhoff's law states that $\alpha = \varepsilon$, and since both values vary with the radiation wavelength, the formula can take the form:

$$\alpha(\lambda) = \varepsilon(\lambda) \tag{6.3}$$

where λ denotes the wavelength. The total radiation law can thus take the mathematical form:

$$1 = \varepsilon + \rho + \tau \tag{6.4}$$

The radiative properties of a perfect blackbody can also be described by Planck's Law, which is generally reassumed as a series of curves (Fig. 6.2). These curves show the radiation per wavelength unit and area unit (W/cm^2 microns) as a function of the temperature. Also known as the *spectral radiant emittance of the blackbody*, the said graph expressed a simple concept: the higher the temperature, the more intense the emitted radiation.

From Planck's law, the total radiated energy from a blackbody can be calculated. This is expressed by the so-called Stefan-Bolzmann law:

$$W = \sigma \cdot T^4 \quad \left[W/m^2\right] \tag{6.5}$$

where σ is the Stefan-Bolzmann's constant (5.67×10^{-8} W/m^2 K^4). The equations described above provide important relationships between emitted radiation and temperature of a perfect blackbody.

6.2.1 The Emissivity and the Temperature Measurements

The radiative properties of objects are usually described in relation to a perfect blackbody ($\varepsilon = 1$). If the emitted energy from a blackbody is denoted as W_{bb}, and that of a normal object at the same temperature as W_{obj}, then the ratio between these two values describes the emissivity (ε_{obj}) of the normal object:

$$\varepsilon = W_{obj}/W_{bb} \tag{6.6}$$

Thus, emissivity is a number between 0 and 1.

Although the emissivity ε of all the materials varies with wavelength, it is often possible to consider it as a fixed coefficient dependent on the characteristics of the object. Thus, Stefan-Bolzmann's law takes the form:

$$W = \varepsilon \cdot \sigma \cdot T^4 \quad \left[W/m^2\right] \tag{6.7}$$

It is important to emphasise that the radiation that impinges on the IR camera lens comes from three different sources. Moreover the sensor receives radiation from the target object, plus radiation from its surroundings that has been reflected onto the object's surface. Both of these radiation components become attenuated when they pass through the atmosphere. Since the atmosphere absorbs part of the radiation, it will also radiate some itself (Kirchhoff's law). Given this situation, a mathematical formula for the calculation of the object's temperature from a calibrated camera's output can be defined. The three components of received radiation power are:

- *Emission from the object* $= \varepsilon \cdot \tau \cdot W_{obj}$, where ε is the emissivity of the object and τ is the transmittance of the atmosphere.
- *Reflected emission from ambient sources* $= (1 - \varepsilon) \cdot \tau \cdot W_{amb}$, where $(1 - \varepsilon)$ is the reflectance of the object (it is assumed that the temperature T_{amb} is the same for all emitting surfaces within the half sphere seen from a point on the object's surface).

- *Emission from the atmosphere* $= (1 - \tau) \cdot W_{\mathrm{atm}}$, where $(1 - \tau)$ is the emissivity of the atmosphere.

The total radiation power received by the camera can now be written as:

$$W_{\mathrm{tot}} = (1 - \tau) \cdot W_{\mathrm{obj}} + (1 - \varepsilon) \cdot \tau \cdot W_{\mathrm{amb}} + (1 - \tau) \cdot W_{\mathrm{atm}} [W/m^2] \quad (6.8)$$

where ε is the object emissivity, τ is the transmission through the atmosphere, W_{amb} is thermal radiation power coming from the object's surroundings, and W_{atm} is thermal radiation from the atmosphere. To correctly measure the object's temperature, IR camera software requires inputs for the emissivity of the object, atmospheric attenuation and temperature and radiant temperature of the ambient surroundings (reflected temperature). Table 6.1 shows the emissivity of the main materials used in building construction.

Table 6.1 Emissivity of the main materials used in building construction

Material	Roughness	Emissivity	Temperature (K)
Aluminium	Polished	0.04–0.06	300–900
	Heavily oxidised	0.20–0.33	400–800
	Anodised	0.80	300
Brass	Highly polished	0.03–0.04	500–600
	Polished	0.09	350
	Oxidised	0.60	450–800
Copper	Highly polished	0.02	300
	Polished	0.04–0.05	300–500
	Oxidised	0.50–0.80	600–1,000
Iron	Highly polished	0.05–0.07	300–500
	Oxidised	0.64–0.78	500–900
	Rusted	0.61	300
	Case iron	0.44	300
	Wrought	0.28	300–500
Stainless Steel	Polished	0.17–0.30	300–1,000
	Lightly oxidised	0.30–0.40	600–1,000
	Highly oxidised	0.70–0.80	600–1,000
Tar	–	0.85–0.96	300
Brick	Common	0.93–0.96	300
	Firebrick	0.75	1,200
Stonework	–	0.80	300
Concrete	–	0.88–0.94	300
Glass	Float	0.90–0.95	300
Soil	–	0.93–0.96	300
Wood	Beech	0.94	300
	Oak	0.90	300
Water	–	0.95–0.96	273–373

Ref. [2]

6.3 Infrared Camera

Although infrared radiation (IR) is not detectable by the human eye, an IR camera can convert it to a photogram that shows the temperature contours of an object or scene. IR covers a portion of the electromagnetic spectrum from approximately 900–14,000 nm (0.9–14 μm) and it is emitted by all objects at temperatures above 0 K (absolute zero). Therefore, thermography is the science that maps the temperatures of objects without contact by using an IR camera.

It is important to emphasise that IR camera does not display the real temperature of the analysed object but a fictitious value, given by the addition of the effective temperature of the object with the reflected ambient temperature and the temperature of the atmosphere.

IR camera construction is similar to that of a digital video camera. The main components are a lens that focuses IR on to a detector, plus electronics and software for processing and displaying the signals and images. The lens is not made from glass but using other materials, for example germanium. Instead of a charge coupled device that video and digital still cameras use, the IR camera detector is a focal plane array (FPA) of micrometer size pixels made of various materials sensitive to IR wavelengths. FPA resolution can range from about 160 × 120 pixels up to 1,024 × 1,024 pixels. Figure 6.3 shows two models of infrared camera: a portable IR camera with a low-average resolution and a high performance model with 640 × 480 sensor resolution.

FPA detector technologies are divided into two categories: thermal detectors and quantum detectors. A common type of thermal detector is an uncooled microbolometer made of a metal or semiconductor material. These typically have a lower cost and a broader IR spectral response than quantum detectors. On the other hand, microbolometers react to incident radiant energy and are much slower and less sensitive than quantum detectors, which are based on the change of state of electrons in a crystal structure reacting to incident photons. Generally IR cameras used for thermographic auditing of buildings are those with thermal detectors.

Fig. 6.3 Example of two models of infrared camera: portable model, on the *left*, high performance model, on the *right* (with permission of Flir Corporation)

6.3.1 Criteria for Instrument Choice

The main requisite to look at for the choice of an IR camera is the definition of the effective spectral bandwidth that the detector could capture, through the optical unit transparency to specific IR. It is possible to distinguish between instruments which are sensitive to *Short Waves* (SW) for high temperature applications and those which are sensitive to *Long Waves* (LW) for the low temperature ones. In building and construction audits LW cameras are generally used.

A second choice requisite is related to the optical characteristics of the imaging system, which quantify its capacity to detect the temperature of small objects and are described by the following specifications. The *Field of View* (FOV) is the area that camera can visualize at a set distance, usually described in horizontal by vertical degrees (e.g., 20° × 10°). The *Instantaneous Field of View* (IFOV), typically specified in milliradians, is the measure of the spatial resolution of the imaging system and is defined as the angle subtended by a single detector element on the axis of the optical system. The *Instantaneous Field of View Measurement* (IFOV measurement, also known as Measurement Resolution) is the smallest detail that one can obtain for an accurate temperature measurement at a set distance. All these parameters are a function of the combination of focal length and sensor resolution. Thermographic auditing of buildings requires at least a 160 × 120 (pixels) IR camera sensor for the interior, whereas it is better to use a 320 × 240 (or greater) sensor for the exterior.

A third choice requisite is the thermal resolution, which is the minimum discard of the temperature that can be measured. This feature is defined by multiple parameters. The most important is the *Noise Equivalent Temperature Difference* (NETD), which is the minimum temperature variation to produce a signal equal to the noise standard deviation of the system. Thermograpic auditing of buildings requires NETD values of 0.05–0.08 °C (Fig. 6.4). For mechanical and electrical applications, this parameter is non-essential because of the system's high Delta-T.

The measurement accuracy is the precision of the temperature measurement. Most of the available systems has values comprised between ± 2 °C below 100 °C (± 2 % above this limit).

Another important requirement to be evaluated during IR camera choice is the thermal range, which is the radiance (or temperature) scale within the limits of which the instrument is calibrated. It must be dependent on the specific use (between −20 °C and 100 °C for thermographic auditing of buildings).

A final requisite to be taken into account is the elaboration frequency, being the number of thermal maps generated in unit time by the thermographic system or by its own device (9 Hz is sufficient for thermographic auditing of buildings).

Furthermore, it is possible to manage all the parameters directly connected to generated thermal images through proprietary software, which allows the modification of: the thermal range, the weather data, the emissivity value of a specific material, the temperature scale etc.

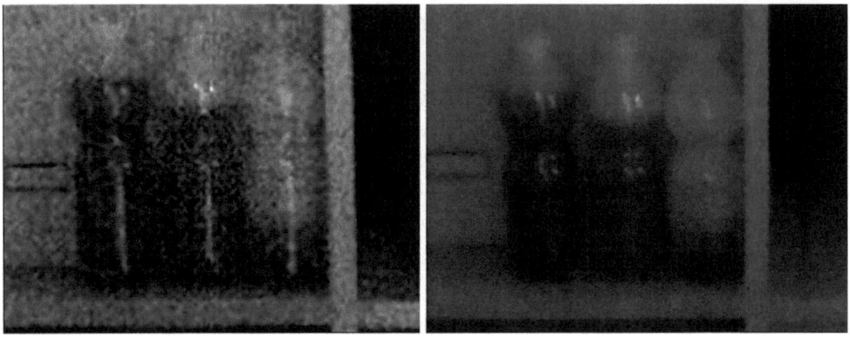

Fig. 6.4 Same scene, different IR camera Thermal Resolution. The coarse-grained thermal image on the *left* was taken by an IR camera with a NETD value lower than the one on the *right*. Therefore, the instrument choice directly depends on the objective needs of the professional

6.4 Skills and Activities of a Thermographic Auditor

The necessary skills of a Thermographic Auditor are defined, at international level, by the ISO 18436-7 standard [3]. This additionally specifies the procedures and the sort of analysis to be conducted. The cited standard introduces three different levels of specialisation, acquirable through specific courses, at the end of which it is compulsory to sit the related examination. Table 6.2 shows the requirements of Thermographic Auditors for each level described in [3].

Table 6.2 Requirements of thermographic auditors for each level of specialisation as described in the ISO 18436-7

Level of specialisation	Requirements
First level	The professional operator is able to conduct a Thermographic Audit using well-defined procedures, to evaluate the analysis results and to process elementary post-elaboration activity of thermal images. He needs 32 h of training and 12 certified months (or 400 h) of professional activity.
Second level	The professional operator is able to conduct an advanced Thermographic Audit through the usage of many operating instruments, and to identify the right corrective action for the observed problem. He needs 64 h of training and 24 certified months (or 1,200 h) of professional activity.
Third level	The professional operator is able to develop procedures, software and advanced instructions for the correct use of Thermography. He is also capable to identify complicated conditions and to recommend thermodynamics corrective actions. He needs 96 h of training and 48 certified months (or 1,920 h) of professional activity.

6.5 Thermography and Energy Audit

Thermographic analysis is very useful for evaluating building energy performance, both for envelope and facilities. In fact it could lead to the correct identification of many energy problems related to bad final design, construction, installation or, generally speaking, to building malfunctions.

For example, the knowledge of structural and energy performance of existing buildings could render feasible the respective retrofitting activities, in order to solve specific critical situations. Moreover, in new buildings it is possible to check the accuracy of building construction compared to the technical specifications defined during the design stage. Therefore, thermography gives the professional a qualitative knowledge about the performances of building envelopes and facilities. It is also used to verify the presence of air seepage, humidity or water leaks.

6.5.1 Infrared Audit of Building Envelope

In order to correctly define the actual performances of the building envelope, both energetically and from a structural standpoint, it is very important to know, in addition to the characteristics of building materials, the reflected ambient temperature as well as the weather conditions (trends of outside temperatures, relative humidity, rainfall intensity, wind direction and velocity, solar radiation). A good theoretical knowledge coupled with long-standing practical experience are required to be able to solve properly problems that, job-to-job, are brought to the attention of the professional auditor. In any case, Table 6.3 provides qualitative information on how to interpret thermal images in order to identify the most common building envelope problems.

6.5.2 Infrared Audit of Facilities

Thermography is very useful also to evaluate the actual condition of both thermal and electrical facilities. In fact, the professional operator can thus verify, without contact, the correct operation of all the related sub-systems and/or determine their "under-wall" distribution. Therefore, in the case of system malfunction, it is possible to act locally to the defect, without proceeding by trial and error. Table 6.4 identifies the most common applications of thermography for facilities management.

6.6 Operational Procedures of Infrared Audit

The operational procedure of building Thermographic Analysis is internationally well-defined by the ISO 6781 standard [4]. It can be used both for new and existing buildings to check global conditions of structures and envelopes, the objective result of a building renovation or the simple laying of a technological system.

Table 6.3 Qualitative information on how to interpret the most common building envelope problems

Thermal Bridges	
	Thermal bridges are easily identifiable through infrared thermography: the different characteristics of materials give to each structural element a different temperature.
Air Seepages	
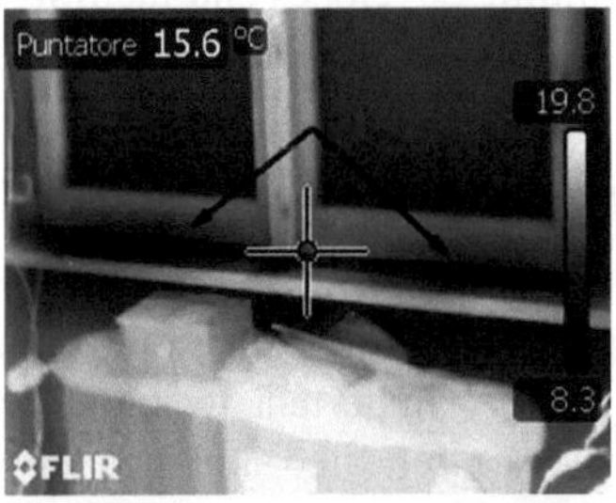	The arrows on the thermal image on the left render evident an air seepage through window frame. It is clearly identifiable because of the lower temperature of the windowsill next to the seepage.
Wall Texture	
	The different thermal resistance of mortar and bricks allows the professional operator to clearly recognise the wall texture, despite the covering plaster.
Thermal Coating	
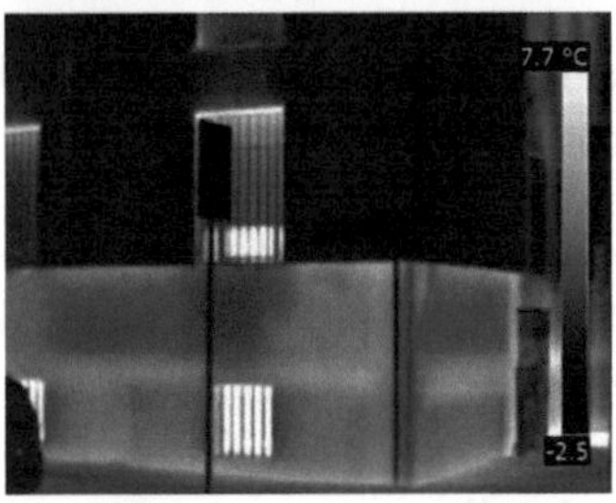	The image on the left is emblematic for understanding the way to recognise thermal coating: the lower part of the wall is clearly hotter than the upper part, owing to the absence of thermal insulation. Structural elements are also easily identifiable, while the coated part of the wall has very uniform temperature.
Humidity	

(continued)

Table 6.3 (continued)

	The thermal image on the left shows a water stain on a ceiling: humidity and/or water infiltrations are generally colder than the rest of the technological item under consideration, owing to the cooling effects of evaporation.

In Europe, the EN 13187 standard [5] has been introduced. It resumes the ISO 6781, but with some changes in order to take account of the evolution both of instruments performances and thermographic analysis methods. The main modification regards the addition of a simplified method usable to verify installation/laying-out a material, a production cycle and many other routine verifications.

6.6.1 Infrared Audit on Field

The first important step of a thermographic analysis is the definition of the aim of the relief. It is essential to plan requirements, duration and costs of all the activities. The procedure of the infrared audit is elaborated on the basis of the available information which needs to be integrated by necessary building surveys, both qualitative and quantitative.

In order to properly evaluate the actual conditions of a building, the professional auditor must acquire information about its architectural, technological and plant characteristics: the situation and conditions of the surroundings (climate included) must also be considered.

The standard EN 13187:1998—Appendix D describes an example for the best weather conditions for infrared audit on-site in Scandinavian countries. The aim is to guarantee an approximate stable profile of surface temperatures. Those conditions, which are to be evaluated case-to-case in function of the climate peculiarities of the relevant country, are as follows:

- For at least the 24 h before the beginning of the infrared audit and during its execution the following relationship must be respected:

$$(T_{\text{in}} - T_{\text{ext}}) > 3/U$$

where T_{in} is the interior temperature, T_{ext} the ambient temperature and U is the thermal transmittance (W/m^2 K) of the element being considered. However, ΔT has to be at least 5 K.

Table 6.4 The most common applications of thermography for facilities management

Outside Radiators Heat Loss

Through thermographic analysis it is possible to identify the external heat loss from radiators. In fact, it is visible because of a localised higher temperature on the exterior surface of the wall being considered (see the thermal image on the left, just under the window).

Uninsulated pipes

The thermal image on the left shows the uninsulated inlet and the outlet pipes of a radiator. They are clearly identifiable from the localised temperature rise on the exterior surface of the wall.

Radiators

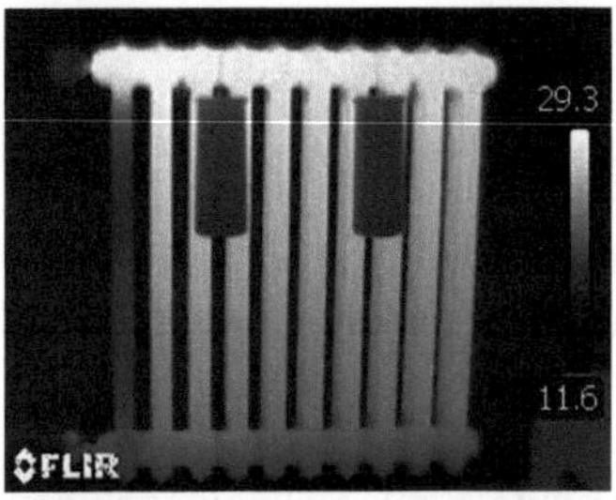

Infrared analysis can be efficiently used to verify the correct operation of a radiator. Looking at the thermal image, one notices a malfunctioning of the first element on the left which is still cold. This could be caused by over-pressurised inlet water.

Air Vent

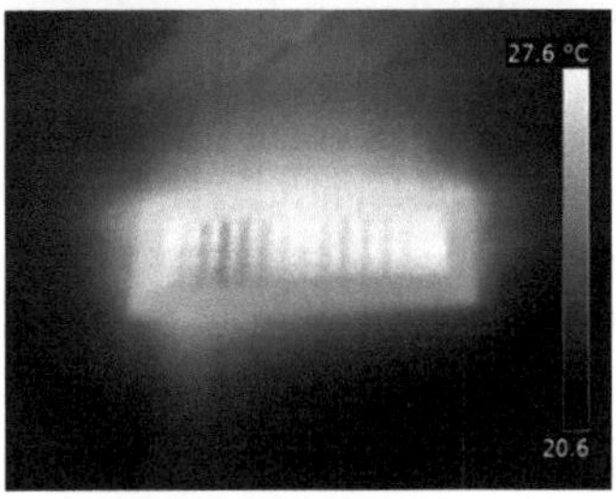

Infrared analysis could also be used to take the temperature of outgoing air from a vent.

Floor Radiant Systems

(continued)

Table 6.4 (continued)

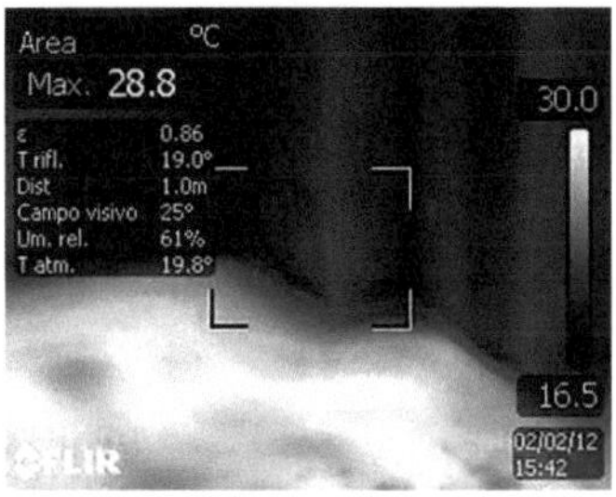	Through thermography the professional operator can verify, without contact, both the actual distribution of a floor radiant system and its correct operation.
Electrical Plants	
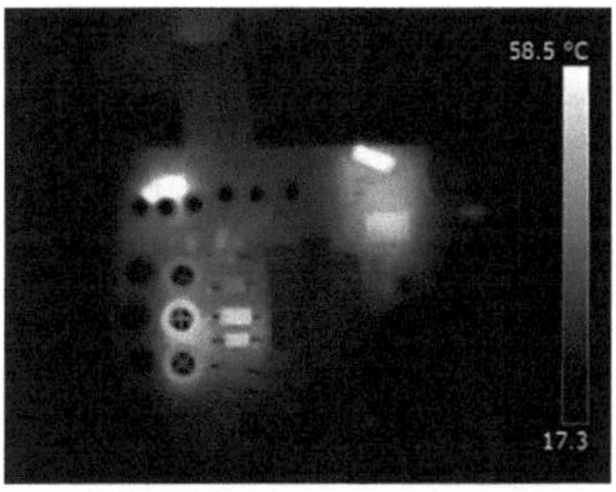	Infrared analysis permits clear recognition of which components of an electrical plants could potentially break down: looking at the thermal image on the left, of the three condenser in series, the central one has a higher temperature than the other two: clearly caused by a greater flow of electric current.
Tracking Pipes	
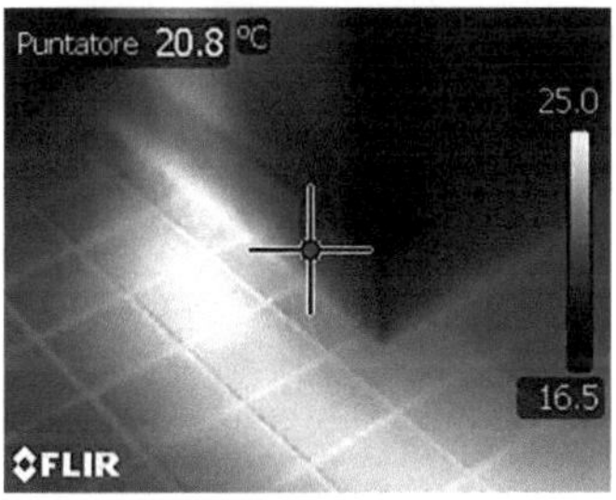	Through thermography, the professional operator can easily track water pipes, in order to act locally to a potential problem area and not by trial and error.

- For at least the 12 h before the beginning of the infrared audit and anyway during its execution, the analyzed building surfaces should not be exposed to direct solar radiation.
- During the thermographic analysis the exterior and the interior temperature must not vary by more than ± 5 °C and ± 2 °C respectively.

If thermographic analysis is to be conducted deviating from the conditions above, it is necessary to explain the motivations for doing so on the thermographic report, which has to be congruous, in its general layout, with the well-defined indications of the EN 13187.

The professional operator could complement the thermographic report with the most correct energetic retrofits in order to solve the registered problems.

Table 6.5 Some common mistakes of infrared audit

Emissivity estimation

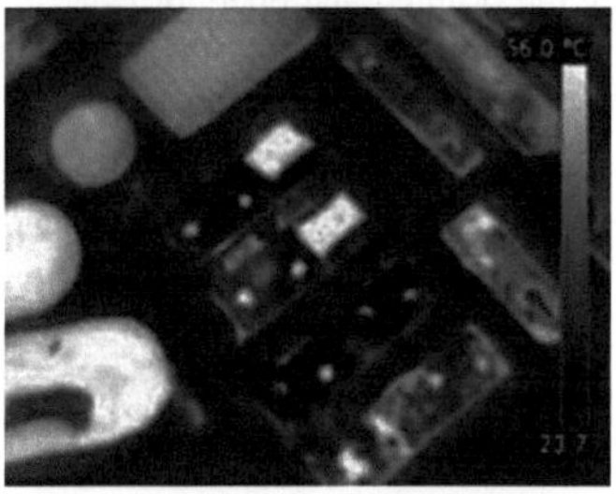

Different objects at the same temperature but with different emissivity coefficients are shown with a different temperature contour by the thermal image.

Effect of the solar radiation

The effect of the solar radiation on the building envelope has to be evaluated case by case. For example, the thermal image on the left shows a thermal coated envelope with a small thermal inertia under the influence of the solar radiation: the "shadowed" part of the surface is colder than the irradiated part, owing to the projection of the balcony.

Reflection

Not choosing the correct angle to minimise the reflections on the thermal image could lead to an erroneous temperature acquisition. The image on the left shows the reflection of IR radiation from three people on the window being analysed.

6.6.2 Common Mistakes

In order to bring a thermographic analysis to a satisfactory conclusion, the professional auditor should pay attention to several very important aspects: the transition from theory to practice could be insidious and the main risk that can be incurred is wasting time and money. Here are some common mistakes which are easily avoidable with a little attention.

The first one is related to the correct usage of the infrared camera, which requires calibration of the internal sensor in function of the exterior temperature. Usually this periodic operation occurs automatically every minute. However, it takes some minutes and several calibration processes before correct measurements are returned (dependant upon the thermal changes to which the instrument is subjected). So, not waiting until the thermal sensor is at temperature could represent a real problem, in terms of gaining correct quantitative information about an object's temperature.

The second one depends on the modality used by the professional to acquire all the necessary data. Moreover, the calculation of temperatures requires the correct acquisition of the reflected ambient temperature as well as the weather conditions (exterior temperatures, relative humidity, rainfall intensity, wind direction and velocity, solar radiation) and the thermal characteristics of analyzed materials. For example, the fictitious rises in temperature owing to the reflections of the surroundings is directly related to the uncorrected evaluation of the reflected ambient temperature.

Not to consider the effect of the solar radiation on the building envelope is also a well-known mistake, which could lead to an incorrect evaluation of temperatures arising from the thermal inertia of the structures present.

Then the correct estimation of emissivity ε: different materials with different emissivity coefficients at the same temperature will be shown by the infrared camera with different temperatures. However, it is necessary to input emissivity on a case-to-case basis in function of the material being analysed (Table 6.5).

Lastly, in the attempt to obtain the temperature of a reflecting object, it is very important to choose an angle that minimises the reflections on the thermal image. Reflections could come from every heat source located next to the object under analysis and may lead the auditor to an incorrect temperature acquisition.

References

1. FLIR, The Ultimate Infrared Handbook for R&D Professionals. FLIR AB, Boston. Available for free on the website: www.flir.com
2. Cengel Y (2008) Introduction to thermodynamics and heat transfer. McGraw-Hill Companies Inc., New York
3. ISO 18436-7:2008 Condition monitoring and diagnostics of Machines. Requirements for qualification and assessment of Personnel. Part 7: Thermography
4. ISO 6781:1983 Thermal Insulation. Qualitative Detection of Thermal Irregularities in building Envelopes. Infrared method First edition
5. EN 13187:1998 Thermal performance of buildings. Qualitative detection of thermal irregularities in building envelopes. Infrared method

Chapter 7
Evaluation of the Performance Improvement

Remedial actions and restructuration works enhancing the energy performance of a building should be evaluated by comparing the baseline and the situation improved, i.e., that which is achieved after the retrofit measures will be implemented. This comparison may be done, as a function of the operating level of the audit, using simplified methods or by evaluating the global effects, through a complete energy assessment by means of simulation software. This chapter examines the issues of the evaluation of the performance improvement and provides the elements of choice to provide an approach consistent with the operational level adopted.

7.1 General Criteria

7.1.1 Approaches to Performance Evaluation

The evaluation of performance improvement of an energy retrofit measure (e.g., the thermal insulation of a wall, the replacement of a window or the substitution of an electric motor, etc.) must compare the baseline with the optimised situation, i.e., after the measure has been implemented. The difference between the energy consumption before and after the retrofit action determines the amount of energy saved which can be transformed into fuel (or electricity) savings, and then into an economic saving and/or or into a reduction in carbon dioxide emissions.

This assessment can be done in a simplified way, considering the effects of a single measure, or in a more complete manner, assessing the effects that the single measure or the set of measures may have on the global energy balance.

If the proposed measures are limited, especially if they do not interfere with each other, the simplified approach can be used. However, this situation is quite rare as it is quite unlikely that a single measure has no effect on the entire system.

The following example clarifies the concept. It is a situation where two apparently very different retrofit measures have been identified:

- thermal insulation of exterior walls;
- replacement lamps for old lamps with high-efficiency lamps.

G. Dall'O', *Green Energy Audit of Buildings*, Green Energy and Technology,
DOI: 10.1007/978-1-4471-5064-0_7, © Springer-Verlag London 2013

The first measure improves the thermal performance of the building envelope: for the part of the wall concerned, and only for that, thermal losses before and after the remedial action are evaluated, as a result of the increasing of the thermal resistance and obviously the reduction of the *U* value (thermal transmittance). The difference between the two energy balances determines the energy savings that can be achieved.

- However, the thermal energy required is provided by an HVAC system which is characterised by an average seasonal efficiency.[1] The net energy saving to be considered in the economic evaluation cannot neglect this aspect: the primary energy savings, which are what lead to an evaluation of the saving in fuel, should be calculated not only considering the reduced thermal dispersion of the building envelope, but also in terms of the necessary heat generated, using the average seasonal efficiency of the system that the auditor must estimate.
- If, for example, the best thermal insulation implies a reduction of energy losses of the building envelope of 500 kWh/year and the seasonal average efficiency of the plant is 0.65, then the primary energy savings generated by the retrofit measure will rise to 500/0.65 = 769 kWh/year.
- In this example, one has simply considered a retrofit measure that improves the energy performance of the building envelope: the heating system is not modified, so it can be taken that its efficiency remains the same. But what happens if the actions undertaken also affect the plant?
- The improvement of the energy performance of the building envelope will reduce the thermal load; hence, a lower heat output is required from the heating system. For this reason, the replacement of the existing boiler with a new more efficient boiler, of lower thermal capacity, is often convenient. When replacing the boiler, for example, the average seasonal efficiency would increase from 0.65 to 0.75, and so the energy savings for the thermal insulation of the building envelope would be reduced to 500/0.75 = 666 kWh/year.
- It may seem paradoxical that an improvement in the plant performance, which however requires greater investment, has the effect of reducing overall energy saved through the improvement required for the thermal insulation of the building envelope. That is indeed exactly what happens in a case like this where retrofit measures are combined, affecting each other.
- A cost-benefit analysis rewards more retrofit actions starting from an inefficient baseline. On the other hand, the aim of the energy audit is to define a strategy that may improve the overall energy efficiency of the building-plant system; the replacement of a heating boiler reduces the cost-effectiveness of the performance improvement of the building envelope. However, an improvement in the average seasonal efficiency generates an improvement in the overall performance and this is precisely the goal one wants to achieve.

[1] The average seasonal efficiency of a HVAC system considers the energy losses of the different subsystems: heat generation, pipes and ducts distribution, regulation or control and emission of the terminals

- Returning to the previous example, other benefits should also be evaluated, such as the better internal thermal comfort owing to the increase in the operating temperature[2] in winter season and the reduction in energy consumption even in the summer season, since better insulation reduces thermal loads.
- In the second example, which involves the replacement of the lamps, the performance improvement results in a reduction in the consumption of electricity (up to 80 % for replacement of incandescent lamps). In the evaluation of the energy advantages, it should however be considered that high-efficiency lamps, providing the same lighting, produce less heat. In the summer season, therefore, one obtains a reduction in energy consumption for air conditioning.
- In the two examples, it has been seen how simple measures are, however, related to other factors which should be taken into account for a more correct assessment of their effects. If the retrofit actions are limited then the approach to performance evaluation through the simplified calculation can be accepted, provided that the auditor is aware of its limitations. The auditor must especially bear in mind the interactions, and so the effects that the measure taken for a system or a component can have over the entire building.

The diagram of Fig. 7.1 shows the correlations between the various energy retrofit actions for a building. From the diagram, although simplified, emerges clearly the great complexity of the building-plant system.

If the proposed measures are different and their effects cannot be evaluated independently, the right approach, although more expensive, is that provided by a global simulation of the entire building-plant system. In this case, it is necessary to model the building in the initial situation, then defining the baseline, and subsequently in the new situation by providing different scenarios.

The choice between a static simulation and a transient simulation depends upon the complexity of the building. Transient simulation requires more time in the definition of the model, but it provides more flexibility and it allows a more reliable assessment of the effects of different measures.

7.1.2 Degree-Days Method

When the modes of usage of the building are constant and the efficiencies of plant systems do not change, the use of methods of assessment of energy consumption based upon heating degree days (DD_H) is the simplest approach. If, however, the internal temperature does not remain constant, but is subject to fluctuations, or the internal thermal loads vary, the use of the transient calculation models is suggested.

---

[2] In design, operative temperature can be defined as the average of the mean radiant and ambient air temperatures, weighted by their respective heat transfer coefficients [1].

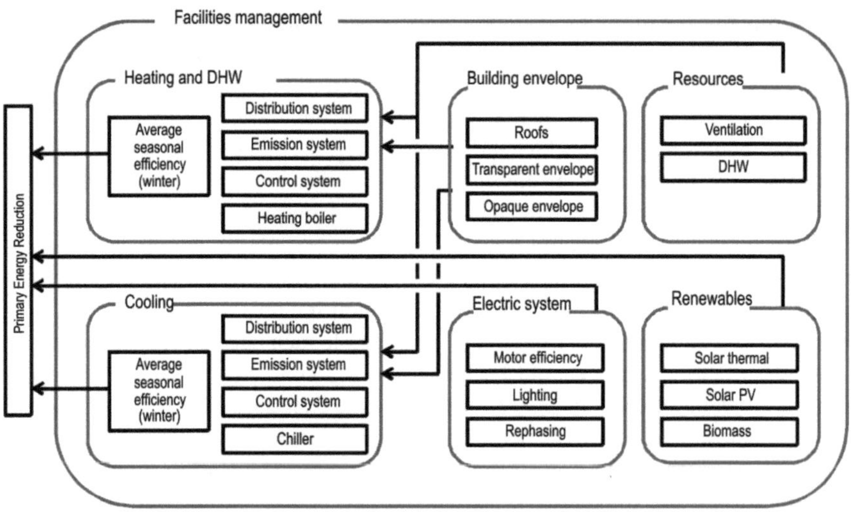

Fig. 7.1 Correlation between the various retrofit measures to improve the energy performance of a building

Before defining the *degree days* one should define the balance temperature of the building θ_{bal} as the value of the outside temperature θ_o at which the heat losses of the building are compensated by contributions from the internal gains q_{gain} (occupants, lighting, equipment, etc.) and external gains (solar radiation), thus the situation in which

$$q_{\mathrm{gain}} = H_t(\theta_i - \theta_{\mathrm{bal}}) \tag{7.1}$$

where

H_t is the overall coefficient of heat losses which takes into account both the losses for transmission and ventilation (W/K);

θ_i is the set point temperature (°C).

From Eq. (7.1) one obtains that θ_{bal} can be calculated by the following equation:

$$\theta_{\mathrm{bal}} = \theta_i - \frac{q_{\mathrm{gain}}}{H_t} \tag{7.2}$$

The winter heating therefore will only be necessary when the outdoor temperature θ_o falls below the value of the balance temperature θ_{bal}.

The heat gains in the winter season contribute to increase the internal temperature of 2 °C, so where there is an average internal temperature by 20 °C, it could therefore be considered that the balance temperature θ_{bal} is on average equal to 18 °C. It should be noted that for well-insulated buildings or those with high internal loads this temperature can be much lower.

The instantaneous value of the energy requirement q_h for winter heating can be evaluated with the following equation:

$$q_h = H_t[(\theta_{\text{bal}} - \theta_o)]^+ \quad (7.3)$$

where the sign + indicates that within the square brackets should be considered only positive values.

If one wants to calculate the energy demand for the winter heating season extended to the whole, the formula to be used is as follows:

$$Q_H = 24\sum_{i=1}^{N_H} q_{H,i} = 24H_t\sum_{i=1}^{N_H}\left(\theta_{\text{bal}} - \theta_{o,i}\right)^+ \quad (7.4)$$

where N_H is the number of days of the winter heating season. From (7.4), winter degrees days can be defined as a function of temperature difference between the balance and the outside temperature according to the equation

$$\text{DD}_H(\theta_{\text{bal}}) = \sum_{i=1}^{N_H}\left(\theta_{\text{bal}} - \theta_{o,i}\right)^+ \quad (7.5)$$

The degree days are thus the sum, extended to the whole winter season considered, of the differences between the temperature of balance and the outside temperature for each of the days considered. Knowing the value of degree days for a given location, primary energy needs E_H can be calculated using the equation

$$E_H = \frac{Q_H}{\eta_H} = \frac{24H_t\,\text{DD}_H(\theta_{\text{bal}})}{\eta_H} \quad (7.6)$$

where

η_H is the average seasonal efficiency of the heating plant (its value may vary on average from 0.50 to 0.85 depending on the age of the plant)

The degree days method expressed in Eq. (7.6) can also be used to determine the thermal summer load, by calculating the cooling degree days (DD_C) using an equation similar to (7.5)

$$\text{DD}_C(\theta_{\text{bal}}) = \sum_{i=1}^{N_C}\left(\theta_{o,1} - \theta_{\text{bal}}\right)^+ \quad (7.7)$$

where N_C is the number of days of cooling in the summer season (the balance temperature of course, is not the same) [2].

The degree-days method can make estimates of energy use in heating season, especially for those buildings characterised by high losses by conduction and ventilation. For the calculation of the primary energy requirement in the summer season; however, this method does have limitations since it does not take into account, some important factors such as that of the thermal mass that retards the effects of the internal loads and the effect of solar radiation that in summer could be relevant. The topic of the criteria for the construction and use of summer degree days remains of great interest and the subject of several publications of interest in this regard are the work of Day and Karayiannis [3, 4] and Claridge et al. [5].

With the degree-days method, we can make an assessment of the energy saved as a result of measures that reduce fuel consumption and improve energy performance. Assuming that the set-point temperature value and the internal gains remains constant, the balance temperature decreases in consistently with what is expressed by the Eq. (7.2). The performance improvement of the building envelope reduces the winter thermal load since H_t and $\mathrm{DD}_H(\theta_{\mathrm{bal}})$ decreases. Energy savings can be calculated using the equation

$$\Delta E_{H,R} = \frac{24\left(H_{T,E}\,\mathrm{DD}_H\left(\theta_{\mathrm{bal},E}\right) - H_{T,R}\,\mathrm{DD}_H\left(\theta_{\mathrm{bal},R}\right)\right)}{\eta_H} \tag{7.8}$$

Equation (7.8) assumes the same value of the average seasonal efficiency of the system while subscripts E and R refer, respectively, to the current situation (existing) and the optimised one (retrofit). In some cases, the measures do not affect appreciably the temperature values of the balance temperatures, so (7.8) can be simplified as follows:

$$\Delta E_{H,R} = \frac{24\left(H_{T,E} - H_{T,R}\right)\mathrm{DD}_H\left(\theta_{\mathrm{bal},E}\right)}{\eta_H} \tag{7.9}$$

If only one of the elements of the envelope is modified (for example the external wall or the roof), the difference $\left(H_{T,E} - H_{T,R}\right)$ is equivalent to the difference between the products of the U transmittances for the respective surfaces A before and after the remedial action taken, thus to the difference $(UA_E - UA_R)$.

The values of H_T can be calculated directly or indirectly. The direct calculation of H_T implies the definition of a comprehensive energy balance of the building, thus the auditor should acquire all the technical information. The evaluation of $H_T$in an indirect way can be done, through a monthly monitoring of the outside temperature and fuel or energy consumption, using the method of energy signature (see paragraph 5.10).

The values of degree days for most locations (summer and winter) can be obtained from the website www.degreedays.net which contains an application online which permits their generation as a function of internal reference temperature. In the same website, there is more information about degree days and the correct way to use them. For a more detailed discussion about this method, refer to ASHRAE [6].

7.1.3 Bin Method

The calculation method is based on the statistical grouping of hours of the year that have similar climatic conditions: each group is called *bin*. These tables (bin tables) can be constructed for different items, such as dry bulb temperature, wet bulb temperature, wind speed, solar radiation, etc. The application of the method, therefore, can be extended to various applications.

For thermal load evaluations in HVAC systems, instead of calculating the thermal load on an average seasonal, monthly or hourly basis, the load is determined for each *bin*. The energy consumed at the seasonal level is then obtained by multiplying the values obtained by the number of hours of occurrence.

The advantages in the accuracy of the calculation are highlighted when there is not a linear dependence between climate and given load, or when the load can be, at certain different times, positive, negative or zero. In HVAC systems, for example, the load can be considered null in a certain range generally between 20 and 26 °C, an indoor temperature interval which does not require either heating or cooling.

This method provides more accurate results than others, particularly in systems with air-to-air or air–water heat pumps in which the coefficient of performance (COP) depends upon the climatic conditions. The following example, shown in Table 7.1, will clarify the simplicity and potential of this method.

The heating energy requirement is to be evaluated for a building for which the thermal dispersion coefficient has been evaluated in 70 kW/K and the internal design temperature is 20 °C. The heat required is provided by an air-source heat pump sized to cover 50 % of the maximum thermal capacity requirements (at design conditions) integrated with an auxiliary heating system.

For simplicity, one considers only the climate parameter of outside temperature whose values have been grouped into 12 bin and thus count the values falling in each class, calculating in this way the temperature values referred to the typical year for the location in question.

In Table 7.1, the second column shows the number of hours during which the outside temperature falls into the range of values reported in the first column.

Table 7.1 Example of application of bin method to a heating system with air heat pump

Temperature (°C)	Hours	Q (kW)	CAP (kW)	COP	P_{EL} (kW)	E_{ELI} (kWh)	P_{EL} (kW)	E_{INT} (kWh)
28 ÷ 31	2							
26 ÷ 28	14							
22 ÷ 26	38							
20 ÷ 22	89							
16 ÷ 20	115	140	1,027	2.3	61	7.00		
14 ÷ 16	128	350	967	2.4	146	18.67		
10 ÷ 14	106	560	906	2.6	215	22.83		
8 ÷ 10	88	770	850	2.7	285	25.10		
4 ÷ 8	75	980	789	2.8	282	21.13	191	14.33
2 ÷ 4	43	1,190	732	2.7	271	11.65	458	19.69
−2 ÷ 2	18	1,400	672	2.5	269	4.84	728	13.10
−4 ÷ −2	4	1,610	614	2.3	267	1.07	996	3.98
Total	720					112.29		51.11

The fourth column (CAP) contains the values of the capacity available from the heat pump, deduced from technical documentation, as a function of the outdoor temperature, whilst the fifth column shows the performance COP as a function of both the temperature and the load factor.

With these data, it is possible to estimate the power consumption P_{EL} and multiplying by the hour, the energy consumed in each bin. The table shows that from the bin 4 ÷ 8 °C, it is necessary to integrate with the auxiliary heating system, for which is calculated, in the same way, power and energy.

With this procedure, it is possible to determine that the integration heating system must be active for 140 h out of 577. Thanks to this method, one can evaluate the correct sizing of the heat pump in such a way as to optimise the initial investment cost. In the example, the total coverage of the needs with the heat pump would entail an investment of almost double the amount, with little benefit in terms savings for the low efficiency operation at low temperatures.

For a more detailed discussion about this method, refer to ASHRAE [6].

7.2 Simplified Calculations

7.2.1 Improvement of the Building Envelope

7.2.1.1 Thermal Insulation of Opaque Walls

The reduction of energy consumption of opaque walls of the building envelope (or shell), both in summer and in winter, can be done by reducing the U value. In the existing buildings, the problem can be solved by increasing the thermal resistance R_T (m^2K/W) assuming that:

$$R_T = \frac{1}{U_E} \tag{7.10}$$

The thermal resistance R can be increased through a higher insulation of the walls, and therefore through the addition of an insulating layer which, depending on the particular situation, can be applied on the outer face (e.g., ETICS[3]), of the inner one (counter wall) or within the same structure filling with suitable insulating materials any air gaps.

By defining U_{E} the U value of the wall before the insulation (i.e., existing wall) and U_{R} the U value of the wall after the insulation, the thermal resistance ΔR_I to add to the wall can be calculated from the equation:

$$\Delta R_I = \frac{1}{U_R} - \frac{1}{U_E} \tag{7.11}$$

[3] ETICS—External Thermal Insulation Composite System.

Table 7.2 Indicative values of thermal conductivity of some insulating materials (commercial values, source ANIT—National Association of Thermal and Acoustic Insulation)

Insulating material	Thermal conductivity λ (W/mK)
Expanded polystyrene (EPS)	0.034 ÷ 0.048
Extruded polystyrene (XPS)	0.032 ÷ 0.036
Expanded polyurethane (PUR)	0.024 ÷ 0.034
Expanded toasted cork	0.040 ÷ 0.045
Wood fibre	0.040 ÷ 0.060
Fibreglass	0.037 ÷ 0.048
Mineral wool	0.037 ÷ 0.045
Expanded perlite	0.066
Expanded clay	0.09 ÷ 0.12
Plaster insulation	0.040

and the thickness of the insulating material to be added can be easily deduced from the equation

$$s = \Delta R_I \cdot \lambda \tag{7.12}$$

where λ is the thermal conductivity of the insulating material expressed in W/mK. Table 7.2 shows some indicative values of the thermal conductivity of insulating materials.

The energy performance improvement of a wall can be evaluated through the following sequence of steps:

- assessment of the U value of the wall in the basic situation U_E;
- determination of the target U value U_R;
- calculation of the thermal resistance ΔR_I to add to the wall, using Eq. (7.11);
- choice of the most appropriate retrofit technology;
- definition of the insulating material in accordance with the technology chosen;
- calculation of the thickness of the insulating material using Eq. (7.12).

The primary energy savings $\Delta E_{H,R}$ can be estimated, using the Degree-days method, with the equation:

$$\Delta E_{H,R} = \frac{24\left((U \cdot A)_E - (U \cdot A)_R\right) \cdot \mathrm{DD}_H\left(\theta_{\mathrm{bal},E}\right)}{\eta_H} \tag{7.13}$$

where η_H is the average seasonal efficiency of the system that should take into account not only the heat losses due to inefficiencies of the system but also the energy losses due to the electrical equipment (i.e., pumps, fans, regulation, burners, etc.).

In the Eq. (7.13), the number 24 indicates the daily operation hours of the system. If the operation is not continuous but there are interruptions, for example during the night, it is possible to reduce this value by applying a reductive coefficient that, however, is not proportional to the actual number of operating hours.

One should take into account of the thermal inertia of the building structure: considering that a precise calculation is not easy, for a first estimation of energy consumption, and then the saved energy, one considers the parameter of 24 h, in any case, consistent with the residential buildings in which a reduction in hours of operation system involves limited energy savings.

For the average seasonal efficiency of the system, for a rough estimate values between 0.55 and 0.85 depending on the status of the plant, can be used.

Example A retrofit measure concerning the thermal insulation of an external wall with an overall area of 50 m^2. To the existing wall, U value = 1.2 W/m^2K, is added an insulating layer (thickness 15 cm, $\lambda = 0.030$ W/m^2K). The location where the intervention is scheduled has 2.400 DD_H, with the internal balance temperature of 18 °C.

If the average seasonal efficiency of the system η_H is 0.6, calculate the amount of primary energy savings in the heating season.

From Eq. (7.12), one obtains that the value of the added thermal resistance is equal to

$$\Delta R_I = s/\lambda = 0.15/0.030 = 5 \ \text{m}^2\text{K/W}$$

The total resistance of the insulated structure will be therefore

$$R_R = (1/U_E + \Delta R_I) = (1/1.2 + 5) = 5.83 \ \text{m}^2\text{K/W}$$

from which it appears that $U_R = 1/RR = 1/5.83 = 0.17\,\text{W/m}^2\text{K}$

From (7.13), one obtains that

$$\Delta E_{H,R} = (24 \times ((1.20 \times 50) - (0.17 \times 50)) \times 2400)/0.6 = 4,944,000 \ \text{Wh}$$
$$= 4.944 \ \text{kWh}$$

7.2.1.2 Energy Performance Improvement of Windows

The calculation procedure described above can also be used for the assessment of window performance (e.g., replacement of glass, replacement of the window or application of an additional frame). Even in this case one must make a comparison between U values before and after the retrofit in order to make appropriate technical and economic evaluations.

In the case of retrofit actions on windows, however, it should be noted that the increase of the thermal performance of the transparent surface (reduction of transmittance) can lead to a reduction of the transparency (see Table 7.3), and therefore a decrease in solar gains. A more accurate assessment of the net benefits that can be achieved requires a global energy balance.

Table 7.3 Values of the solar transmission factor g and U value for some types of glass

Glass type	Solar transmission factor g	U value U_G (W/m^2K)
Single glazing (6)	0.74	5.7
Double glazing (6-8-6)	0.64	3.1
Double glazing (6-12-6)	0.65	2.8
Low-emission insulating glass with air (4-16-4)	0.52	1.4
Low-emission insulating glass with argon (4-16-4)	0.52	1.1

Example In a building all the windows are replaced. Existing ones have an average transmittance value (frame and glass) U_E of 4.8 W/m^2K, while new ones will have an average U_R of 2.3 W/m^2K. The total gross area of the windows is 60 m^2 and the location has 2,400 DD_H and 260 DD_C. The average seasonal efficiency for winter heating η_H is 0.7, while the average seasonal efficiency of the summer cooling is 1.8.

Calculate energy savings both in winter and in summer.

Using Eq. (1.13), one can calculate energy saving in winter and energy saving in summer:

$$\Delta E_{H,R} = (24 \times ((4.8 \times 60) - (2.3 \times 60)) \times 2,400)/0.7 = 12,342,857\,\text{Wh} \\ = 12.343\,\text{kWh}$$

and

$$\Delta E_{C,R} = (24 \times ((4.8 \times 60) - (2.3 \times 60)) \times 260)/1.8 = 520,000\,\text{Wh} = 520\,\text{kWh}$$

In assessing the overall energy, it is necessary to distinguish between the calculated energy in winter (thermal primary energy), from which the fuel consumption can be easily estimated, and the calculated energy in the summer (electric energy) which must be evaluated separately, unless one is employing non-electric chillers (e.g., gas absorption chillers).

So far as the energy evaluation of windows, or more generally of the transparent surfaces of the envelope, are concerned, the calculation based on the degree-days method is approximated (in particular for summer conditions.), so useful for a rough estimate of the effects. A more precise assessment should be made through a simulation that takes into account solar gains.

7.2.2 Ventilation Control

In many existing buildings uncontrolled infiltration, besides causing energy wastage, may generate considerable discomfort, due to cold and hot air currents.

The solution in this case is to reduce as much as possible these infiltrations, for example through joint sealing of the closures of the door and window frames, sealing the joints wall-door or window, the sealing of the bins or, even better, the complete replacement of the frames. However, this measure should be accompanied by another one: the installation of controlled mechanical ventilation system, in order to ensure the air changes in accordance with air quality requirements.

The analytical evaluation of the uncontrolled infiltration is quite complex and unreliable, but it can be done directly through the blower door test (see Sect. 5.3.4), while the air changes on the basis of which it is necessary to size the system can be taken from technical standards of hygiene according to the intended use.

An approximate evaluation of the energy savings achievable can be made with the equation:

$$\Delta E_{HV,R} = \frac{24(\rho_a \cdot C_a \cdot V)(n_E - n_R(1 - \eta_{\mathrm{REC}})) \cdot \mathrm{DD}_H(\theta_{\mathrm{bal},E})}{\eta_H} \tag{7.14}$$

where:

$\rho_a C_a$	is the volumetric heat capacity of the air which may be assumed equal to 0.34 Wh/m^{3}K;
V	is the net volume of the space, expressed in m^3;
n_E	are the estimated or measured air changes due to infiltration in vol/h;
n_R	are the air changes guaranteed by the installation of mechanical ventilation system expressed in vol/h;
η_{REC}	is the average seasonal efficiency of the heat recovery unit (if applicable);
$\mathrm{DD}_H(\theta_{\mathrm{bal},E})$	are the winter degree days of the location calculated for the balance temperature;
η_H	is the average seasonal efficiency of the heating system.

Example In a building with a net volume of 1,200 m^3, uncontrolled air infiltration generates 0.8 vol/h. Infiltrations are sealed, and the building will be equipped with a control mechanical ventilation system that provides an air change of 0.5 vol/h of fresh air. The mechanical ventilation system is equipped with a heat recovery unit with a seasonal average efficiency of 0.75. The location where the building is situated has 3,600 $\mathrm{DD_H}$. Considering that the average seasonal efficiency of the heating system is 0.75, estimate the annual savings of primary energy.

Applying the (7.14), and substituting the values, one obtains that:

$$\begin{aligned}\Delta E_{HV,R} &= 24 \times (0.34 \times 1{,}200) \times (0.8 - 0.5 \times (1 - 0.75)) \times 3{,}600 \\ &= 23{,}794{,}560\,\mathrm{Wh} \; = 23{,}795\,\mathrm{kWh}.\end{aligned}$$

For a full evaluation of the energy saved it would be appropriate to deduct from the value of $\Delta E_{HV,R}$, the amount of electrical power absorbed by the fan of the ventilation system.

7.2.3 *Actions on HVAC Systems*

If the purpose of the retrofit of the building envelope is to reduce the energy requirements, the scope of the actions on HVAC systems is to produce the energy required maximizing the efficiency and therefore reducing energy wastage.

A good approximation of the average seasonal efficiency η_H (related only to thermal energy, i.e., so without considering the energy consumption of electrical auxiliaries as pumps, fans, etc.) can be evaluated from the relationship

$$\eta_H = \eta_E \cdot \eta_C \cdot \eta_D \cdot \eta_P \tag{7.15}$$

where

η_E is the emission system efficiency,
η_C is the regulation system efficiency,
η_D is the distribution system efficiency, and
η_P is the seasonal average production system (e.g., heating boiler) efficiency.

7.2.3.1 Improvement of the Emission System Efficiency

The emission efficiency η_E is the ratio between the useful energy requirements for space heating with a reference emission that can provide a temperature which is perfectly uniform and of the same value in the various heated spaces, and the real emission system under the same conditions of temperature internal reference and external temperature.

The emission efficiency, depending on the type of heating terminal, has values ranging from 0.99 (radiant panels integrated in the structure) to 0.95 (convectors). So-called "radiators" have an emission efficiency of 0.95.

There are no strategies to improve the emission efficiency unless than redevelopment involving the replacement of the heating terminals.

7.2.3.2 Improvement of the Regulation System Efficiency

A regulation system that does not respond accurately and quickly to the energy demand, generates fluctuations of temperature within the heated spaces which cause increases in heat exchanges by transmission and ventilation.

The regulation or control efficiency η_C is a parameter that expresses the deviation between the amount of energy required in real conditions compared to that under ideal ones, and is given by the ratio between the useful energy requirements of space heating with a theoretical perfect regulation and that required for the heating of the same spaces but with the real control system.

The efficiency values of the regulation systems, if we exclude the really rare case of manual adjustment which can attain 0.78, with a centralised climatic control system ranges from 0.82 to 0.88 depending upon the associated heating terminal.

The values of maximum performance, up to 0.99, are obtained with control systems for single space with pre-set and adjustable modulating control.

7.2.3.3 Improvement of the Distribution System Efficiency

The distribution system efficiency η_D is the ratio between the energy requirements of a heated space and the heat supplied by the heating production system. It depends on the heat losses and thus the quality of the thermal insulation of the piping network (or ducts if the thermal transport fluid is air).

In the case of heating system pipes, the distribution system efficiency takes values ranging from a minimum of 0.87 to a maximum of 0.99.

The distribution system efficiency of an existing plant can be improved through the insulation of the pipes, at least for those portions still accessible.

A correct evaluation of the energy saved by thermally insulating the pipes can be made through the analytical calculation of heat losses from the pipes. Table 7.4 shows the values of the pre-calculated transmittance per unit of length, expressed in W/mK depending on the diameter of the pipe, the conductivity of the thermal insulation λ used and the thickness of the insulating layer. The values were calculated assuming that the pipes currently run outside.

7.2.3.4 Improvement of the Production System Efficiency

The average seasonal efficiency of production system η_P is the ratio of the heat supplied from the production system in the heating season to the primary energy demand in the season.

The production efficiency depends on the type and the status of the heating boiler type. In old inefficient boilers the average seasonal efficiency could reach very low values, for example 0.75, whereas in modern boilers it can reach values even above 0.95.

In condensing boilers the efficiency, referred to the *lower calorific value* (LCV) of the fuel, can be greater than 1 if the temperature of return water from the plant is maintained at a value lower than 40 $\div$ 50 °C. The replacement of a heat generator with a traditional-type condensation boiler allows one therefore to obtain a considerable improvement in the overall efficiency.

The best performances are obtained by replacing a boiler with a heat pump: the values of the seasonal production efficiency in this case are around 1.5 $\div$ 1.7.

For a rough estimate of the energy (fuel) saving thanks to the replacement of an existing heat generator with a new heat generator efficiently (for example, a condensing boiler) it is possible to use the following equation:

Table 7.4 Values of the pre-calculated thermal transmittance of pipes insulation per length unit, expressed in W/mK [7]

Thickness (mm)	λ (W/mK)	External diameter of pipe (mm and inches)								
		21.35 1/2″	26.75 3/4″	33.6 1″	42.3 1″ 1/4	48.2 1″ 1/2	60.15 2″	75.75 2″ 1/2	88.65 3″	113.95 4″
10	0.030	0.241	0.283	0.336	0.403	0.448	0.539	0.658	0.756	4.309
	0.035	0.268	0.314	0.373	0.447	0.497	0.597	0.728	0.836	4.305
	0.040	0.306	0.359	0.425	0.509	0.565	0.679	0.827	0.949	4.313
	0.045	0.336	0.393	0.466	0.557	0.619	0.743	0.904	1.038	4.318
15	0.030	0.193	0.224	0.263	0.311	0.344	0.410	0.495	0.565	3.998
	0.035	0.216	0.251	0.294	0.348	0.384	0.457	0.552	0.630	4.292
	0.040	0.249	0.289	0.338	0.400	0.442	0.525	0.634	0.723	4.298
	0.045	0.276	0.320	0.374	0.442	0.488	0.580	0.699	0.797	4.303
20	0.030	0.166	0.191	0.221	0.260	0.285	0.337	0.403	0.458	3.613
	0.035	0.186	0.214	0.248	0.291	0.320	0.377	0.452	0.513	4.284
	0.040	0.216	0.248	0.287	0.337	0.370	0.436	0.522	0.592	4.290
	0.045	0.240	0.275	0.319	0.374	0.410	0.484	0.578	0.656	4.294
30	0.030	0.135	0.153	0.175	0.202	0.221	0.257	0.304	0.342	2.864
	0.035	0.152	0.172	0.197	0.228	0.248	0.289	0.342	0.385	4.276
	0.040	0.177	0.201	0.229	0.265	0.289	0.336	0.397	0.447	4.280
	0.045	0.198	0.224	0.256	0.296	0.322	0.375	0.443	0.498	4.283
40	0.030	0.117	0.132	0.149	0.171	0.185	0.214	0.250	0.280	4.269
	0.035	0.132	0.149	0.168	0.193	0.209	0.241	0.282	0.316	4.271
	0.040	0.155	0.174	0.197	0.225	0.244	0.281	0.329	0.368	4.275
	0.045	0.173	0.194	0.220	0.252	0.273	0.314	0.368	0.411	4.277

$$\Delta FU = \frac{\eta_{\text{eff}} - \eta_{\text{old}}}{\eta_{\text{eff}}} FU_{\text{old}} \tag{7.16}$$

where:

η_{eff} is the average seasonal efficiency of the new boiler;
η_{old} is the average seasonal efficiency of the existing boiler;
FU_{old} is the actual fuel (or energy) consumption.

7.2.4 Renewable Energy Sources

Renewable energy sources can provide a significant contribution to the reduction in energy consumption. Their application, most common in new buildings, can also be assessed for existing buildings undergoing renovation. A comprehensive framework guide to the technology and the potential of renewable energy sources is proposed by Michaelides [8].

7.2.4.1 Solar Thermal Systems

The absorbing surface of the solar system can be calculated from the ratio between the energy required by the users, considering also the thermal losses of the plant, and the energy that the solar system is able to provide per unit area.

The variability in climatic data requires the running of an energy balance at least monthly. If the more favourable months, characterised by the maximum incident solar radiation, are considered, then the system is undersised for the remaining months during which the integrative system remains activated. On the other hand, if the system is sized for the less favourable months, the collecting surface is oversised and in periods characterised by maximum solar radiation, the solar thermal system provides a quantity of energy which, in fact, is not employed because it is in excess of requirements.

An optimum design is one that exploits the system during the period in which solar radiation is sufficient, which would be in the period between April and October (obviously for the locations in the northern hemisphere).

Owing to its complexity, the sizing of a solar thermal system requires the use of a calculation model and a software package. The calculation model most widely used for solar thermal systems is the F-Chart, elaborated by Beckman et al. [9].

The availability of a well-exposed surface for the installation of solar collectors often constitutes a constraint to this choice. Table 7.5 shows the corrective coefficients to be applied to surfaces variously tilted and orientated, starting from the value of solar radiation on the horizontal surface (latitude 45° N). From the table it can be seen that even for non-optimal orientations there are no particular penalties of incident solar radiation: this demonstrate the flexibility of solar thermal systems [but also for solar photovoltaic (PV) systems] for an integrated installation on buildings.

For a preliminary sizing of a thermal solar system Table 7.6 can be used. This refers to three locations respectively characterised in three latitudes: 45° N, 40° N and 35° N (the analytical calculation using calculation procedures between standard is however considered indispensable).

The table has been designed with the following parameters:

Table 7.5 Corrective coefficients to be applied to surfaces variously tilted and orientates starting from the value of solar radiation on the horizontal surface (latitude 45° N) [7]

Solar collectors orientation	Solar collectors tilt angle				
	20°	30°	45°	60°	90°
0°	1.11	1.13	1.11	1.03	0.75
±15°	1.10	1.12	1.11	1.03	0.76
±30°	1.09	1.11	1.10	1.03	0.78
±45°	1.07	1.09	1.08	1.02	0.79
±60°	1.05	1.06	1.04	0.99	0.78
±90°	0.99	0.97	0.94	0.88	0.70

- solar collector type: flat plate selective
- collectors orientation: South ± 15°
- collectors tilt: 25 ÷ 30°
- water supply temperature: 12 °C
- DHW temperature: 40 °C
- Heat losses from solar circuit: 5 %

Once the location, for Latitude, is chosen and identified the ratio between the solar collector surface and the daily DHW demand, from the table we can read directly the following values:

- annual solar integration (thermal energy demand/energy supplied from the solar system);
- solar integration in the period between April and October;
- utilisation coefficient of solar system (i.e., the ratio between of the energy required for DHW and the energy supplied and used by the solar system).

The system is economical as much as the value of the utilisation coefficient of the solar system is close to 100 %.

The system is economical wherever the value of the utilisation coefficient of the solar system is close to 100 %.

From the table it can be observed that it is not convenient practically, even in locations characterised by a high solar radiation, to size the system for a maximum annual integration. It might be more convenient to ensure good coverage in the period from April to October.

The installation of a solar heating system in an existing building can be cost-effective if:

- there is enough space for the installation of solar collectors, and this space is conveniently exposed to the solar radiation;
- there is a space for the installation of the thermal storage;
- that the building is already equipped with a centralised DHW system.

Without meeting the third condition, the solar heating system can nonetheless be supplied but the installation cost is higher since it is necessary to supply all the users with a new hot water hydraulic distribution network.

Example One has to size the solar collecting surface for a solar thermal system for a location with a latitude of 45° N with a daily DHW requirement of 2,000 L.

From Table 7.6 it can be seen that with a ratio between the solar collector surface and the daily DHW demand of 12.5 m^2/m^3 a quite high (87 %) solar integration in the summer period is obtained, maintaining however a good utilisation coefficient (93 %).

Confirming this choice the solar collector surface of the plant will be equal to: 12.5 (m^2/m^3) × 2 (m^3) = 25 m^2; the solar system sized in this way will cover 58 % of the annual energy requirements.

7.2.4.2 Solar Photovoltaic Systems

The availability of some forms of incentives, such as those adopted in some countries such as Germany, Spain and Italy, has rendered convenient the installation of solar PV systems that can produce electricity.

In a collective building, a solar PV system can be designed to meet the electrical requirements of the equipment in the communal areas [staircase lighting, outdoor lighting, electricity for lifts (elevators), etc.].

For these systems, the major constraint is the availability of space for the installation of PV modules (an average of 8 m^2 per peak kW). PV systems give less problems than thermal solar systems because only electrical connections are required, those being simpler to execute than hydraulic connections.

Table 7.7 shows the productivity per square metre of average electrical panel and power unit installed for a few different locations in different latitudes. (the analytical calculation also in this case is needful).

In Table 7.7 with the symbol balance of system (BOS) one considers the losses due to equipment and circuits needed to transfer the electricity produced by the photovoltaic field to the network or users.

7.2.5 Remedial Action/Works on Electrical Installations

7.2.5.1 Power Factor Correction

We define *rephasing* (or power factor correction) any measure to improve the power factor ($\cos \varphi$) of a given electrical load, in order to reduce, with the same value of active power consumption, the value of the current that circulates in the electrical circuit.

Table 7.6 Annual and seasonal (April to October) solar integration and utilisation coefficient for thermal solar plants as a function of the ratio between solar collector surface (m^2) and DHW demand (m^3) for three locations [7]

Latitude (°N)		Solar collecting surface/daily DHW demand (m^2/m^3)								
		5	7.5	10	12.5	15	17.5	20	22.5	25
45	Annual (%)	25	37	50	58	63	65	67	69	71
	April–October (%)	38	57	75	87	93	95	96	97	98
	Utilisation coefficient (%)	100	100	99	93	84	74	67	62	57
40	Annual (%)	34	52	66	74	79	83	85	88	90
	April–October (%)	45	70	90	97	100	100	100	100	100
	Utilisation coefficient (%)	100	100	96	86	77	69	62	57	52
35	Annual (%)	40	60	75	83	88	92	95	96	97
	April–October (%)	51	76	95	99	100	100	100	100	100
	Utilisation coefficient (%)	100	100	95	83	74	66	59	54	49

Table 7.7 Production of electricity from a PV system. Hypothesis: tilt of panels surfaces 30°, facing south, module efficiency 12.5 % BOS efficiency of 85 %, 1 kWp of 8 m^2 of panels [7]

Latitude (°N)	Annual solar radiation	Electricity (AC current)	
	kWh/m^2	kWh_{el}/m^2	kWh_{el}/kWp
45	1,372.4	145.8	1,167
40	1,737.4	184.6	1,477
35	1,963.7	208.6	1,669

The scope of the power factor correction is above all to reduce energy losses in distribution grids and to reduce the absorption of the "apparent" or reactive power. Authorities normally require of electricity distributors, users, and non-residential consumers, contract clauses that require the user to keep the value of the power factor above a certain threshold or pay a penalty. The power factor correction of an electrical system represents one of the most convenient technological measures from the economic point of view.

In circuits with resistive loads such as filament of lamps, water heaters, certain types of ovens, the power consumption is the whole active power. In circuits whose users that have installed inductive and/or capacitive devices such as motors, ballasts of fluorescent lamps, transformers, a part of the apparent power, said *reactive power*, is stored and then transferred, while only the part of the said active power is actually used.

A purely reactive load, therefore, does not absorb active power but only reactive power. Then it does not absorb energy from the grid but generates energy losses in the cables and in the devices constituting the system as an active load of the same power.

With industrial users, generally, most of the loads are constituted by motors and transformers and are therefore of the inductive type, which, especially when run unloaded, absorb reactive energy (expressed in kVA_{RH}). Both the reactive power and the active power, the only type useful and transformable, for example, into mechanical work, contribute to the total current flowing in the cables and components of the network. This additional current, linked to reactive loads, causes a decrease in the ability of transporting useful energy on a cable, since (when assimilating the electric cable to a hypothetical tube) its presence "steals" space from a certain amount of active energy.

The reactive power, therefore, constitutes an additional load for the generators, transformers and transport and distribution lines, obliging the electrical utilities to oversize their generators at the expense of efficiency and also causing a greater voltage drop in the line, which results in further losses of active power.

To overcome this problem, batteries of capacitors (capacitive loads) are inserted in parallel with the motors in order to contrast the effect of inductive loads, tending to bring back to "phase" voltage and current. For this reason this operation is called "power factor".

Normally, for electrical installations with low voltage and power input greater than 15 kW:

- when the average monthly power factor is less than 0.7, the user is required to correct the power factor to the facility;
- when the average monthly power factor is between 0.7 and 0.9 there is no obligation for the correction of the system but the user pays a penalty for reactive energy;
- when the average monthly power factor is greater than 0.9 and less than 1 there is no obligation for the correction of the system and no fee payment is due for reactive energy.

To improve the power factor one can take action with a series of technical measures, such as:

- use properly sized motors and transformers;
- do not leave motors and transformers without load.

Il the good practices do not give sufficient results it is necessary to re-phase the system:

1. by the provision of parallel motors of capacitor (capacitive loads);
2. through the use of automatic static compensators.

To know the capacity Q_c of the capacitor required to reach the target, we need to know the reactive power to be compensated, the value of which is obtained with the following equation:

$$Q_c = P \cdot (tg\varphi - tg\varphi') \quad (7.17)$$

where $\varphi\prime$ is the value of phasing that to be achieved and P is the power. Generally it is not possible to obtain a standard value capacitor for a power factor correction which is exactly equal to the desired one, for this reason it is appropriate to perform the calculation for the total power factor correction ($\varphi\prime = 0$) and choose a product that falls between the two values calculated. A capacitor bank of excess capacity would even cause an increase in the reactive power.

To know the value of the capacitors to be used simply apply another equation:

$$C = \frac{Q_c}{n_c \cdot \pi \cdot 2f \cdot V^2} \quad (7.18)$$

where:

n_c corresponds to the number of capacitors, in three-phase systems three capacitors are used (connected in star or delta), in single-phase systems only one (in series or parallel);

f is the frequency of the electricity grid;

V^2 is the square of the voltage to which the capacitors are subject, in the case of single-phase systems, corresponds to the voltage of the system, as well as in the case of three-phase systems with a delta connection of capacitors, while in the case of star connection is must divide the system voltage to $\sqrt{3}$.

The evaluation of cost savings due to the increase of the power factor cannot be generalised but depends on the structure of the supply contract.

7.2.5.2 Replacement of Inefficient Electric Motors

The replacement of existing electric motors with high-efficiency electric motors generates a saving of energy that can be assessed in different ways.

The simplified method provides that the motor is always operated at full load and that the speed change is negligible. The power savings ΔP_R owning to the replacement of the motor can be calculated with the following equation:

$$\Delta P_R = P_M \left(\frac{1}{\eta_E} - \frac{1}{\eta_R} \right) \tag{7.19}$$

where:

P_M is the mechanical power generated by the motor;
η_E is the full-load efficiency of the existing motor;
η_R is the full-load efficiency of the new motor.

The electricity saving due to replacement of the motor can be calculated by the equation:

$$\Delta E_{EL} = \Delta P_R \cdot N_h \cdot LF_M \tag{7.20}$$

where:

N_h is the number of hours per year during which the motor is operating;
LF_M is the average load factor of the motor during the year.

7.2.6 Energy Efficiency of Lighting Systems

7.2.6.1 Efficient Management of Lighting Systems

A reduction of the energy consumption from lighting can be achieved using devices able to turn off the lamps if there is no presence of people or if natural lighting is adequate.

Thanks to the progress in electronics sector, nowadays one can readily adjust the luminous flux produced by lamps: the adjustment can be manual or automatic, related to the presence of persons or to the maintenance of predetermined lighting levels, or according to the natural light variation.

Table 7.8 Potential energy savings with retrofits of occupancy sensors [11]

Space application	Range of energy saving (%)
Offices (private)	25 ÷ 50
Offices (open space)	20 ÷ 25
Rest rooms (WC)	30 ÷ 75
Conference rooms	45 ÷ 65
Corridors	30 ÷ 40
Storage areas	45 ÷ 65
Warehouses	50 ÷ 75

The continuous adjustment of the lighting level in a space permits the user to always have the maximum visual comfort with the minimum energy consumption. The energy savings in this case cannot be generalised but must be assessed as a function of the intended use and the mode of occupation of the building.

The European standard EN 15193 [10] provides a method of assessment of energy consumption of lighting systems as a function of the types of light fixtures used and systems of control and management (centralised switching-off, flow control, presence sensors etc.). Table 7.8 provides some range of potential energy savings for lighting systems that can be obtained in various types of space using presence sensors.

7.2.6.2 Installation of High-Efficiency Lamps

The replacement of existing lamps with high-efficiency lamps generates an energy saving ΔE_{LI}, expressed in kWh, which can be evaluated through the following equation:

$$\Delta E_{LI} = \frac{N_{\text{lum}}(WR_E - WR_R)N_h}{1{,}000} \tag{7.21}$$

where:

N_{lum} is the number of replaced lamps;
WR_E is the power, in W, of the existing lamps;
WR_R is the power, in W, of the new lamps;
N_h is the number of operating hours per year of the lamps.

In a lighting system with fluorescent lamps, but with a traditional electromechanical power system, improved energy efficiency can be achieved without changing the lighting, by installing a new electronic ballast and possibly replacing the lamps. The electronic ballasts also ensure a more rapid switch-on, longer service life of the lamps and the ability to adjust the luminous flux.

In the assessment of energy benefits it must be considered that, in summer, the lower power consumption of the lamps corresponds to a reduction of the share of internal heat loads that must no longer be rejected from the installation by the HVAC systems.

In the winter period, however, the reduction of the load causes a reduction in the internal gains that are instead useful in this season. However, the delivery of the same amount of heat via the air conditioning system is more convenient since that amount of heat emitted by the lamps would be provided using electricity.

7.3 Overall Evaluation of Thermal Performance

A more accurate and precise evaluation of the energy performance of a building should be done through a comprehensive assessment, which means a modeling of the entire building.

The inclusion of technical and performance characteristics of the building in the original state i.e., before the renovation is useful to define the baseline, that represents the starting point from which the improvements due to different scenarios of retrofit actions can be estimated. The performance improvement is global, in other words it will not be possible, in this case, to evaluate the effect of the single action.

The overall assessment of the performance can be made through a *steady state simulation* or through a *transient simulation* with dynamic models.

7.3.1 Steady State Simulation of Buildings

The European reference for the calculation of the energy balance energy is the standard EN ISO 13790:2008 [12] that gives calculation methods for assessment of the annual energy use for space heating and cooling of a residential or a non-residential building, or a part of it, referred to as "the building".

The method includes the calculation of:

- the heat transfer by transmission and ventilation of the building zone when heated or cooled to constant internal temperature;
- the contribution of internal and solar heat gains to the building heat balance;
- the annual energy needs for heating and cooling, to maintain the specified set-point temperatures in the building—latent heat not included;
- the annual energy use for heating and cooling of the building, using input from the relevant system standards referred to in ISO 13790:2008 and specified in Annex A.

ISO 13790:2008 has been developed for buildings that are, or are assumed to be, heated and/or cooled for the thermal comfort of people, but can be used for other types of building or other types of use (e.g., industrial, agricultural, swimming pool), as long as appropriate input data are chosen and the impact of special physical conditions on the accuracy is taken into consideration.

Energy requirements are calculated at standard conditions, i.e., normalising all the information regarding the ways in which the user uses the equipment (such as maintaining a constant internal air temperature) and assuming that the external weather conditions, in winter or in summer, remain constant for the same site.

In the simplest case of an heating system, energy demand represents the amount of energy that must be provided to ensure that the building has a constant inside temperature throughout the heating season. The energy required to meet the needs of the building is provided by a heating system which, owing to its inefficiencies, consumes a greater amount of energy that is precisely the primary energy.

The concept is well highlighted in Fig. 7.2, which shows the diagram of the energy balance for the assessment of the global energy requirements of a building and the numbers have the following meanings [13]:

(1) represents the energy needed to fulfill the user's requirements for heating, cooling, lighting etc., according to levels that are specified for the purposes of the calculation;
(2) represents the "natural" energy gains—passive solar heating, passive cooling, natural ventilation, illumination using daylighting—together with internal gains (occupants, lighting, electrical equipment, etc.);

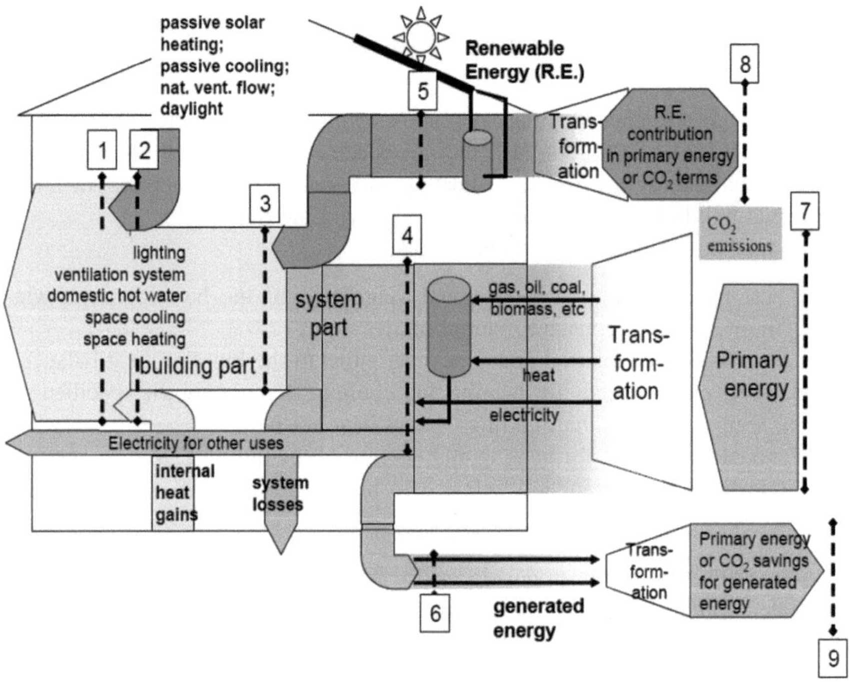

Fig. 7.2 Schematic illustration of the calculation scheme [13]

(3) represents the building's energy needs, obtained from (1) and (2) along with the characteristics of the building itself;
(4) represents the delivered energy, recorded separately for each energy carrier and inclusive of auxiliary energy, used by space heating, cooling, ventilation, domestic hot water and lighting systems, taking into account renewable energy sources and CHP/co-generation. This may be expressed in energy units or in units of the energyware (kg, m^3, kWh, etc.);
(5) represents renewable energy produced on the building premises;
(6) represents generated energy, produced on the premises and exported to the market; this can include part of (5);
(7) represents the primary energy usage or the CO_2 emissions associated with the building;
(8) represents the primary energy or CO_2 emissions associated with on-site generation which is actually used on-site and thus is not subtracted from (7);
(9) represents the primary energy or CO_2 savings associated with energy exported to the market, which is thus subtracted from (7).

Figure 7.2 is a schematic illustration and is not intended to cover all possible combinations of energy supply, on-site energy production and energy use. For example, a ground-source heat pump uses both electricity and renewable energy from the ground; and electricity generated on site by photovoltaic's could be used entirely within the building, or it could be exported entirely, or a combination of the two. Renewable energy wares like biomass are included in (7), but are distinguished from non-renewable energy wares by low CO_2 emissions. In the case of cooling, the direction of energy flow is from the building to the system.

For the calculation of the energy performance for heating and cooling, the process involves the following steps:

1. Definition of the boundaries of all the air-conditioned and non air-conditioned areas of the building.
2. If required, definition of the boundaries of the different zones.
3. Definition of the internal and external conditions and input data relating to the outside climate.
4. Calculation for each month and for each area of the building, the net energy requirements for heating and cooling.
5. Aggregation of the results for the different months and in different areas served by the same plant.

The primary energy demand of the building must take into account the inefficiencies of the systems. The HVAC system is divided into five subsystems:

- emission subsystem;
- regulating or control subsystem;
- distribution subsystem;
- thermal storage subsystem (if any);
- generation subsystem.

For each of these subsystems, thermal energy losses and electrical energy required by the auxiliary electric devices of the plant (for example electric pumps, fans, etc.) are calculated. The primary energy demand of the building is calculated by summing the energy requirements of the building, the heat lost through inefficiencies of the plants and the primary electrical energy supply needed to run the auxiliaries.

7.3.2 Dynamic Simulation of Buildings

In the steady state simulation of buildings, the calculation model considers the boundary conditions, such as the outdoor temperature and the indoor air temperature, or the solar radiation, constant for the calculation period. If the evaluation of the thermal balance is monthly, for example, it is assumed that the external and internal air temperature and the solar radiation are constant for the entire month.

The limits of this model are that it does not consider the dynamic effects of heat transmission due to the variation of the external environmental parameters (temperatures, solar radiation, wind speed, etc.) and does not consider the thermal inertia of the building structures.

The dynamic simulation models, on the contrary, make a calculation in transient conditions so the results of the building simulation are much closer to the real situation in which the environmental climatic parameters change continuously.

The dynamic calculation models are certainly more complex, as they require much more data concerning the climatic data, the thermo-physical characteristics of the building components (walls, windows, roofs, etc.), the characteristics of the facility equipment and the space management.

If a climatic data file of a steady state model contains a maximum of 12 sets of values (e.g., temperature, solar radiation, humidity, etc.), i.e., a set once per month, a climate data file of a transient model contains at least one complete set per hour, so 8,760 sets.

The dynamic simulation models or building simulation programmes (BSP) require much more effort in modeling the building, which must be three-dimensional in order to evaluate the effects of shadows on the building facades, in the description of technical systems and in the description of usage profiles.

In addition, the time required for the processing increases considerably (although the increase in the speed of calculation of a PC tends to solve this problem) since there is a verification, always with the step of calculation defined by the user, not only of the interactions between the internal and the external environment but also of those between different environments, characterised by different profiles of use and by different models of plant management, and the interactions between different plant systems. With a dynamic simulation model the air conditioning system and the lighting are not two separate entities but are systems which interact with each other, simulating what is the real behaviour of the building.

This increased commitment in the use of dynamics simulation, compared with a greater effort guarantees to the designer or, more particularly, the auditor, a great deal of flexibility in the modeling of buildings. If the inputs to the model are correct, and the user is skilled, using the dynamic simulation the effects that will lead different scenarios of retrofit measures can be predicted very accurately. For this reason the use of dynamic simulation of buildings is cost-effective.

The main advantages of using a dynamic simulation model are as follows:

- a more realistic energy balance for heating and cooling;
- an accurate predictive assessment of primary energy consumption;
- a careful evaluation of indoor environmental parameters (temperature, relative humidity, air speed, etc.);
- a proper assessment of the thermal component both in winter and in summer;
- a more correct assessment of the effects of solar radiation, because the solar path is simulated hour by hour;
- a more correct assessment of the effects of shading systems (shelves, overhangs, etc.) or the effects of shadows from other buildings;
- a correct evaluation of the energy components due to bioclimatic greenhouses or direct gain systems;
- an assessment of the energy performance of special enclosures such as double skin facades and ventilated walls.

Some simulation software can be implemented with computational fluid dynamics (CFD) in order to verify systems with natural ventilation or to better control the indoor environmental conditions of spaces of considerable height such as gallcrics.

The use of dynamic simulation of buildings has become essential to simulate the behaviour of complex buildings with large window areas, or to simulate the behaviour, and then verify the performance, of buildings with high or very high energy efficiency (e.g., Zero Energy Buildings).

The USA department of energy (DOE) has established and maintains a building energy software tools directory (BESTD), that is a database of the main software used in the energy field, including those for the dynamic simulation, some of which are free (http://apps1.eere.energy.gov/buildings/tools_directory/). The cost of the simulation software itself is relatively low whereas higher is the cost, in terms of the time, necessary to learn how to make a correct use of the models.

Some of the softwares presented have the calculation engine inside them. This is the case for example of TRNSYS, ESP-r or EnergyPlus. The first of these, developed over 30 years ago, is sold for a fee while the other two are distributed free of charge. The problem lies in the interface of these softwares which is not user-friendly: the description of the building must be done manually by filling in an ASCII file that contains all the information, both those relating to the building envelope, and those that relate to facilities. However, these are programmes that accept open interfaces developed by third-party software houses. Two of the programmes presented, DesignBuilder and eQuest, were born as interfaces and maintain EnergyPlus as the calculation engine. The availability of user-friendly interfaces contributed to the diffusion of these instruments to a wide range of users with no particular skills on modeling. The graphical user interfaces greatly facilitate the description of the building.

For a computer simulation of a building to be of value in evaluating energy retrofit measures, it must be accurate. To be "accurate," the model should account for essentially all of the sources and uses of energy in a building.

Accuracy in computer simulation of buildings, according to the experience of Thumann and Younger [14] is founded upon three basic areas:

1. an intimate understanding of the simulation tool being used, including its various idiosyncrasies and nuances;
2. an intimate understanding of the building being simulated, vis-à-vis its physical and operational characteristics—in essence, in existing buildings, the quality of the survey or "audit" determines the quality of the simulation;
3. careful analysis and critique of output data (just because it is carefully prepared and computer generated does not mean it is correct)—our comments elsewhere herein generally apply to "mainframe" programs, though they also apply to other simulation tools.

By utilising the above techniques, the above mentioned authors have found it possible to regularly model buildings within 5 % of their actual annual energy use with a high degree of confidence in the simulation of each energy-using system and functional use of energy in the building. It should be noted that, in buildings where weather is a strong energy use factor, modeling to less than 10 % variance from the actual energy use may be of limited value as our ability to predict weather for a given future year may not even be that accurate [14].

7.4 Software for Building Simulation

In the paragraphs below are some brief descriptions, taken from BESTD, of the most commonly software for dynamic simulation of buildings used worldwide.

7.4.1 TRNSYS

An energy simulation programme whose modular system approach makes it one of the most flexible tools available. TRaNsient SYstem simulation program (TRNSYS) includes a graphical interface, a simulation engine, and a library of components that range from various building models to standard HVAC equipment to renewable energy and emerging technologies.

TRNSYS also includes a method for creating new components that do not exist in the standard package. This simulation package has been used for more than 30 years for HVAC analysis and sizing, multi-zone airflow analyses, electric power simulation, solar design, building thermal performance, analysis of control schemes, etc.

The TRNSYS input file, including building input description, characteristics of system components, manner in which components are interconnected, and separate weather data (supplied with program) are all ASCII files. All input files can be generated with a graphical user interface known as Simulation Studio. The envelope for the TRNSYS building can now be three-dimensionally created in the TRNSYS3D plugin for Google SketchUp™ and edited in the TRNBuild interface.

Basic output format is ASCII. The data included in those files can be life cycle costs; monthly summaries; annual results; histograms; plotting of desired variables (by time unit). It is also possible to plot variables online (as the simulation progresses).

The computer platform for TRNSYS are Windows 98 or higher (NT, 2000, ME, XP, Vista, Windows 7, etc.) (Distributed source code will compile and run on any FORTRAN platform.)

Due to its modular approach, TRNSYS is extremely flexible for modeling a variety of energy systems in differing levels of complexity. Supplied source code and documentation provide an easy method for users to modify or add components not in the standard library; extensive documentation on component routines, including explanation, background, typical uses and governing equations; supplied time step, starting and stopping times allowing choice of modeling periods. The TRNSYS programme includes a graphical interface to drag-and-drop components for creating input files (Simulation Studio), a utility for easily creating a building input file (TRNBuild), and a program for building TRNSYS-based applications for distribution to non-users (TRNEdit). Web-based library of additional components and frequent downloadable updates are also available to users. Extensive libraries of non-standard components for TRNSYS are available commercially from TRNSYS distributors. TRNSYS also interfaces with various other simulation packages such as COMIS, CONTAM, EES, Excel, FLUENT, GenOpt and MATLAB.

The commercial version of TRNSYS is sold at €4,500 whilst the "educational" version has a price reduced to €2,500. Demo versions and information are available by contacting the local distributors (the official website is (http://energyplus.gov).

7.4.2 EnergyPlus

One of the most common dynamic simulation models in the world, developed by the DOE in the USA, EnergyPlus, has its origins in two simulation models: building loads analysis and system thermodynamics (BLAST) and DOE-2, developed between the late 70 s and early 80 s. The purpose of the two models was to become useful tools for engineers and architects for the correct sizing of air conditioning systems, the development of studies for upgrading the energy efficiency, performance optimisation, etc.

EnergyPlus is a simulation model that allows for an energy balance under dynamic building-plant system conditions and to evaluate the heat load in summer and winter. Based on a description of the thermo-physical characteristics of a building by the user, the programme performs the calculation of the heat load in summer and winter necessary to maintain thermal conditions of well-defined setpoint through the use of a HVAC system.

The programme calculates the primary energy requirement (considering the performance behaviour of the main components and plant systems), and performs other simulations of detail necessary to verify that the simulation is consistent with the required performance.

EnergyPlus includes innovative simulation capabilities including time steps of less than an hour, modular systems simulation modules that are integrated with a heat balance-based zone simulation, and input and output data structures tailored to facilitate third-party interface development. Recent additions include multi-zone airflow, electric power simulation including fuel cells and other distributed energy systems, and water manager that controls and report water use throughout the building systems, rainfall, groundwater, and zone water usage.

EnergyPlus accepts interfaces developed by third-parties. Open studio Plug-in for Google SketchUp allows us to use the simulation of EnergyPlus simulations starting from the graphical user-friendly 3-D Google SketchUp. It was tested by IEA BESTEST for the evaluation of the thermal load of the building and for HVAC systems. Amongst the users of the software, there are designers (architects and consulting engineers), operators, owners and energy consultants. EnergyPlus is also used by professionals, universities and researchers. From the first release, which was in April 2001 85,000 copies of the programme were distributed.

EnergyPlus uses as input a simple ASCII file. The interfaces are made available by developers and greatly facilitate data entry. As output the programme generates ASCII files that can be imported directly from spreadsheets for further processing.

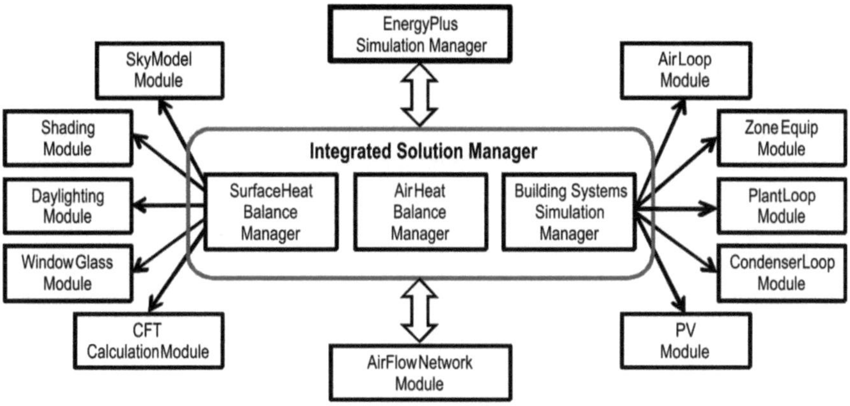

Fig. 7.3 The internal structure of EnergyPlus (*Source* Technical documentation EnergyPlus)

The programming language used is FORTRAN 2003, the programme works with Windows XP/Vista, Mac OS, and Linux. Figure 7.3 shows the internal structure of EnergyPlus.

Accurate, detailed simulation capabilities through complex modeling capabilities. Input is geared to the 'object' model way of thinking. Successful interfacing using IFC standard architectural model available for obtaining geometry from CAD programs. Extensive testing (comparing to available test suites) is completed for each version and results are available on the web site. Weather data for more than 2,000 locations worldwide available on the web site (http://energyplus.gov).

7.4.3 ESP-r

Environmental system performance, research version (ESP-r) is an energy simulation code able to simulate the behaviour of buildings in terms of thermal loads, lighting and sound, and to assess the flows of energy and carbon dioxide emissions: related systems control indoor environment and construction materials also in the transient.

Considering all aspects simultaneously, ESP-r allows the designer to explore and manage the complexity of the relationship between shape of the building, airflow systems, and control systems.

It is a deterministic model, based on the solution of the governing equations of physical phenomena analysed starting from the boundary conditions determined by the input data (thermo-physical characteristics of the building, internal gains, climatic data). For the solution of the balance equations of mass and energy of the structure is used the method of finite differences. The programme basically consists of three modules:

- Project Manager;
- Simulator;
- Results Analyser.

The definition of the project is accomplished by accessing the satellite modules, such as the database (sequences of different climatic characteristics, wind profiles, components, properties of window frames, etc.) and the modules for the definition of shading systems, factors view, etc. The interaction between the three modules can be realised continuously, so as to support the work of the designer: the programme user analyses the results, change some parameters of the problem and performs simulations in an iterative cycle until reaching the desired solution.

Template files generated with ESP-r can be exported in a format compatible with EnergyPlus complete with all the attributes of the components (materials, surfaces, etc.).

There are hundreds of ESP-r users, mainly distributed throughout Europe and Asia: these include consulting engineers, architects, energy consultants and researchers. To use the software skills are required in building thermal physics and environmental control systems.

The geometry of the buildings can be defined using external CAD, or using the tool internal to the programme. ESP-r is compatible with AutoCAD and ECOTECT that can be used to build the model of the building. The models generated can in turn be exported for processing with other software such as Radiance and TSBI3.

ESP-r has been developed in C and FORTRAN (F77 or F90) and compiled with compilers UNIX and Linux. It can work with different operating systems: Sun-Solaris, Silicon Graphics: Sparc5, Linux, Mac OS X 10.1 or newer.

The ESP-r software has been developed between 1977 and 1984 by the Energy Systems Research Unit, University of Strathclyde and is a freely available software and is released under the GNU General Public License (Gnu), downloadable from the official site (http://www.esru.strath.ac.uk/).

7.4.4 eQuest

This is the dynamic simulation software widely used in the U.S. (it is estimated that there are about 10,000 downloads per year) and tested to evaluate the energy performance of buildings. The interactive graphics, the parametric analysis and the calculation speed make this software tool capable of performing simulations of the performance of the entire building in the different phases of the project, from that preliminary to executive. The calculation engine is eQuest the DOE 2.2, model tested for several years and reliable. eQUEST has been tested with the ASHRAE Standard 140.

For a basic usage are not theoretically necessary expertise to energy analysis. To use the detailed interface of the software, however, requires knowledge in building technologies. Experience with other simulation models, particularly those that are based on DOE-2, is useful.

Among the users of the software, there are designers (architects and consulting engineers), operators, owners and LEED energy consultants. eQuest is also used by professionals—quali ??, universities and researchers. EQuest can be simulated using various measures of energy retrofit, so it can be used as part of the Energy Audit.

The input data can be provided in three ways: through an outline of the project through a project developed in more detail and through a detailed interface (DOE-2.2).

As regards output, the programme provides a graphical report synthetic or a comparison between the results of different simulations, as well as a parametric relationship. The software has been developed through a C++ interface for the calculation engine DOE-2.2 in FORTRAN. It runs with Microsoft Windows operating systems 98/Me/NT/2000/XP/Vista.

eQUEST is supported primarily through public funding from California's Savings By Design (www.savingsbydesign.com) and Energy Design Resources (www.energydesignresources.com), and is available at no cost from www.Energy DesignResources.com and www.doe2.com. Long-term average weather data (TMY, TMY2, TMY3, etc.) for 1,000+ locations in North America are available via automatic download from within eQUEST (requiring an Internet connection).

7.4.5 DesignBuilder

DesignBuilder is a graphics modeller with user-friendly interface for buildings and building facilities connected to them. The software provides a variety of services ranging from verification of energy consumption in summer and in winter, the comfort control and sizing of HVAC systems. The output is based on a dynamic simulation sub-hourly which uses as the calculation engine platform EnergyPlus. DesignBuilder can be used to simulate:

- many of the most common HVAC systems;
- buildings with natural ventilation systems;
- buildings with natural lighting (from daylighting);
- double skin facades;
- advanced shading systems;
- greenhouses and bioclimatic systems in general.

So far as the skills to use this programme are concerned, it is worth mentioning that DesignBuilder has been specially developed to facilitate the use of a simulation model by architects and engineers. For this reason DesignBuilder requires minimal basic experience, free tutorials are made available to accelerate learning.

DesignBuilder is used by engineers, architects, researchers and academics. The construction of the building model can be made through a 3D graphical interface inside the software but you can also import the model built by external software (e.g., AchiCAD). It is possible to create internal libraries and templates.

The windows and other transparent openings in the building can also be defined parametrically: This allows one to quickly build simple models to be used for simulations in early design phases. DesignBuilder not only simplifies the interface but also facilitates the management of output. Through simulations which can be made at annual, monthly, daily, hourly or sub-hourly intervals the following information can be obtained:

- Energy consumption broken down by fuel and end-use.
- Internal air, mean radiant and operative temperatures and humidity.
- Comfort output including temperature distribution curves, ASHRAE 55 comfort criteria, Fanger PMV, Pierce PMV ET, Pierce PMV SET, Pierce discomfort index (DISC), Pierce thermal sens index (TSENS), Kansas Uni TSV.
- Site weather data.
- Heat transmission through building fabric including walls, roofs, infiltration, ventilation etc.
- Heating and cooling loads.
- CO_2 generation.
- Heating and cooling plant sizes can be calculated using design weather data.

Data can be displayed graphically or in tabular form to be exported in a range of formats to spreadsheet and custom reports DesignBuilder has been developed on the C++ platform for the graphics engine, platform Visual Basic and runs on Microsoft Windows XP and 2000.

DesignBuilder comes with a free version fully functional for a period of 1 month (the official website http://www.designbuilder.co.uk) while the license at a cost of around €1,500. Local developers as well as providing the program in different languages provide technical support (website: http://www.designbuilder.co.uk).

References

1. ASHRAE Handbook, Fundamentals (2009) Thermal comfort, Chap. 9. Atlanta, p 3
2. CRES (2000) Energy audit guide, part B: system retrofits for energy efficiency. Project financed by the European Commission, Athens, p 5
3. Day AR, Karayiannis TG (1999) Identification of the uncertainties in degree-day based energy estimates, Accepted BSERT

4. Day AR, Karayiannis TG (1999) A new degree-day model for estimating energy demands in buildings. Accepted BSERT
5. Claridge DE, Krarti M, Bida M (1987) A validation study of variable base degree-day cooling calculations. ASHRAE Trans Part 2 93:90–104
6. ASHRAE Handbook, Fundamentals (2009) Degree day and bin method, Chap. 31. Atlanta, p 17
7. Dall'O' G, Gamberale M, Silvestrini G (2010) Manuale della certificazione energetic degli edifici. Edizioni Ambiente, Milan
8. Michaelides EE (2012) Alternative energy sources. Springer, London
9. Beckman WA, Klein SA, Duffie JA (1977) Solar hating design: by the f-chart method. Wiley, New York
10. CSN EN 15193 (2007) Energy performance of buildings: energy requirements for lighting
11. CRES (2000) Energy audit guide, part B: system retrofits for energy efficiency. Project financed by the European Commission, Athens. p 12
12. EN ISO 13790 (2008) Energy performance of buildings: calculation of energy use for space heating and cooling
13. prCEN/TR 15615 (2006) Explanation of the general relationship between various CEN standards and the energy performance of buildings directive (EPBD) (Umbrella document)
14. Thumann A, Younger WJ (2007) Handbook of energy audit, 7th edn. Fairmont Press, Lilburn

Chapter 8
Definition of the Green Energy Plan

The Green Energy Plan outlines strategies for improving the sustainability of a building and its related facilities. In the author's vision of energy audit, the technical report is not the final document of the process, but the starting point to promote the implementation of the retrofit measures. For this reason, its structure must be complete, rigorous and balanced between technical and economic aspects but also aimed at communicating well the results to the client/owner. This chapter provides the reader with guidelines for organising and structuring the green energy plan in accordance with these goals.

8.1 General Criteria

8.1.1 Clarity in the Objectives

The Green Energy Audit is realised with an operational plan, described in the *technical report*, that defines the strategies and the actions that result in measures for the limitation in the usage of resources.

Whereas the on-site audit phase, with further processing, allows the auditor to define what it is technically possible to achieve, the Green Energy Plan explicitly states what should be done.

In defining and then proposing to the client resource containment measures, the auditor should consider that the objective of the Green Energy Audit is to improve the sustainability of the building.

Since some measures (energy retrofit) result in a cost-effectiveness while others do not, before developing the operational plan of action, or green energy plan, it is important that the auditor understands not only what the goals are, but also what are the expectations of the owner/client.

In Table 8.1, there are some questions that the auditor may ask the owner and, depending on the possible responses, some interpretations.

G. Dall'O', *Green Energy Audit of Buildings*, Green Energy and Technology,
DOI: 10.1007/978-1-4471-5064-0_8, © Springer-Verlag London 2013

Table 8.1 Some questions that the auditor may ask the client and, depending on the possible responses, some interpretations [1]

Possible questions	Interpretations of responses
Is reducing the cost of management the only goal of the audit?	If so, few opportunities remain to suggest other motives and one must understand what financial commitment the customer is ready to sustain and for how long. A negative response provides an opening to other possible reasons; it is for the auditor to define a broader framework
Is the heightened value of the building as a result of the retrofit work a factor to consider?	A positive response can lead to prediction of the time of return on investment, which may exceed the period of use of the building, since all that is not recovered from the improved performance can be recovered from the consequent increased value of the building when it is sold
If the client is already planning redevelopment of the opaque building envelope, is it for technological reasons or to improve the image?	The measures applied to improve opaque building envelope performance are those that require larger investments and longer payback times An affirmative answer to this question gives the auditor the opportunity to offer these remedial actions. The economic evaluation of the investment should reasonably take into account only the value added related to improved energy performance
Is the client's goal to support domestic economic resources, or is the use of external financing possible?	The availability of external resources allows for more substantial investments, and therefore, for the containment of resources to be amplified. On the other hand, the availability of internal resources can reduce the costs of financing. The auditor must obtain this information in order to properly process the economic evaluation
Does the client-company usually prepare an environmental sustainability report?	Many companies prepare and publish an annual report on environmental sustainability that highlights the efforts made to reduce the company's impact on the environment. This choice, which is not normally required, shows that the client pays attention to these aspects If this is the case, the auditor can be more confident in proposing to the client remedial actions that improve the sustainability of the building or its facilities whilst not guaranteeing energy-savings and an income that can offset the expenses

This consultation should be scheduled, obviously, at a meeting preceding the drafting of the Green Energy Plan. With the objectives clarified, the next step concerns the definition of the retrofit measures.

Chapter 13 of the book gives the reader a set of 97 retrofit measure sheets describing retrofit measures that can be taken. Such measures are grouped into five sections:

1. *Building envelope or shell* (roofs, basements, walls, transparent envelope, shading, illumination using daylighting, etc.).
2. *Mechanical systems* (heating, summer cooling, ventilation, DHW, water services, etc.).
3. *Electrical systems* (generation, distribution and use of energy, lighting).

4. *Renewable energy sources* (solar thermal, solar photovoltaic, biomass, etc.).
5. *Improving management* (improving the management, maintenance and energy accounting, etc.).

The list of measures can be integrated as possible intervention strategies will also take into account particular situations and what the market offers with new technologies. The measures chosen, however, can cover much of the actions in the civil sector (residences, offices, commercial buildings, schools, etc.).

8.1.2 The Retrofit Measure Sheets

The *measure sheets* are structured to facilitate both quick reading and a more detailed examination. The information that allows the reader to understand in a synthetic way the contents of each sheet are shown at the top section as shown in the example of Table 8.2 which concerns a measure concerning replacement of windows.

The measure sheets provide the auditor the main parameters for a preliminary conscious choice, but at the same time contain references to general insights. The measures may include

- Actions on parts of the building (building envelope or facilities) that although "operative" and not subject to urgent replacement or upgrading, still cannot guarantee the best performance.
- Actions on parts of the building (building envelope or facilities) that are rather obsolete and may be subject to urgent replacement or upgrading.
- Actions concerning new installations in order to improve the sustainability of building.
- Actions to improve the management and maintenance.
- Each proposed measure is identified by a five-character alphanumeric coding.
- The first digit covering the area of interest (1 building envelope, 2 mechanical systems, 3 electrical systems, 4 renewable energy sources and 5 improved management).
- Two letters concern the subcategory (e.g., LI lighting, RS renewable energy sources, etc.).
- Two progressive numbers for the various subcategories.
- Code 3.RS.01, for example, identifies the intervention of installation of a new solar plant for DHW.

Table 8.2 Example of the top section of a retrofit measure sheet containing the synthetic data of the retrofit measure

Working	**C**	Obsolete	**A**	New	-	O&M	-
Saving potential			□■■■	Payback			■■
Reliability			□■■■	Feasibility			■■■
Environmental effects			**EC**				

Each sheet is structured into three sections:

- the heading;
- the scores section;
- the descriptive part;

The heading includes

- the identification code;
- the title of the measure;
- the assessments (rating) which show the feasibility of the measure.

Regarding the assessments there are three situations that do not necessarily relate to the building but can refer to a component or system:

- *Working/Operative*: in this case, retrofit actions are not planned (for example, it is proposed a thermal insulation in a facade that, despite being inefficient from the point of view of energy, does not present situations of degradation).
- *Obsolete*: in this case, the proposed measure is the result of an energy audit which takes into account the fact that there may be synergies with technological redevelopment or regulatory compliance.
- *New*: in this case, the installation of a component or an additional system is provided.
- *O&M*: in this case, the measure does not provide replacements of components or installation of new components, but an improvement of management strategies (considering the real needs of users) or more attention to maintenance.

For these three situations the assessments have the following meanings:

(a) The application of the measure is very convenient;
(b) The application of the measure is cost-effective;
(c) The application of this measure is feasible even if the margins of convenience are not very high;
(d) The application of this measure is not very convenient because complex or unprofitable, in the case of maintenance or management measures are required of expertise particularly costly.

The *scores section*, placed in the upper right part of the sheet, considers a series of indicators, expressed as a rating ranging from 1 to 4 that we describe below: the indicators refer to cases where the application of the measure is very convenient (A) or cost-effective (B).

Savings potential, expressed as annual percentage reduction of primary energy consumption refers to the extent implemented

■■■■	>70 %
■■■	40 to 70 %
■■	20 to 40 %
■	<20 %)

In the assessment of the energy-savings are considered only the benefits obtained by analysing the measure indicated in the sheet (i.e., not the benefits referred to the building as a whole). In the calculation of the ratings, for each measure, technical and economic evaluations were made considering the maximum achievable for that measure.[1]

Payback of the investment expressed as simple payback (SPB):

■■■■	<5 years
■■■	5 to 10 years
■■	10 to 20 years
■	>20 years

Financial returns are always referred to the benefits related to the single retrofit measure.

Reliability of the retrofit measure, indicates how the measure will be effective and reliable over time

■■■■	*high reliability* (the measure maintains its performance for a period of time equal to the useful lifetime of the building)
■■■	*good reliability* (the measure guarantees its performance with low maintenance)
■■	*mediocre reliability* (measuring its performance guarantees in time but with frequent maintenance)
■	*poor reliability* (it is very difficult to maintain performance over time because of a considerable commitment and/or need for expensive technical skills).

Feasibility of the remedial action, provides guidance on how is easy or not to implement the measure:

■■■■	*maximum ease* (the implementation of this measure requires minimal effort and not high expertise)
■■■	*ease normal* (no special efforts are needed since it is routine procedure)
■■	*difficulty* (the implementation of the measure requires a special effort and expertise with high levels of qualification)
■	*high difficulty* (the implementation of this measure is particularly difficult and may have contraindications)

Improvement of sustainability: the positive effects on the improvement of sustainability (reduction of energy consumption, reducing environmental impact, improving comfort) generated by the implementation of the measure are assessed and summarised:

---

[1] In the case of the thermal insulation of a wall, for example, it is obvious that the potential savings depend on the quality of the insulation added (thermophysical characteristics of the insulating material and its thickness).

A	*reduction of energy consumption* (fuel, electricity, etc.)
E	*reduction of resources consumption* (non-energy as water, etc.)
I	*reduction of the environmental impact* (we consider only the direct effects, such as improved thermal insulation certainly reduces fuel consumption but has no direct effect on the environmental impact whereas the replacement of a heat generator reduces emissions and hence has a direct effect on the environmental impact)
C	*comfort improvement* (thermal comfort, lighting comfort, IAQ)

In assessing the ratings, other symbols are used: □□ indicates a possible extension of the rating criteria (□□■■ means ranging from 2 to 4), NA indicates that the assessment of the specific evaluation criteria for the specific measure is not applicable.The descriptive part includes:

- a *general description* section that describes the measure synthetically from the technical point of view, but mainly focuses on the reasons that can make its implementation very convenient in the overall context as well as any interdependency (or alternatives offered) with other measures that may be cited. In this section, reference is made to issues related to the application of the measure described, as well as the selection criteria;
- a *tips and warning section* in which we provide tips to increase the effectiveness of the measure, but at the same time highlight possible weaknesses.

8.2 Retrofit Measures on the Building Envelope

Retrofit measures on the building envelope (or shell for U.S. technicians) have the purpose of making more sustainable the relationship between the indoor space (or confined environment) in which it is necessary to ensure optimal comfort conditions, and the external environment.

Through the opaque and/or transparent walls of the envelope, heat fluxes must be controlled considering that the strategies in summer and in winter may change according to the different external climatic conditions. The general concepts of the strategies to be adopted can be summarised as follows:

- improvement of the thermal quality of the opaque and transparent envelopes by means of increasing of the thermal resistance (reduction of the U value);
- control of solar radiation with the aim to making the most of the free heat contributions in winter, while ensuring protection from the sun in summer;
- control of the lighting component of the solar radiation, favouring the technical solutions that use natural lighting but, at the same time, protecting the interior spaces from the effects of glare;
- reduction of uncontrolled air infiltration.

To better understand how the retrofit strategies may be defined, it is important to refer to the energy balance on the building as illustrated in a simple schematic way, Fig. 8.1 fulfils that purpose for a winter energy balance.

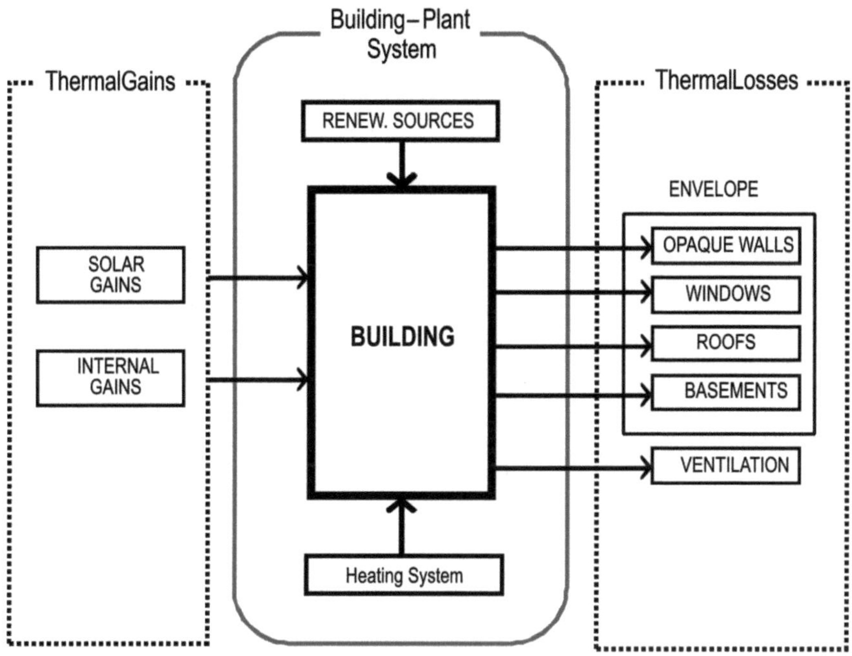

Fig. 8.1 Simplified energy balance for a building during heating season

The objective of an HVAC system is to maintain within the indoor spaces the design operating conditions, such as the air temperature constant at 20 °C in the winter season. The building interacts with the external environment, where the climatic conditions may vary: through the walls that define the envelope, when the outside temperature is lower it generates a heat flux to the outside. The thermal losses from the envelope increase as the outside temperature decreases, with a dependency upon the values of the thermal resistance of the individual components (opaque vertical walls, window, roofs, basements, etc.). Better insulation, increasing the thermal resistance to the passage of heat, therefore helps to reduce heat losses.

To ensure the healthy environment of the internal spaces, it is necessary to maintain suitable IAQ. A correct air ventilation of internal spaces is therefore necessary but, on the other hand, it does contribute significantly to increase thermal dispersion, since a part of the inside air must be continuously replaced with external fresh air which must be heated. The global energy losses for heating are the sum of the losses due to thermal dispersions through the envelope and the losses due to ventilation of spaces.

In the energy balance of a building, however, not only the heat losses but also the heat gains should be considered. Internal free gains are provided not only by people and equipment within the building (e.g., lighting systems, appliances, etc.),

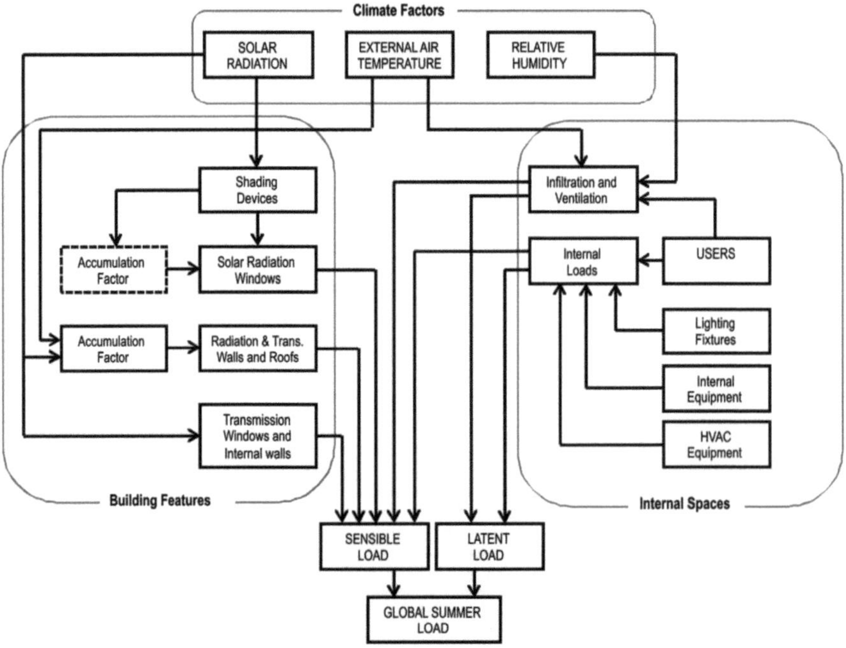

Fig. 8.2 Simplified calculation diagram for the summer cooling energy balance on a building

but also as natural heat coming from the contributions of solar radiation through the transparent and opaque components of the building envelope.

If the objective is to maintain a constant internal comfort temperature, the difference between energy losses and energy gains must be compensated by the installation of an HVAC system: its contribution to the energy balance could be dramatically reduced by employing renewable energy sources (e.g., solar thermal, solar PV, biomass, etc.).

The energy balance shown in Fig. 8.1 considers only winter heating. In the overall energy assessment in the winter season, we should consider also other types of energy uses, such as those related to the production of domestic hot water, as in summer we must consider energy uses related to air conditioning.

The energy balance for summer cooling is more complex, as shown in the simplified diagram of Fig. 8.2. The summer cooling load is greatly influenced by the building components of the envelope and in particular by the transparent surfaces that interface with the solar radiation: the shielding shading devices and the optical characteristics of glazing (in particular the characteristics of reflection) are very important to minimise the heat intake due to solar radiation.

During the winter heating all free heat contributions, whether internal or external gains, are useful in the energy balance; during summer cooling, the opposite applies since all the incoming heat (solar radiation) and generated heat (people, lighting, equipment, etc.) must be removed by the HVAC system. For this reason, the retrofit measures that aim to reduce these loads are very useful to save energy besides increasing thermal comfort.

8.2.1 Thermal Insulation of Roofs and Basements

Flat and pitched roofs of a building surfaces are the more exposed surfaces to solar radiation. When choosing retrofit measures, even this aspect should be considered, by providing solutions that not only increase the thermal resistance to the heat passage, but reduce the negative effects, in summer, of the solar radiation. The techniques that can be used are essentially three:

- ventilation (for example through a ventilated roof);
- reflection (using reflective materials);
- damping of the thermal wave through a layer of material (for example, with the green roof technology).

As far as the roofs themselves are concerned, it is necessary to make a distinction between those which are pitched and those which are flat. For pitched roofs, it must be verified whether below the roofing cover there is a living space (or a space that as a result of planned restructuring is expected to be used) or not. If the space is occupied (or otherwise used), the roof should be directly insulated, otherwise this measure is not convenient since the heat-dispersing element is the base slab which delimits the spaces below and, therefore, it is useful to thermally insulate this element.

For the thermal insulation of a pitched roof, one can consider two retrofit measures:

- action effected from the outside, which provides for the removal of the tiles and the replacement of the existing roof with a new one: this measure is particularly convenient if renovation works on the roof are planned for maintenance reasons;
- action effected from inside which involves the application of one or more insulating panels (in this case, it is essential to check that the existing roof is waterproof and that its maintenance status is good).

If the spaces under the roof are not used, action will have to be taken to improve the energy performance of the last slab. The addition of the insulating layer can be made from the outside, by placing the insulation on the attic floor (see Fig. 8.3 c), or inside the room under the attic, through a ceiling. In the first case, the retrofit can be very simple and effective if the spaces are not used at all, for example stretching insulating fibrous materials (e.g., rock wool, fibreglass, wood fibres, etc.) if the attics are partially used (e.g., for storage of materials), it is necessary to choose insulating materials resistant underfoot or to protect the floor surface with a layer of resistant material.

The insulation of the slab from the space below could reduce the height of the room: this could be positive or negative depending upon the original height of the room.

Even the flat roofs can be insulated from the inside, practically with the same technique described above, and from the outside. This retrofit measure, however, is more difficult in economic terms as it includes a complete recovering of the roofing. If provision is made for extraordinary maintenance of the roof, the auditor may propose this solution using the technology that best fits with the existing situation.

Concerning basements (slabs on cellars, slabs on ground, pilotis, etc.), the improvement of the energy performance involves the addition of an insulating layer that can occur only from the outside or from inside depending on the existing situation.

The improvement of the energy performance of the envelope has the effect of reducing the heat losses and, therefore, the thermal loads both in winter and summer. The retrofit measures are therefore less effective if no provision is made to modify the control system so that the HVAC system can provide less heat at constant environmental conditions.

The thermal insulation of pitched roofs sometimes permits the owner to acquire additional living spaces; the HVAC system, in this case, should be extended to meet the needs of the new areas.

8.2.2 Thermal Insulation of Walls

The choice of the most appropriate measure or measures to improve the energy performance depends on the existing situation. The measures only aimed at improving the energy performance pay for themselves in a very long time if they are made outside of a maintenance plan which does, however, provide for the renovation of facades. In this case, the technique of isolation from the outside through the External thermal insulation composite systems (ETICS) solution is the most effective one.

The insertion by blowing of the thermal insulation into the wall cavities, if present, is configured instead as an feasible remedial action, relatively inexpensive and minimally invasive. The works can be carried out from the inside in the cases where it is not possible to intervene from the outside, for example because of historic or environmental constraints, or to correct specific situations.

The additions of insulation of the vertical walls involve a reduction of the thermal load (both in winter and summer) and it is therefore necessary, even in this case, check whether the control system of the HVAC system is capable of allowing the plant to supply the amount of heating or cooling strictly necessary to guarantee the internal comfort condition.

8.2.3 Increasing of the Thermal Performance of Fenestration

In the energy balance of a building the transparent envelope, if compared to opaque one, is more complex. In order to better understand the complexity of the issues relating to energy performance of transparent envelope, Table 8.3 shows some terms indicated in the "window energy glossary" published by US DOE [2].

The transparent separation between interior and exterior environment involves energy flows in both directions:

- energy fluxes in the infrared spectrum (heat);
- energy fluxes in the visible spectrum (solar radiation);
- luminous energy fluxes.

Considering the thermal fluxes, a transparent surface has a selective effect, allowing the passage of the solar radiation (visible spectrum) and retaining, instead, the heat emitted from the internal environment (infrared spectrum). In selecting a particular type of glazing, it is therefore necessary to consider both the aspects that define the energy balance, in relation to the climatic characteristics of the location and to the reference season. Whereas for opaque walls, neglecting the economic aspects, the more thermally insulated they are, the better, the analysis for choosing a transparent surface, or fenestration (windows and skylights), is more complex.

In the energy balance, the effect of solar radiation can be positive in certain periods of the year and negative in others.

Solar transmission through windows and skylights can provide free heating during the heating season, but it can cause a building overheating during the cooling season. Depending upon orientation, shading and climate, solar-induced cooling costs can be greater than heating benefits in many regions. In fact, solar transmission through windows and skylights may account for 30 % or more of the cooling requirements in a residence in some climates [2].

As the sun's position in the sky changes during the day and seasonally, window orientation has a strong bearing on solar heat gains.

Therefore, solutions capable of favouring the exploitation of solar radiation (external gains) should be chosen but only when this is useful to the indoor environment. On the other hand, the indoor environment must be protected from the solar radiation when an excess thereof would simply cause an increase of the thermal load.

Through the technical elements that best represent the transparent envelope, windows or skylights, it is possible to control the effects of natural lighting and, in some cases, the ventilation. Energy losses in winter, summer cooling loads, lighting and ventilation, are all factors that must be considered when selecting a fenestration: the choice, therefore, is not simple.

In order to limit the complexity of the problem and to provide the reader with more straightforward information, the retrofit measures have been divided-up considering the functions separately. The measure sheets were therefore divided into three groups:

Table 8.3 Window energy glossary published by US DOE (continue) [2]

Item	Description
Air leakage rating	A measure of the rate of infiltration around a window or skylight in the presence of a strong wind. It is expressed by the relationship between the air flow and the window perimeter length. The lower a window's air leakage rating, the better its airtightness
Fenestration	A window or skylight and its associated interior or exterior elements, such as shades or blinds. The placement of window openings in a building wall is one of the important elements in determining the exterior appearance of a building
Gas fill	A gas other than air placed between window or skylight glazing panes to reduce the U-value by suppressing conduction and convection
Glazing	The glass or plastic panes in a window or skylight
Infiltration	The inadvertent flow of air into a building through breaks in the exterior surfaces of the building. It can occur through joints and cracks around window and skylight frames, sash, and glazings.
Low-emittance (Low-E) coating	Microscopically thin, virtually invisible, metal or metallic oxide layers deposited on a window or skylight glazing surface primarily to reduce the U-value by suppressing radiative heat flow through the window or skylight
Shading coefficient (SC)	A measure of the ability of a window or skylight to transmit solar heat, relative to that ability for 1/8-inch clear, double-strength, single glass. It is equal to the Solar Heat Gain Coefficient multiplied by 1.15 and is expressed as a dimensionless number with a value between 0 and 1. A window with a lower Shading Coefficient transmits less solar heat, and provides better shading
Solar heat gain coefficient (SHGC)	The fraction of solar radiation admitted through a window or skylight, both directly transmitted, and absorbed and subsequently released inward. The Solar Heat Gain Coefficient has replaced the Shading Coefficient as the standard indicator of a window's shading ability. It is expressed as a dimensionless number with a value between 0 and 1. A window with a lower Solar Heat Gain Coefficient transmits less solar heat, and provides better shading
Spectrally selective glazing	A specially engineered low-E coated or tinted glazingthat blocks out much of the sun's heat whiletransmitting substantial daylight.
U-Value (U-Factor)	A measure of the rate of heat flow through a material or assembly. It is expressed in units of W/m^2 °C. Window manufacturers and engineers commonly use the U-value to describe the rate of non-solar heat loss or gain through a window or skylight. A lower window U-value means greater resistance to heat flow and better insulating value
Visible transmittance	The percentage or fraction of visible light transmitted by a window or skylight

- measures to improve the thermal performance where we rank with the energy aspects, passive and active, of the different technologies;
- measures aimed at protection from the sun in which particular attention was paid to the summer season;
- measures for illumination using daylight, in order to increase lighting performance.

Retrofit measures improving the energy performance of a transparent envelope involve direct or indirect correlations with other measures which should be adopted:

- *improved energy performance*: implies a check of the local control system in order to regulate the output of HVAC systems;
- *sun protection*: in the case of automated mobile screenings is essential to provide a control system able to efficiently manage the shields;
- *illumination using daylight*: it implies an adaptation of artificial lighting through control systems that favour natural lighting whilst avoiding glare.

8.3 Retrofit Measures on Mechanical Systems

The energy requirements for climate control inside buildings is satisfied by means of mechanical equipment: the heating system in winter and the cooling systems in summer. The acronym HVAC (Heating Ventilation & Air Conditioning) indicates a system able to satisfy both these requirements, including also the ventilation necessary to ensure the IAQ.

Figure 8.3 shows the schematic diagram of an ideal HVAC system, integrated with a DHW system: the diagram highlights all possible combinations of plant components including renewable energy sources.

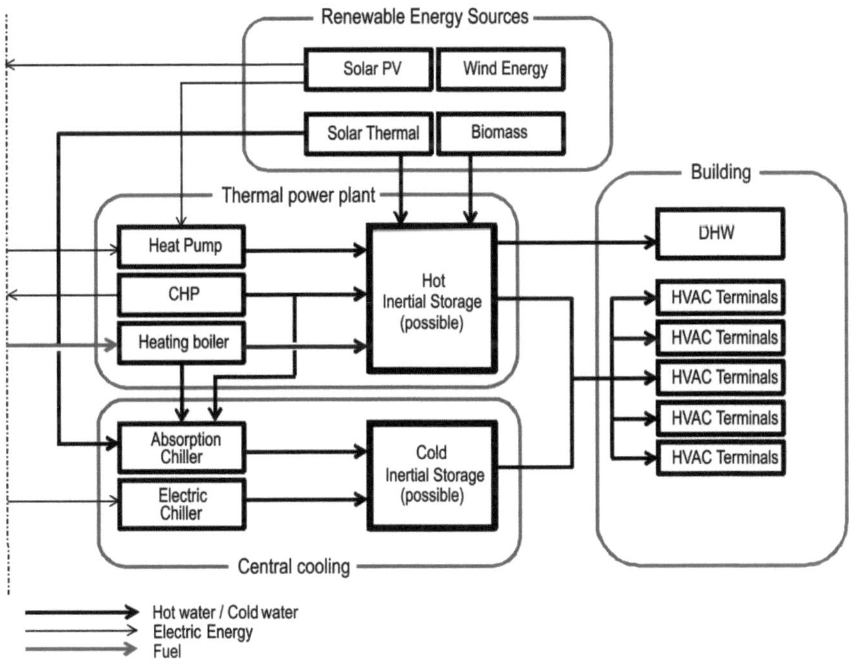

Fig. 8.3 Schematic diagram of a HVAC and DHW integrated system

The diagram identifies three specific areas where the plant components are located: thermal power plant (heating generation), central cooling (cooling generation) and the building (HVAC terminals).

The diagram in Fig. 8.3, as mentioned previously, is complete and includes all possible options. In existing buildings, where energy audits are performed, the plant schemes are much simpler: heat generation systems are normally unique (by type, summer and winter), no storage systems are installed and renewable energy sources are rarely used.

From the functional point of view, each system may be divided into five subsystems:

- *Generation subsystem* (heat generator, chiller, heat pump, etc.), whose task is to generate heat and cold, which will then be transferred to the heat transfer fluid;
- *Distribution subsystem* consisting of a hydraulic and/or aeraulic network network (piping, ducts, pumps, fans, etc.) whose function is to transport thermal vectors to HVAC terminals installed in the conditioned spaces of the building;
- *Storage subsystem* (if any) consisting of a water buffer tank, hot and/or cold, whose function is to ensure a thermal stabiliser between generation and distribution;
- *Regulating or controlling subsystem* defined as the set of devices that allow the system to deliver the required quantity of heat, or cooling, that is strictly necessary to control the climate, avoiding wastage;
- *Emission subsystem* consisting of HVAC terminals (such as radiators, fan-coils, radiant systems, air grille, etc.) whose function is to transfer the heat or the cooling (heat removal) to the conditioned spaces.

The various subsystems are interconnected according to a sequential scheme such as that shown in Fig. 8.4. Each subsystem has thermal losses due to its inefficient to operate, and it needs its auxiliary equipment powered by electricity.

To provide a certain amount of energy, the building (energy needs) plants consume more energy, primary energy demand, therefore, be derived from the algebraic sum of the following energy quantities:

- the net building energy requirement;
- the sum of the heat losses due to the inefficiencies of the different subsystems excluding possible heat recovery;
- the electricity demand of auxiliary converted into primary energy, taking into account the average efficiency of the national electricity grid.

The retrofit measures of mechanical systems on the basis of a careful analysis of the existing plant and components (through field surveys and monitoring) have the objectives of reducing the primary energy (thermal and electrical) of plants, so as to reduce energy losses by improving overall performance.

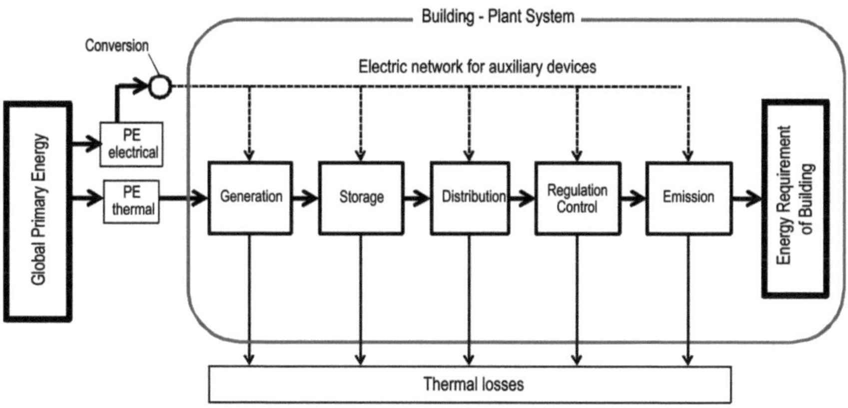

Fig. 8.4 Schematic diagram the subsystem sequence in a HVAC system

In the following sections, the retrofit measures suggested are discussed, divided for each of the above-mentioned subsystems.

8.3.1 Increasing of the Energy Performance in Heat Generation Systems

The first step to improve the energy performance of the heat generation sub-system is, of course, to enhance the management and maintenance activities (for example through a frequent check on the combustion efficiency).

The field survey, however, may highlight the need and also the convenience of replacing the existing heat generator with a high performance system (for example a condensing boiler or a heat pump). Diagram in Fig. 8.6 shows how important it is, considering the plant chain, to ensure that the heat generation subsystem can operate with maximum performance and, therefore, with minimal heat losses.

Before replacing a the existing heat generator with a new one, an evaluation of the effective heating power demand of the building, by means of a precise calculation of thermal losses, should be performed in order to avoid a possible oversizing. On this point, several aspects must be considered

- the thermal capacities of boilers in old buildings often were not calculated in an analytically way but rather were estimated approximately: for this reason, often the installed heating system has a thermal capacity of at least twice the real needs;
- if the thermal inertia of the buildings is high (typical in existing buildings), the thermal power requirement can be reduced by increasing the hours of operation of the heating system (the building can be used as a thermal flywheel avoiding the need to cover power peaks);

- if retrofit measures to improve the thermal insulation of the building envelope are planned, the thermal power of the heating system required to compensate thermal losses can be reduced dramatically;
- a precise calculation of the real heating capacity of the system is important also because a reduction of it involves a reduction of the other components of the heating plant (e.g., sizing of pipes, pumps, terminals, etc.).

In replacement of a heat generator, the most efficient technology available should be chosen: that is currently condensing technology. In this case, however, it is advisable to check the conditions of the circuit to prevent deposits and sediments contained in the water which could damage the new generator in a very short time.

As an alternative to the heating boiler it is possible to install an electric heat pump or a gas-fired heat pump. The installation of a Combined Heat Power (CHP) unit, thermal machine which simultaneously produces electricity and heat, must be evaluated in terms of both the technical and the economical pros and cons.

For all these possible remedial works, one must take into account that the greater cost is not only linked to new equipment but also to installation costs (included the disposal of the old generator) the accessories that must be provided (safety and control devices, modification and adaptation of existing circuits, etc.).

8.3.2 Increasing of the Energy Performance in Cooling Generation Systems

Cooling generation systems (chillers) should be replaced because inefficient or because inconsistent with the new regulations relating to refrigerant gases. It can be cost-effective to replace an existing chiller by a new and more efficient one. In recent years, significant improvements in the overall efficiency of mechanical chillers have been achieved by the introduction of two-compressor chillers, variable-speed centrifugal chillers and scroll compressor chillers.

So far as the problem of refrigerant gases is concerned, it is worth emphasising that the choice of using chillers that use environmentally compatible refrigerants should be carefully considered, even if not compulsory, since it is a choice which improves the sustainability of the building. Prerequisite 3EA of LEED NC protocol (CFC Reduction in HVAC&R Equipment) states that "Zero use of CFC-based refrigerants in new base building HVAC&R systems. When reusing existing base building HVAC equipment, complete a comprehensive CFC phase-out conversion" [2]. As regard the potential technologies and strategies, the same document states that "When reusing existing HVAC systems, conduct an inventory to identify equipment that uses CFC refrigerants and adopt a replacement schedule for these refrigerants. For new buildings, specify new HVAC equipment that uses no CFC refrigerants" (see Chap. 9 for a discussion on this topic).

If the existing chiller is relatively new (less than 10 years old), it may not be cost-effective to replace it entirely with a new non-CFC one. Just the conversion of the chiller to operate with non-CFCs may probably be the most economical option. However, the non-CFC refrigerants (e.g., R-134a and R-717) may reduce the chiller cooling capacity owing to their inherent properties. Fortunately, this loss in energy efficiency can be limited by upgrading some components of the cooling system, including the impellers, orifice plates and gaskets, even the compressors themselves.

In some cases, the conversion with upgrade may actually improve the chiller performance. Some of the strategies that can be used to improve the efficiency of existing chillers include [4]:

- increment of the evaporator and the condenser surface area for more effective heat transfer;
- improvements in the compressor efficiency and control;
- enlargement of the internal refrigerant pipes for lower friction;
- ozonation of the condenser water to avoid scaling and biological contamination.

The energy auditor should be faced mainly with two technical aspects: the type of new chiller and its cooling capacity.

As regard the first aspect, there are many possibilities, a description of the various types of central chillers, taken from [4] is provided below.

Electric chillers use a mechanical vapour compression cycle. There are currently three major types of electric chillers available in the market, defined according to the type of the compressor they use:

- *Centrifugal compressors* use rotating impellers to increase the refrigerant gas pressure and temperature. Chillers with centrifugal compressors have capacities in the range of 300–25,000 kW. For capacities above 4,500 kW, the centrifugal compressors are typically field erected.
- *Reciprocating compressors* use pistons to raise the pressure and the temperature of refrigerant gases. Two or more compressors can be used under part-load conditions to achieve higher operating efficiencies. Capacities of 35–700 kW are typical for chillers with reciprocating compressors.
- *Rotary compressors* use revolving motions to increase refrigerant gas pressure. One of the most ingenious rotary compressors is the scroll type, while the most conventional are the screw compressors, which have several configurations. The capacity of rotary chillers can range from 3 to 1750 kW.

Engine-driven chillers, like those electrically driven, use reciprocating, rotary or centrifugal compressors to provide mechanical refrigeration: they can be powered by turbines or gas-fired engines. Engine-driven chillers can have capacities up to 15,000 kW, but usually have a high first cost.

Absorption chillers operate using a concentration-dilution cycle to change the energy level of refrigerant (water) by using lithium bromide to alternately absorb heat at low temperatures and reject heat at high temperatures. A typical absorption chiller includes an evaporator, a concentrator, a condenser and one absorber. These chillers can be direct fired (using natural gas or fuel oil) or indirect fired:

- *Direct-fired absorption chillers* can be cost-effective when the price of natural gas is low and are available on the market in two types: single-effect and double-effect. Some of them can be used to produce both chilled and hot water, known as chillers/heaters, and can be cost-effective especially when heating needs still exist during the cooling season (e.g., in buildings with large service hot water needs). Capacities ranging from 100 to 5,000 kW are available for direct-fired chillers.
- *Indirect-fired absorption chillers* operate with steam or hot water (with temperature as low as 140 °C) from a boiler, a district heating network, an industrial process, etc. Small absorption chillers using solar energy have been developed and analysed. Cooling capacities from 15 to 425 kW are available, although typical sizes range from 200 to 5000 kW. Double-effect chillers can be considered only for high temperatures.

When examining the cooling capacity of chillers in existing buildings, it is also true for cooling generation systems that previous oversizing is another problem that may warrant the replacement of the old cooling systems. Indeed, several existing chillers may well have a capacity that is significantly higher than their peak-cooling load and operate exclusively under part-load conditions, with reduced efficiency and hence increased operating and maintenance costs. When the oversized chillers are more than 10 years old, it may be cost-effective to replace them with smaller and more energy efficient units, operating with non-CFC refrigerants [4].

If retrofit measures to improve the summer energy performance of the building envelope (increasing of the thermal insulation, implementation of sun protection strategies, etc.) are planned, the capacity of the cooling system can be reduced very significantly. A precise calculation of the real cooling demand of the system is important also because its reduction involves a reduction in size of the other components of the related plant (e.g., sizing of pipes, pumps, terminals, cooling tower, etc.).

A final consideration concerns the general strategies to improving the overall energy performance of the building: since the market-place proposes reversible chillers (i.e., refrigerant machines able to operate for cooling as chillers and for heating as heat pumps) practically without additional costs, why not consider a solution with a single generating system (reversible chiller) for winter heating and summer cooling?

8.3.3 *Increasing of the Hydronic and Air Distribution System Performance*

8.3.3.1 Hydronic Distribution Systems

The aim of distribution systems is to transfer the heat or the cooling generated from the associated plant (heating boilers, heat pumps, CHP, chillers, etc.) up to the HVAC terminals. Refrigerants are also used as a means of energy transfer between components of refrigeration equipment (condensers and evaporators).

Regarding hydronic distribution systems, the pumping of heated or cooled water is commonly used in many buildings. In order to avoid freezing in extremely cold conditions, water is often diluted with an antifreeze fluid. The water temperature for heating depends on the type of heating terminal installed: the range varies from a minimum of 35–45 °C (low temperature radiant floor or ceiling panels) to 70–80 °C (radiators).

Chilled water systems distribute cold water to terminal cooling coils to provide dehumidification and cooling of HVAC systems. Low temperature radiant floor or ceiling panels in occupied spaces are also connected to chilled water distribution systems so as to reduce the sensible loads (a control of the air humidity is required, in this case, to prevent an increase in the relative humidity RH and vapour condensation on the cold surfaces).

Condenser water systems connect mechanical refrigeration equipment to outdoor heat rejection devices (e.g., cooling towers or air cooled condensers).

Hydronic systems are also used in thermal solar systems to connect solar collectors with the storage tank or, in the DHW systems, to supply water to the users.

Examining the energy performance of hydronic distribution systems, one is confronted with two technical aspects:

- the energy required for pumping the fluid (water or mixture water-refrigerant);
- the heat losses (positive or negative depending on the system, heating or cooling circuit, and the environment in which the pipes are installed).

In existing buildings, energy required for pumping can be a significant portion of the energy used. In fact, pumping energy is roughly inversely proportional to the fifth power of pipe diameter, so a small increase in pipe size has a dramatic effect on lowering pumping energy [5].

Flow rate can easily be reduced, so reducing greatly the energy consumption due to pumping, by reducing the difference between outlet and inlet temperature of the fluid. The temperature difference selected depends upon the ability of the system to operate with lower return water temperatures: in any case, the auditor should always check on this possibility. If the existing boiler is replaced with a condensing boiler, a lower return temperature of the fluid is required in order to operate properly. This is because too high a value of the water return temperature does not allow the boiler to reach the conditions for the condensation from the flue-gas.

Once the conditions of the distribution system have been checked, and there are no other options for its optimisation, it is possible to replace the existing pumps with high efficiency pumps (e.g., variable-speed pumps equipped with inverter): the reduction of the energy required for pumping is significant and the auditor should always consider this option.

In existing buildings, sometimes the distribution system does not fit with the new requirements of the users. The modification of the activities within the living spaces over time can require a complete check-up on the hydronic distribution system, in order to avoid, for example:

- maintaining activated the heating or cooling system for the whole time that a small part of the building requires heating and/or cooling (in this case, an independent circuit is the best solution);
- maintaining oversized pipe dimensions neglecting the fact that some HVAC terminals have been replaced over time (a re-design of the distribution circuit should be made).

The thermal losses of the hydraulic distribution systems are due to the heat or cold dispersions through the outer surface of the pipes when the insulation of these is not adequate. The verification of the validity of insulation can be done by measuring the thickness, and comparing this with the minimum thickness required by the regulations, but also by resorting to a measure of the surface temperature using an infrared thermometer or better yet, a infrared camera.

8.3.3.2 Air Distribution Systems

In HVAC systems, air distribution systems are used essentially for three different purposes:

- providing distributing heating and cooling from the air handling units to spaces that need them;
- maintaining a IAQ within the air-conditioned spaces by means of air ventilation in accordance with the users requirements;
- ensuring the complete internal air conditions which satisfy both thermal requirements and IAQ.

Air distribution systems are not only a means of energy distribution, since their essential role is that of providing fresh and purified ait to occupied spaces.

On purely energy aspects, the general concepts applicable to air distribution systems are the same as those for hydronic distribution systems, discussed above: again in this case two technical aspects must be examined:

- the energy required for moving (blowing) the fluid (air);
- the heat losses (positive or negative depending on the system, heating or cooling ducts and the environment in which the ducts are installed).

The inefficiencies of the aeraulic distribution system can be determined on the basis of the thermal losses through any non-optimally insulated enclosing panels of the ducts but also on the basis of air leaks that may occur at the joints between the sections of air ducts or corresponding to the duct junctions.

In cases where examination of the air ducts shows-up evident signs of obsolescence, it may be convenient to arrange for their partial or total replacement. The thermal insulation of ducts is a retrofit measure that can significantly reduce losses in both winter and summer season.

The electrical energy input of the fans can be reduced by using fans with variable-speed drive motors. This type of remedial action still requires a recalculation of the pressure losses of the circuit.

Another aspect of air system design is that there are temperature limitations on supply air since that air is the energy medium that directly impacts space occupants. While care must always be taken in how air is introduced into an occupied space, this is especially critical the colder the supply temperature becomes [5].

8.3.4 Increasing of the Air Handling Units Performance

In air conditioning systems or in ventilation systems, air is treated before being introduced in the conditioned spaces. The types of HVAC systems can vary greatly depending on the needs of users, the diagram in Fig. 8.5 represent a typical HVAC configuration: it is useful to understand the different plant components.

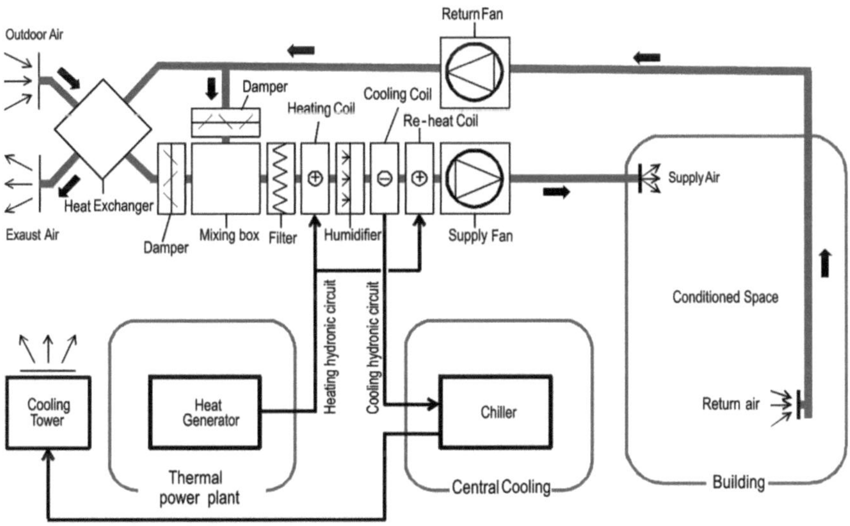

Fig. 8.5 Schematic diagram of a full air HVAC system

The core of the system is the air handling unit where the air taken partially from outside (outdoor air) and partially from the return duct (recycled air), is mixed, filtered and thermally treated. Typically, there are three treatment coils: heating coil and re-heat coil, connected to the hydronic heating circuit and the cooling coil connected to the hydronic cooling circuit.

The supply fan provides the air with the pressure necessary to move it thorough the supply ducts up to the HVAC terminals (e.g., air grille) with a residual pressure, whilst the return fan has the function of drawing the air out of the conditioned spaces through a separate duct (return duct).

The diagram of Fig. 8.5 also shows a heat recovery unit (air to air heat exchanger), in which the air extracted yields part of its heat, which is then used to preheat the incoming outdoor air.

If the air handling unit is obsolete or appears to be inadequate after an extension of the air-conditioned spaces, its complete replacement is the most suitable solution.

The energy required to condition ventilation air can be significant. The auditor should estimate the existing volume of fresh air and compare this estimation with the amount of ventilation air required by the appropriate standards and codes. Excess in air ventilation should be reduced if it can lead to increases in heating and/or cooling loads. However in some climates and periods of the year or parts of the day, providing more ventilation can be beneficial and might actually reduce cooling and heating loads through the use of airside economiser cycles [4].

There are two ways to determine the switchover point and decide when it is better to use more than the minimum required amount of the outdoor air to cool a building. The first method is based on dry-bulb temperatures and the other on enthalpies, known respectively as temperature and enthalpy air-side economisers: they are described briefly below.

- *Temperature Economiser Cycle*: in this cycle, the outside air intake damper opens beyond the minimum position whenever the outside air temperature is colder than the return air temperature. However, when the outdoor air temperature is either too cold or too hot, the outside air intake damper is set back to its minimum position. Therefore, there are outdoor air temperature limits beyond which the economiser cycle should not operate.
- *Enthalpy Economiser Cycle*: this cycle is similar to that based on temperature, except that it is the enthalpy of the air streams which is used as the controlling property instead of the temperature. Consequently two parameters are typically measured for each air stream in order to estimate its enthalpy (e.g., dry and wet bulb temperatures). The enthalpy economisers are thus less common, since they are more expensive to implement and less robust to use, despite their potential to achieve greater savings if properly operated [4].

Where the ventilation system does not exist, air changes cannot be guaranteed effectively. In fact only a ventilation system ensures the proper flow rate as a function of the real needs of the users. The installation of a controlled mechanical ventilation systems still requires a verification of the permeability of the walls that delimit the building envelope.

8.3.5 Increasing of the Performance of Control/Regulation Systems

The indoor temperature control during both heating and cooling seasons has significant impacts on the thermal comfort within the occupied spaces and on the energy use of the HVAC systems. It is therefore important for the auditor to assess, through measurements and monitoring, the existing indoor air temperature controls within the facility to evaluate the potential for reducing energy use and/or improving indoor thermal comfort without any substantial initial investment.

The manually imposed set-point of the air temperature, does not in itself guarantee the performance over time, since the set-point values could be modified by the users. The control system, properly chosen, properly installed and properly adjusted, is the best solution.

The functions of the control system are critical to any objective of improving the energy performance and, at the same time, the indoor environmental comfort.

The installation of zone control adjustment systems (e.g., zone valves or thermostatic valves) allows modulation of the supply of heating or cooling to an area in function of the actual needs, but also allow efficient usage of free energy contributions from solar radiation and internal gains (occupants, equipment, etc.).

The local control systems become even more effective if the control is associated with consumption meters.

8.3.6 Increasing of the Performance of DHW Systems

Normally energy auditors focus their interest on heating and cooling systems and electrical systems, believing that these facilities are responsible for the most considerable usages of energy. Domestic Hot Water (DHV) systems are often neglected, since their consumption of energy are relatively marginal.

DHW systems, on the contrary, are responsible for high energy consumption in many types of users such as residential buildings, hotels, sports centres, gyms and hospitals.

As regards residential buildings, it can be observed how the energy performance for heating and cooling, thanks to the better thermal insulation of the building envelope and the better performance of equipment tends to improve reductions in energy consumption but the same cannot be said for DHW systems which nowadays in green buildings represent a major cause of energy consumption.

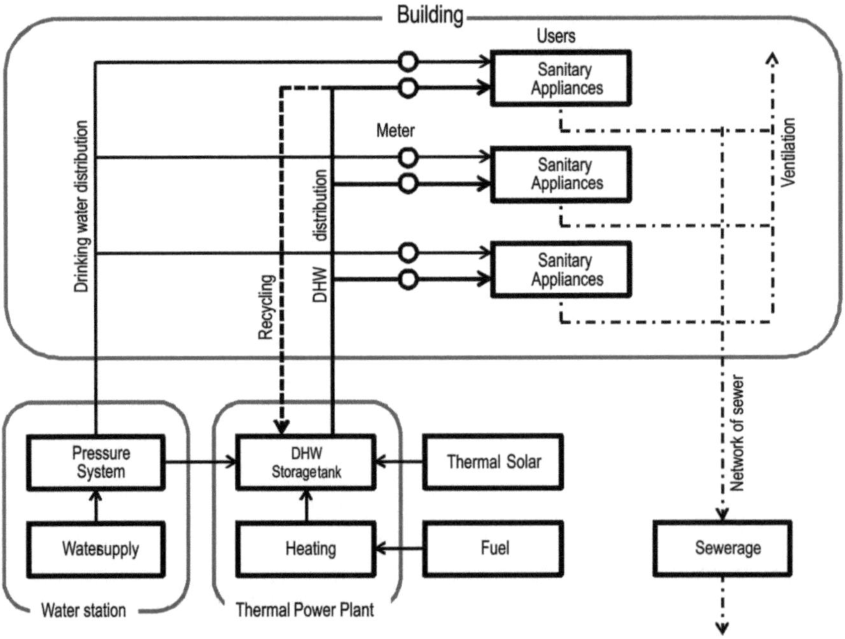

Fig. 8.6 Schematic diagram of a DHW system

The energy audit of DHV facilities is important and useful in helping to improve the overall sustainability of a building: the savings measures are diverse and easy to achieve.

The diagram in Fig. 8.6 represent a typical DHW system configuration: it is useful in order to understand the different plant components.

The water coming municipal water supply must have sufficient pressure to reach the most unfavourable users and exit the dispensing tap with a certain residual pressure. If the pressure is insufficient (common situation for tall buildings) it is necessary to provide for the installation of a suitable hydraulic compensation unit (overpressure group). The distribution circuits may be one or two depending on how the hot water is produced. The network is unique (only cold water) if the heating takes place locally (domestic water heaters) or is double if the heating of DHW, as shown in the diagram of Fig. 8.6, is centralised.

Drinking water is the heated using a heating boiler, a heating exchanger but also using a thermal solar system. A hydraulic distribution network shall deliver hot water to the users. In correspondence of each user (e.g., flat), two meters are normally installed (one for drinking water, one for hot water), so any user can pay for his own consumption.

A recycling circuit guarantees that the availability of hot water is immediate also related to the furthest.

Energy retrofit measures that can be foreseen are as follow:

- check that the water temperature is not too high compared to the values required;
- verify the possibility to turn off the circulating pump at night or in the hours of non-use of the plant;
- replace the heat generator with a more efficient one (for example condensing boiler or heat pump);
- install an independent heat generator if DHW system and heating system are connected to a single heat generated too big;
- replace any electrical kettles with gas-fired heaters or heat pumps;
- verify the level of thermal insulation of the distribution pipes;
- check if there is the process heat at a low temperature that can be used for this purpose;
- verify if it is possible to install a solar thermal system.

8.3.7 Reduction in the Usage of Drinking Water

The basic resources are obviously essential in every building, but what has changed over the last several decades is the realisation that water is rapidly becoming a precious resource. While the total amount of water in its various forms on the planet is finite, the amount of fresh water, of quality suitable for the purposes for which it may be used, is not uniformly distributed [5].

Drinking water is a precious resource and its conservation should be one of the objectives of the green energy auditor. The reduction of the use of drinking water can be obtained in two ways:

- with a more conscious use of it (for example installing saving devices as aerator valves, taps timers, etc.);
- by replacing drinking water with non-potable water for the uses in which the sanitary drinking water is not needed (e.g., for toilets, irrigation, etc.).

The first strategy does not require particular commitment, since one simply changes the existing plant by introducing devices that make sure that less water is used, whilst still maintaining the quality of service, but the water used remains always drinking water. In the second case, instead, it is necessary to provide a dual supply system, one for drinking water and one for non-potable water.

We define *drinking water* as water that can be used for human consumption without harmful consequences to health. The potability of the water is defined by law by a series of parameters that are divided into five points: organoleptic parameters, which relate to those characteristics of the water perceptible by the senses, physical–chemical parameters that define the characteristics of natural waters, parameters concerning the substances which are undesirable and over certain thresholds can become toxic, parameters related to the presence of toxic substances and microbiological parameters related to the presence in the water of specific micro-organisms harmful to health.

Non-potable water is water that, while not responding to chemical, physical and biological of drinking water does not contain anything that is polluting or otherwise dangerous for all the people who simply come into contact with it. The distinction between drinking water and non-potable water is defined by the local health regulations whose observance is monitored by the competent authority.

Non-potable water may be used for many applications and the list below gives some examples:

- power urinals or vessels;
- industrial laundries and industrial cleaning in general;
- watering plants;
- supply of fountains, ornamental pools and similar;
- make-up circuits of cooling towers;
- make-up circuits of open type expansion vessels for heating systems;
- make-up circuits for condensate collection trays;
- heating circuits for heating or cooling of other fluids;
- adiabatic cooling and humidifying equipment in HVAC systems;
- indirect cooling circuits of equipment in general;
- fire fighting hydrants, sprinklers, etc.

The distribution of non-potable water must be distinguished in each point of the related piping and terminals from that of drinking water.

No connections are allowed between a drinking water supply and a distribution system of non-drinking water even when equipped with shut-off valves. All components of the distribution networks of non-potable water must eventually be marked prominently and indelibly with the words and symbols in accordance with the local regulations.

8.4 Retrofit Measures on Electrical Systems

Retrofit measures on electrical systems aim to reduce the wastage of a fine form of energy. The conversion factors between electricity and primary energy depend on how the electricity is generated and then the structure of the national electricity grid. Assuming a range between 0.4 and 0.45, to produce 1 kWh of electricity, then 2.5–2.2 kWh, respectively, of primary energy must be used, usually supplied by fossil fuels. For this reason, auditors should pay particular attention to identifying the reasons for electrical wastage and propose appropriate solutions to reduce them.

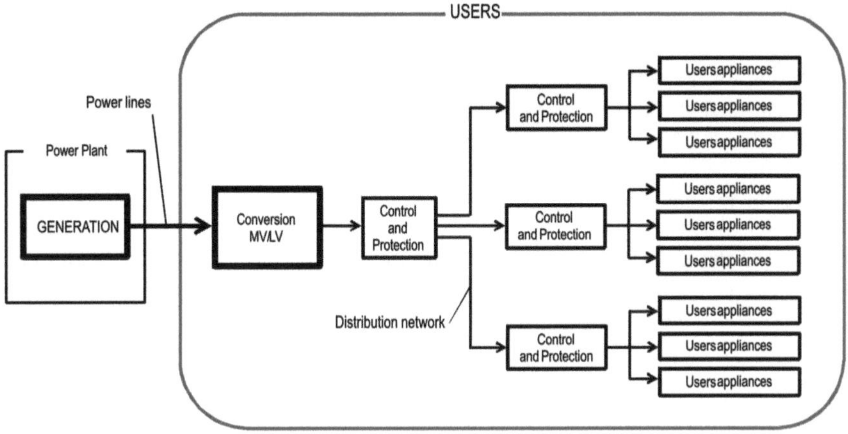

Fig. 8.7 Schematic diagram of an electrical system, from the generation to the user appliances

Figure 8.7 shows a schematic diagram of an electrical system powered from the public grid. The elements that constitute an electric system are the following:

- generation systems;
- transmission grid/network or power lines;
- conversion equipment;
- distribution network;
- protection components and control;
- users appliances.

Telecommunications plant, transport systems (lifts/elevators, escalators, etc.) and lighting systems all belong to the electrical systems of a building.

Electrical energy and costs are saved by managing demand loads, reducing running hours, improving equipment efficiency and properly maintaining distribution systems.

8.4.1 Reducing Energy Consumption on Generation, Distribution and Utilisation Subsystems

The retrofit measures for electrical systems are amongst the most cost-effective and, being minimally invasive, are well suited to be applied in existing buildings.

The retrofit strategies may include:

- the reduction of the power factor (cos ϕ);
- the reduction of wastage, namely the improper use of equipment when not really needed (for example, thanks to the installation of timers, of presence sensors, etc.);

- check on and replacement of wires and cables if the operating conditions have changed (e.g., the connection of devices with absorbed electric power higher than the design value);
- the replacement of appliances with more efficient versions (e.g., high efficiency motors);
- the reduction of the peak power through the installation of control systems able to better manage the electrical loads.

Some of the above-mentioned measures are intended to reduce the consumption of electricity, while others aim to optimise the electricity supply contracts and so reduce running costs for electricity.

8.4.2 Reducing Energy Consumption on Lighting

The electricity consumption of lighting systems represents a significant proportion of the total consumption of electricity of a building. The reasons for this and probable sources of excess are summarised below:

- lighting of spaces when it is not necessary to illuminate them;
- lighting areas in the wrong manner;
- lack of exploitation of natural lighting;
- use of inefficient lighting equipment;
- poor maintenance of the lighting fixtures.

The strategies should consider and solve for all three of the following aspects:—better lighting, illuminate when it is necessary and light efficiently.

Energy audit of lighting systems should provide the Auditor with all the elements in order to propose a global strategy favouring, whenever possible, illumination using daylight.

The following is a list of retrofit measures that can be taken in order to increase the efficiency of lighting systems:

- replacing interior and outdoor lamps;
- installing occupancy sensors;
- installation twilight switches;
- installing daylight sensors;
- installing high-frequency electronic ballasts;
- installing dimmers.

Table 8.4 shows a list of possible retrofit actions for lighting systems and the approximate energy-savings that can be obtained.

Table 8.4 List of possible retrofit actions for lighting systems and the approximate energy-savings [6]

Audit finding	Corresponding retrofit measures	Approximate energy-saving
Lighting level in corridor area at 500 lux, which is rather high, however capital cost not available for retrofit	Disconnect power supply to some lightings and lower illumination to a suitable level, say 100 lux	15–30 % for corridor lighting
Lighting along windows areas turned "ON" during the day time, providing a lux level well over 700 lux	Maintain the lighting at 500 lux by: turning off corresponding perimeter lighting; or if both interior lighting and perimeter lighting share the same control switch, re-wire to facilitate ? independent control switches for each of the 2 zones Alternatively replace the lighting ballasts (only if lighting can suit) at perimeter with dimmable electronic type and control by means of photo sensors	20–30 % for lighting at perimeters
T12/T10 fluorescent tube used in lighting (e.g., exit sign)	Replace with T8 fluorescent tube (not feasible for quick start type)	10 %
T8 fluorescent lighting (fixture & tube) used	Replace with T5 fluorescent lighting	30-40 %
Manual ON/OFF control for lighting	Add occupancy sensor control	>20 %
Electromagnetic ballast used in lighting with T8 fluorescent tube	Replace with electronic ballast	20–40 %
Incandescent lamps are being used	Change to compact fluorescent lamps or retrofit with fluorescent tube lighting	80 %, more if spaces have AC: the extra is cooling energy to offset the higher heat dissipation incandescent lamp

8.5 Use of Renewable Energy Sources

In new high efficiency or green buildings, renewable energy sources are those of reference: their use in existing buildings for retrofit contributes considerably to improve the overall efficiency and, of course, to make the project more sustainable.

Renewable energy sources more suitable for architectural integration with the existing buildings are solar (thermal and PV) energy and biomass.

Solar energy can be used:

- for DHW with or without integrating by means of an heating system;
- for low temperature requests (e.g., swimming pools heating);
- for HVAC systems (Solar Heating and Cooling);
- for electricity production.
- for DHW with or without integrating by means of an heating system;
- for low temperature requirements (e.g., heating of swimming pools);
- for HVAC systems (Solar Heating and Cooling);
- for electricity production.

The use of thermal solar systems for DHW can be applied in all buildings, provided that there is space available for the installation of the solar collector, since this type of system is independent of the others.

The possibility of using the solar heating system for winter heating can provide interesting results, if the thermal heating loads are dramatically reduced by means of retrofit measures on the building envelope and the heating system is equipped with low temperature terminals (e.g., floor or ceiling radiant panels, fan-coil units, etc.).

The use of this system can be still cost-effective for buildings used only occasionally (vacation homes) in which the heat generated by solar is not used at times when the building is not occupied and in such a case can help to play a part in frost protection.

The use of solar thermal for summer air conditioning, utilising an absorption machine, is convenient since there is a good coincidence between the energy demand and the energy supplied by solar.

Installing solar PV systems, thanks to the incentives provided in some countries such as Germany, Spain and Italy, is very widespread: the direct production of electricity from the sun offers a wide range of possibilities to satisfy many electrical needs, lighting and appliances as well as heat. The production of heat must be done not directly using electricity but via a heat pump.

The installation of a biomass boiler is an interesting alternative, with two conditions: the first is that the necessary space for the storage of fuel has been identified, the second is that biomass is readily available on-site.

8.6 Building Control Systems for Energy Efficiency

In energy and environmental management of a building, the human factor is a key element. Among the measures that contribute to promote a better management of the resources, there are actions that increase the awareness of the users who should be trained in how to use energy and avoid its wastage.

The involvement of people does, however, rely on a sense of individual responsibility and therefore the results that can be achieved are subjective. A strategy of this type is interesting and convenient because good results can be obtained with virtually zero costs. However the users, even if highly motivated, can but rely on their own sensory capacities, skills that not always lead an individual to act effectively.

A building control system (BCS) permits the management of energy resources with an objective approach, starting from certain information detected by appropriate devices, such as sensors for the detection of the temperature, carbon dioxide concentration, lighting levels or the presence of people.

A BCS has also the ability to process the information acquired by means of the sensors and to generate actions for regulation, control, optimisation and accounting. All this is done with maximum flexibility and total synergy between different systems. The added value that a building control system can provide compared to a standard regulation and control system is that all energy-using equipment and provision of services can be managed together.

A BCS is able to manage the following items [5]:

- *delivering energy efficiency* (through scheduling, unloading, and fault detection, controls have the capability of reducing building energy usage);
- *delivering water efficiency* (used primarily in landscape irrigation and leak detection, controls can reduce water usage);
- *delivering IAQ* (controls and regulates the quantities of outdoor air brought into the building, zone ventilation, zone temperature, and relative humidity and can monitor the loading on air filters);
- *Commissioning process* (of all the building systems, controls are the most susceptible to problems in installation);
- *obtaining LEED certification*;
- *sustained efficiency* (control systems help ensure continued efficient building operation by enabling Measure & Verification of building performance and serving as a repository of maintenance procedures.

An in-depth explanation of the issues listed above can be found in [5].

The design of a BCS stems from the analysis of customer requirements, that analysis then becomes a gradually more specific definition and results in a set of procedures that will be implemented in the software.

8.7 Management and Maintenance Improvement

Most of wastage of energy and resources is caused by poor management or a lack of maintenance. The measures which lead to an improvement of the management are among the most cost-effective in that they do not require special investments (sometimes zero). On the other hand, proper maintenance of the installations not only maintains the high performance of the individual components, but prevents possible situations in which the plant may not work, owing to unexpected breakdowns.

An initial list of measures that can be applied is as follows:

- the correct setting of control devices (e.g., reduction of the hours of operation of a plant or precise calibration of temperature in the occupied spaces);

- disabling components which are consuming energy unnecessarily;
- implementation of control procedures and monitoring of consumption;
- implementation of maintenance procedures;
- implementation of information strategies and incentives amongst users.

Some of the above measures are sometimes underrated in terms of their effectiveness, in particular those concerning the proper information and awareness, with doubts as to whether the users do become more conscious of the nature of the facilities creating conditions of comfort. Nevertheless, there are case studies in which the implementation of these measures in schools and offices have shown that it is actually possible to obtain great benefits for virtually no cost.

User awareness at various levels, from students to employees, of the appropriate use of energy resources generates a change of approach with positive effects also outside the areas for which they are promoted. It is clear that the awareness of the importance, and the effective reduction of the wastage of energy and resources in the workplace also changes people's general attitude when they are elsewhere, for example at home.

8.8 Economic Assessments

The retrofit measures to be taken to improve the sustainability of a building often require considerable investment, although some of them, mainly those of energy, are directly paid back on the basis of the energy saved over time.

The Green Energy Plan should therefore carefully consider the economic assessment of remedial actions, since the cost-effectiveness can be a discriminating factor in moving forward from the theoretical proposal to the effective implementation.

The economic assessment of retrofit actions is discussed in detail in Chap. 10, whereas in this chapter the objective is to provide some ideas on how to approach these aspects and how to manage them in a proper comparison with the critical issues that the auditor inevitably faces in the definition of the Green Energy Plan.

8.8.1 The Definition of the Parameters for the Economic Assessment

Before going more deeply into the question it is useful to reflect on a few points:

- the cost-effectiveness of energy efficiency measures is characterised by a strong variability: retrofit on the plants generally provides a greater profitability, while retrofit on the building envelope has longer payback time;

- one of the elements that underlie the economic assessment is the length of the evaluation period: for some actions this evaluation is simple, for others instead it is quite complex;
- if the length of the evaluation is necessarily long (e.g., replacement of windows or ETICS) the economic assessment is less reliable, since it is not simple to estimate the probable variation of some parameters influenced by unpredictable external factors (e.g., the cost of energy or the cost of money).

8.8.2 The Definition of Costs

In an economic assessment, the initial costs of the action to be taken must be defined. Whether or not the measure involves enhancement of energy and environmental performance, the definition of the initial costs is not so simple, since the measure could in fact fulfil several functions and some of these are not necessarily related to the increase in energy and environmental performance. In order to make a consistent economic assessment of the measure, it is therefore important to determine the portion of total the initial cost to be considered. To better understand this important aspect, here are some typical examples:-

- A retrofit measure provides *the replacement of an obsolete heating boiler* with a new one (e.g., a condensing boiler or a heat pump). A correct evaluation of the cost to consider for the economic assessment should be made on the basis of the difference between the cost of the new high performance heating boiler (or heat pump) and a new boiler with performance compatible with the minimum required by laws and regulations. As regard the increase in performance (and the energy-saving evaluation), for consistency with the given approach, the new high performance heating boiler should be compared with the new standard heating boiler.
- A retrofit measure provides *the replacement of an existing heating boiler (not obsolete in this case)* with a new one (e.g., a condensing boiler or a heat pump). A correct evaluation of the cost to consider for the economic assessment should be made considering the cost of the new high performance heating boiler. As regard the increase in performance (and the energy-saving evaluation) for consistency with the given approach, the new high performance heating boiler should be compared with the existing boiler.
- A retrofit measure *provides thermal insulation of an external wall* of a building with the ETICS technology (insulation from outside). In this case, two scenarios should be taken into account. If the facade still requires an extraordinary maintenance, the cost to be considered is the difference between the full cost of the measure and that of a simple renovation (for example with restoring plastering and painting). If the facade does not require particular maintenance operations, the cost to be considered is the global cost. In the first case, obviously, the measure is much more profitable than in the second case, since the costs of are much lower.

- A retrofit measure *provides the installation of a solar thermal system for DHW.* In this case, the to consider is the whole cost of the solar system, because the new system does not replace, but rather is added to the existing one (in this case the heating boiler).

On defining the costs of the retrofit measures of facilities component replacement (e.g., the first two example relating to boiler replacement), the auditor must obviously consider also the installation costs to be involved for the disposal in landfills of old equipment and other materials.

8.8.3 Economic Indicators

An economic evaluation is based on the calculation of economic classics indicators, of which the most commonly used are the following:

- Net Present Value (NPV);
- Profit Index (PI);
- Internal Rate of Return (IRR);
- Payback Time (PBT).

Each of these economic indicators, which are discussed in Chap. 10, when calculated, express a characteristic of the investment and then determine its convenience or not from different points of view.

The *net present value* (NPV) takes into account more systematically the time of cash flows, cost of money including interest on the capital cost investment, life time of equipment/installation, etc., which can better reflect the effectiveness of the investment. This method gives a present value to future earnings which are expected to be derived from an investment.

The *profit index* (IP) is equal to the ratio of discounted benefits and discounted costs: if it is greater than 1, the project could be accepted; otherwise, it has no reason for being passed.

The *internal rate of return* (IRR) is a measure of the return in percentage to be expected on a capital investment. This takes into account the similar aspects as for NPV. IRR is often used for comparison of two or more projects and allows for an immediate comparison between the rates of return offered by banks to deposit the same amount of money that would be needed to execute specific remedial action.

The *payback time* (PBT) is the number of years required to recover the capital invested. This method, which is simple in its calculation, normally excludes consideration of the timing of cash flows, inflation rate, interest rate of capital cost, depreciation, opportunity cost, etc. It is useful to obtain a rough estimate of the validity of a project and is easily understood by anyone. However, its accuracy will usually be within a certain, albeit reasonable range.

The most widespread indicators are those that offer the customer a greater ease of understanding, including SPB which reflects the years of payback, and IRR which expresses the hypothetical interest rate that could be provided by a virtual bank in which one could have invested the money used to make the investment.

8.8.4 Life Cycle Cost

A commonly accepted method for assessing the economic benefits of retrofit measure projects over their lifetime is the *Life-Cycle Cost* (LCC) method.

The method is used to compare two or more alternatives of a given project (for example, a replacement of an old compression chiller with a new compression unit or with an absorption machine). Only one alternative will be selected for implementation based on the economic analysis.

The basic LCC procedure is quite simple since it seeks to determine the relative cost-effectiveness of the various alternatives. For each of these, including the base case (corresponding to maintaining the current situation without improvement), the total cost is computed over the project lifetime considering all costs associated with the retrofit measure (e.g., initial costs, maintenance costs, operating costs including energy, fuel escalation rates, inflation, interest on the investment, salvage value and other lifetime expenses). These expenses over the life of the project are multiplied by their present value factor: the sum of all the present values is called the *life cycle cost.*

The cost is commonly determined using one of two approaches: the present worth or the annualised cost estimate. Then typically the alternative with the lowest LCC is selected.

8.8.5 Economic Assessment and Sustainability

In a classic energy audit, measures that do not lead to the direct economic benefits and therefore do not pay for themselves within a reasonable time, say between 4 and 12 years, are rarely taken into account.

The Green Energy Plan proposes a mix of measures, some of which actually show an immediate convenience while others generate long-term advantages. We refer to those measures that improve the sustainability of, and indeed enhance, the building if the market is able to accept this value.

LEED certification, for example, certainly enhances important aspects: the building that achieves high levels of classification, such as Gold or Platinum, is a quality building, a building that ensures the best conditions of comfort, which produces a lower impact on the environment and, obviously also saves energy.

However, it is not only the last mentioned, directly correlated to an economic advantage, which comes into play in the overall assessment, but also the overall performance.

In a market where a high LEED certification brings rewards, the building has an intrinsic value greater than an uncertified building. But how much more is it for a building with these characteristics? In their research, Popescu et al. [7] tried to evaluate the impact of energy efficiency measures on the economic value of buildings.

Their research highlighted what more recent studies indicate, being that energy efficiency and energy certification produce tangible benefits recognised by the real estate market. US and EU studies disapprove the null hypothesis that there is no relationship between the market value of a real estate asset and its energy efficiency. The paper argues that since reliable proofs exist, market driven reasons, such as higher prices for energy performing buildings, can be included in energy policies.

Traditionally, potential energy-savings represent the only parameter taken into consideration in energy audits. In the presented study, the added value due to energy performance was taken into consideration in additional to potential cost reductions from energy-savings.

This type of evaluation, however, is not easy because it is only in recent years that awareness has taken hold and it will take more time before the real estate listings are able to assess a value on an environmental, rather than just income basis. The process, however, has started and this type of development will emerge quickly.

8.9 Assessment of Environmental Effects

The Green Energy Audit, as has previously been stated, aims to improve the sustainability of the building. This means not only a reduction in the use of energy from fossil fuel sources, but also a full improvement of sustainability, taking into account the reduced impact that it may have on the ecosystem, health, the economy and, in the case of buildings that house a workplace, productivity.

The measures taken to promote this improvement must therefore be evaluated also for the other effects, direct or indirect, that they are able to generate or, in other words, there must be an assessment of environmental effects.

8.9.1 Analytical Evaluation of Environmental Effects

A first approach to the assessment of these effects can be analytical, considering the reduced impacts for each action. The improvement in energy efficiency of a

Table 8.5 Emission factors for the main fuels (*source* IPCC–intergovernmental panel on climate change)

Type of fuel	Standard emission factors ($t\ CO_2/MWh$)
Natural gas	0.202
Residual fuel oils	0.279
Kerosene	0.259
Municipal waste (not covered by biomass)	0.330
Motor gasoline	0.249
Diesel	0.267
Coal	0.341
Industrial waste	0.515
Vegetable oils	0.000
Biodiesel	0.000
Bioethanol	0.000
LPG	0.227
Biomass short chain	0.000

plant, for example, reduces emissions in the same way that the optimisation of the water systems reduces the consumption of water.

For this type of evaluation, which the author believes should in any case be made, it is useful to make an estimate of how, for example, the measures can contribute to the reduction in the emission of greenhouse gases. However, it is not able to take into consideration all effects that may contribute to the improvement sustainability in the broadest sense of the term.

The analytical approach to the evaluation of the environmental effects can be applied to all measures which are aimed at reducing the use of resources. If intervention or set of interventions, improve the energy efficiency of a building, the less energy can be translated into a reduction in CO_2 emissions which can be estimated by applying standard emission factors.

The indicated factors are official and are taken from the Intergovernmental Panel on Climate Change (IPCC): all the processes that generate CO_2 are in fact contained in a database available on the site http://www.ipcc-nggip.iges.or.jp/EFDB/find_ef_main.php.

Table 8.5 shows the emission factors, expressed in tonnes of CO_2/MWh for the most common fuels. The table takes into account only the emissions of CO_2, for the case of other gases generated from the combustion processes, the applicable factors are reported as values of CO_2 equivalent. These factors, however, depend on the combustion process in question and their assessment must be conducted in an analytical manner.

The CO_2 equivalent is the unit of measurement used for the GWP (Global Warming Potential) of greenhouse gases. CO_2 is the reference gas used to measure all others, so the GWP of $CO_2 = 1$.

The effect of methane CH_4 for the heating-up of the Earth is equivalent to 21 times that of CO_2, while nitrous oxide N_2O is equivalent to 310 times that of CO_2.

If the energy saved does not result directly from fuels but also from electrical energy, it must be considered that the primary energy requirement can be calculated by dividing the electrical energy consumption for the efficiency of the national electrical system. The value of this depends on the manner in which the energy is generated and the efficiency of the overall system. Then its value, which varies from country to country, comes directly from the competent authority and is regularly updated. The same authority gives the values of emission factors overall.

8.9.2 Life Cycle Assessment

A more comprehensive assessment of a retrofit project is the assessment that considers its application for the entire life cycle, called life cycle assessment (LCA).

Born in the industrial field, LCA has been for many years applied to the building sector, considering both the single component and the building as a whole.

The basis of the methodology is the "life cycle" approach i.e., that approach which brings awareness of the damage or the potential environmental impacts owing to what is happening in each of the phases of the life cycle of a product, production, transportation, use, recycling, reuse or disposal.

LCA allows to understand how each choice made in the design and production has then a fall-back consequence during the distribution, use and disposal.

Through an in-depth knowledge of the object one can make informed choices on how to acquire raw materials and define production processes, about who will use the product and on how to perform maintenance and decommissioning.

The LCA is the scientifically recognised method of quantitative evaluation of the environmental damage caused by a product/building/service.

8.9.3 Certification Protocols for Environmental Sustainability

A more complete approach is more comprehensive than that offered by the certification of environmental sustainability which can be made on the basis of one of the official internationally recognised protocols such as those shown in Table 8.6. Environmental certification is primarily a tool to facilitate dialogue between the building and the environment, to enhance the positive aspects, but at the same time to bring out the critical issues that must be addressed and resolved.

In environmental certification, many aspects are considered: the energy cycle, the water cycle, air, soil, waste and materials. The evaluation parameters can vary depending upon the scheme chosen, but they have a common denominator: a comprehensive assessment of quality.

Table 8.6 Comparison of different rating systems for sustainable buildings [8]

System (country)	Key aspects of assessment and versions	Level of certification
BREEAN (Great Britain) 1990	Management, health and well-being energy, water, material, site ecology, Pollution, transport, land consumption BREEAM for: courts, ecohomes, education, industrial, healthcare, multi- residential, offices, prisons, retail	Pass Good Very good Excellent Outstanding
LEED (USA) 1998	Sustainable sites, water efficiency, energy & atmosphere, material and resources, indoor air quality, innovation & design LEED for: new construction, existing buildings, commercial interiors, core and shell, homes, neighbourhood development, school, retail	LEED Certified LEED Silver LEED Gold LEED Platinum
Green star (Australia) 2003	Management, indoor comfort, energy, transport, water, material, land consumption and ecology, emissions, innovations Green star for:—office—existing buildings, office—interior design, office—design	4 Stars: 'best practice' 5 Stars: 'Australian excellence' 6 Stars: 'world leadership'
CASBEE (Japan) 2001	Certification on the basis of "building environment efficiency factor'' BEE = Q/L Q … quality (Ecological Quality of buildings) Q1—Interior space Q2—Operation Q3—Environment L … loadings (Ecological effects on buildings) L1—Energy L2—Resources L3—Material Main criteria: (1) Energy efficiency (2) Resource consumption efficiency (3) Building environment (4) Building interior	C (poor) B B+ A S (excellent)

In all protocols, rating systems have been developed to measure the sustainability level of the buildings and provide best practice experience in their highest certification level. Using several criteria, compiled in guidelines and checklists, building owners and operators are given a comprehensive measurable impact on their buildings' performance.

The purpose of rating systems is to certify the different aspects of sustainable development during the planning and construction stages. The certification process means quality assurance for building owners and users. Important criteria for successful assessments are convenience, usability and adequate effort during the different stages of the design process [8]. For further details on LEED protocols, please refer to Chap. 9.

8.10 Structure of the Technical Report

8.10.1 Basic Concepts

The Audit Report is the document that informs the Owner of what has been done during the audit and the measures that can be implemented to make the building, facilities or infrastructure more energy efficient and more sustainable.

If the Energy Auditor is a competent professional expert then contained in this document all the strategies proposed will be both technically and economically valid. The purpose of an energy audit, however, is not limited to the preparation of the Report, but goes further: the purpose of the report is to bring about effective remedial actions, by encouraging the Owner to take those actions.

The preparation of the Report is therefore a key phase, a phase in which the Auditor must be able to communicate to the Owner the results of his work, in a rigorous but clear and convincing way.

The Technical Report is intended to be a useful document not for demonstrating the Auditor's skills (those are inferred) but to support the Owner in the decisions to be taken. It may seem a banal observation, but many technical reports seem designed to only to move information between experts rather than to transfer it to those who will effectively it.

A Technical Report should meet several requirements:

- it should be structured in a systematic way so as to facilitate its reading and contain a correct balance of arguments;
- it should be rigorous in content but at the same time understandable even by people not necessarily expert;
- it should ensure flexibility in reading, namely satisfy readers who may or may not be interested in learning about all the report from the technical point of view but at the same time do not want to lose a vision of its global framework;
- it should be concise in clarifying its content: it is not the quantity but the quality that makes for a better report.

There are no general rules for structuring a technical report nor even unique methods to describe its contents. Although the "manuals of style" are good reference texts from which every technician can derive inspiration, the writing of the technical report is a personal matter.

The purpose of this section is to provide suggestions and propose a scheme of the general structure for the Green Energy Technical Report, leaving the Auditors, of course, the freedom to make any changes which appear necessary to determine their personal style.

Figure 8.8 shows a typical structure of a Green Energy Technical Report, explanations will be discussed in the next sections.

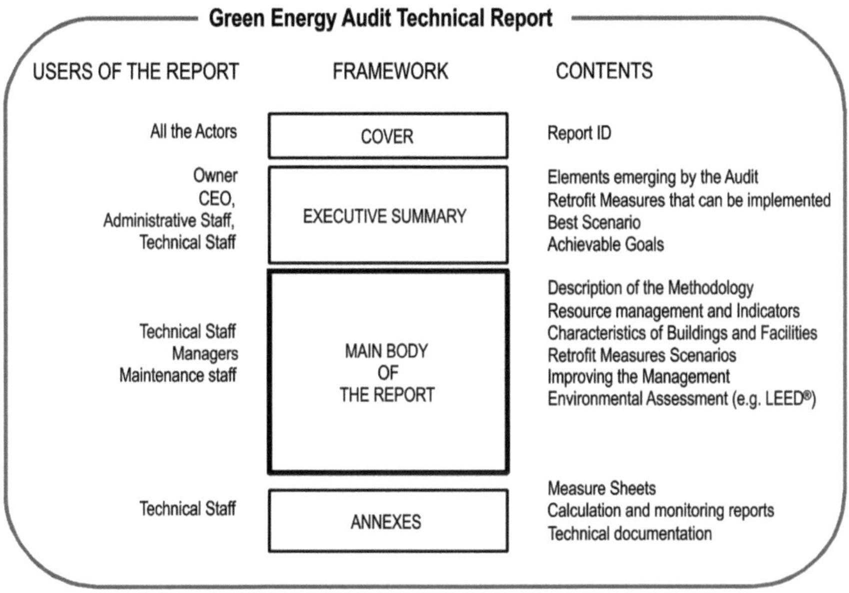

Fig. 8.8 Structure of the Green Energy Audit technical report

8.10.2 Who Needs the Technical Report

The title of this section poses a question whose answer seems obvious: the technical report serves the Owner. But who exactly is the Owner? In the simplest case it is a person, such as a private user or building manager. In some cases, however, that is not so: the term Owner may be a whole organised structure, composed of several people who may have different skills and roles.

Before starting to write the report, the Auditor should reflect on this question: it must be understood who will read the report, because only in this way will it be possible to communicate properly the project.

If the Owner's structure is complex (e.g., a Company), probably the Chief Executive Officer or the Managing Director will be the decision makers, persons without technical expertise to understand the details of the proposed solutions and persons that, probably, do not even have the time or willingness to read many pages of the report.

The structure of the report should include a very brief *executive summary*, (to be an absolute maximum of two pages), within which the proposals and economic evaluations will be listed. The Owner's technical staff, however, will need to go into detail on the topics covered, so a part of the report will describe in more detailed manner the solutions proposed with in-depth technical evaluations.

The Owner's Maintenance staff will be more interested in issues regarding maintenance aspects: for example, it will determine whether the proposals made by the Auditor could cause difficulties or whether, on the contrary, they will solve problems. The structure of the document, then, should be flexible in order to meet the needs of the different persons who will analyse the technical report.

8.10.3 A More Direct Language in Writing the Document

Technical writers normally use technical, very formal language as "insiders". This language may be fine when all the interested parties have a technical background, write articles for (technical) journals or have presented papers at seminars and conferences. In this case the situation is different: the technicians communicate his project to an interlocutor who may not be a technician or, if a technician, may not necessarily be an expert like the author of the report.

A general rule is to avoid the setting of the sentences in a non-involved and debatable manner: "it is believed that the power consumption is too high," for example, is a phrase that reveals to those who read a willingness to turn away from what is says: better to say "the power consumption is much too high," a decisive statement that exudes confidence on the part of the ruling or the writer.

Similarly, the phrase "the replacement of incandescent lamps with energy-saving lamps is to be recommended" can be better and more effectively formulated with the phrase "replace incandescent lamps with energy-saving lamps". In the first sentence there is a recommendation from which the writer distances himself whilst the second is direct, clearly linked to the writer and decisive.

The Auditor makes proposals on the basis of his technical evaluations, and for these proposals that he takes full responsibility. If the proposals are more than one, he will demonstrate the ability to provide the Owner, for each of the solutions, with clear advantages and disadvantages.

8.10.4 The Graphical Display of Information

Wherever possible the technical information should be displayed using charts and graphs, much more explicit in their display than tables of information.

Pie charts can be used for example, to display the distribution of the energy consumption of the various building utilities as parts of the overall consumption, whilst bar charts can be used for the allocation of consumption over time. These charts, as well as line diagrams better highlight the specific consumption trends over time.

The definition of not only the baseline energy balance, but also the optimised energy balances, can be supported by the graphics of Sankey diagrams where the thickness of the arrows is proportional to the amount of energy or flow transported.

8.10.5 The Structure of the Technical Report

The structure of the technical report will change depending on the type of the Audit.

- *Walkthrought Audit* needs a brief report in which facilities, building envelope, maintenance and management and any related inefficiencies are identified. A first lot of retrofit measures and indications on whether or not to further the investigation (by means of Standard or Simulation Audits) are provided.
- For the *Standard Audit* and *Simulation Audit* is normally required an extensive report in which, besides a detailed description of facilities, building envelope, maintenance and management inefficiencies should be defined along with retrofit measures, grouped into scenarios, with technical and economic evaluations.

Both documents, however, regardless of the length of the text, should be clear, concise and well structured. In the following section, a proposals example for a Standard and Simulation Audit is present.

8.10.6 The Structure of the Walkthrough Audit Report

8.10.6.1 Report Cover

The graphic aspect of the cover is normally chosen by the Energy Auditor, in some cases, however, the Owner imposes documentation standards that must be met. Normally the cover page should carry the following information:

- title of the report;
- name of the Owner;
- location of the audited building;
- report date;
- name of the Auditor (or of the person in charge if the audit was executed in team);
- revision reference and date of revision;
- signature of the Report.

If Energy Audit has been performed by a team, the list of consultants and specialists involved can be conveniently reported on either the cover page or the second page.

1. *Analysis of the current situation*

This section contains a brief report needed to define the situation of the status quo, highlighting the problems that emerged as a result of the survey. The contents of this section should include:

- general overview of the building or building complex;
- main dimensional features of the buildings;
- outline description of the facilities (HVAC systems, DHW systems, electrical, lighting, etc.);
- description of the (thermal and electrical) supply contracts with a summary table showing the energy consumption for a period of at least three years;
- general characteristics of management modes (hours of operation, indoor environmental conditions, etc.);
- consumption indicators and comparison with benchmarks (if available);
- results of the site survey;
- initial list of critical aspects emerging.

2. *First list of possible retrofit measures*

This section should contain a first list of possible retrofit measures, the economic assessment will be a rough estimate.

3. *Conclusions*

This section provides a summary assessment of the issues raised by the Audit, highlights the problems identified and, amongst the retrofit measures proposed, identifies those that deserve further investigation, suggesting a possible need to implement a more detailed audit.

8.10.7 The Structure of the Standard and Simulation Audit Report

8.10.7.1 Report Cover

The features are the same as the cover page of the Walkthrough Audit.

1. *Executive Summary*

The Executive Summary is not merely a summary of the results of the audit report, but much more: in a few pages, preferably no more than two, the Auditor shall transmit to the customer the most important information that puts him in a position to decide whether or not to proceed, on the reading of the report. The Executive Summary should include the following points:

- a summary of what the Energy Audit has highlighted;
- a list of measures that can be implemented to improve the energy and environmental sustainability (considering the technical and economic aspects);
- the indication of the most cost-effective scenario;
- a summary of the objectives that can be achieved by implementing the more convenient scenario.

2. *Description of the methodological approach*

The section should include the following points:

- a summary of the contractual agreements (commitments made by the auditor);
- scope of the work (Audit targets agreed with the Owner);
- acquisition of documentation (highlight where documentation has been provided by the Owner);
- description of field surveys;
- description of instrumentation and monitoring;
- calculation procedures adopted.

3. *Resource Management and indicators*

The contents of this section are as follows:

- identification of thermal and electrical loads (e.g., winter heating, summer cooling, ventilation, lighting and other electrical equipment or systems, etc.);
- characteristics of energy supply contracts;
- characteristics of the Owner's structure (number of employees, hours of employment, etc.).
- breakdown of energy consumption (heating and electricity) and water for the type of user;
- indicators of specific consumption (expressed by dividing the energy consumption at the surface or volume for each type of user);
- comparison between the consumption indicators and benchmarks (if available);
- description of the facilities management (internal management, external contracts, availability of a monitoring system, remote control, etc.);
- description of existing management and maintenance strategies (preventive maintenance, scheduled maintenance, etc.);
- a summary of the results of monitoring (the full report can be found in the annexes).

In addition to the descriptions, one can refer to the checklist that will appear in full in the appendices that shall be mentioned in the text.

4. *Data on the characteristics of the building and facilities*

The contents of this section are as follows:

- general layout drawing of the building or complex of buildings covered by the audit with highlighted the elements that could generate shadows (surrounding buildings, trees, etc.).
- identification of individual buildings on a site map showing the location of technological stations, substations and in the increasingly schematic diagram of the major backbone of the plants;

- geometrical characteristics of the buildings (floor area, gross heated volume, gross surface dispersant, shape factor, etc.);
- thermo-physical characteristics of the building envelope of individual buildings (U-values of opaque and transparent surfaces, etc.).
- description of the HVAC systems, DHW systems, etc.;
- description of the electrical equipment (transformer substation, distribution, utilities, etc.);
- description of the lighting systems (type of lighting systems installed in individual rooms);
- description of other plant systems (e.g., water systems, irrigation systems, etc.).

In addition to the descriptions, one can refer to the checklist that appears in full in the appendix that shall be mentioned in the text.

5. *Definition of the retrofit measures*

This section should contain, for each scenario, the following information:

- description of the measure and reasons for its choice;
- cost estimate;
- assessment on the reduction of operating costs;
- description of other non-quantifiable economic benefits (e.g., improved comfort, emission reductions, etc.);
- environmental effects.

For measures that reduce energy consumption, a comparison between the consumption before (baseline) and after (providing for the implementation of the measure) is required.

6. *Definition of measures to enhance facilities management*

In this section, measures related to improvement of maintenance management are discussed. This section should contain, for each proposed measure:

- description of the measure and reasons for its choice;
- cost estimate;
- assessment on the reduction of operating costs;
- description of other non-quantifiable economic benefits (e.g., improved comfort, emission reductions, etc.);
- environmental effects.

7. *Preliminary evaluation of LEED rating*

This section provides a preliminary assessment of the LEED ratings that can be achieved in with different scenarios proposed.

8.10.7.2 Appendices

This section contains all the additional information and technical insights:

- verification calculations and preliminary sizing;
- summary of the monitoring campaign;
- infrared audit report;
- audit checklists;
- detailed description of the proposed retrofit actions;
- technical documentation.

It should be remembered that the pattern of above technical report is intended as a frame of reference: auditors are of course free to develop their own formats. The definition of the structure of the report should in any case be communicated to the customer at the time of the definition of the terms of contract.

References

1. Dall'O' G, Speccher A, Bruni E (2012) The Green Energy Audit, a new procedure for the sustainable auditing of existing buildings integrated with LEED Protocols. Sustain Cities Soc 3:54–65
2. USGBC (2002) Green building rating system for new construction and major renovation (LEED-NC) version 2.1
3. US Department of Energy (1997) Selecting windows for energy efficiency, Merrifield 11. 4
4. CRES (2000) Energy audit guide, part B: system retrofits for energy efficiency, p 19, Project financed by the European Commission, Athens
5. ASHRAE (2003) ASHRAE GreenGuide—The design, construction, and operation of sustainable buildings. Butterworth-Heinemann-Elsevier, Burlington
6. EMSD (2007) Guidelines on energy audit, Hong Kong
7. Popescu D, Bienert S, Schutzenhofer C, Boazu R (2012) Impact of energy efficiency measures on the economic value of buildings. Appl Energ 89(2012):454 463
8. Bauer M, Mösle P, Schwarz M (2010) Green Building, guidebook for sustainable architecture. Springer, Munich

Part II
Tools

Chapter 9
Green Energy Audit Versus LEED® Protocols

This chapter analyses the activities and structures of the building with the aim of identifying the possible contribution that they can make to satisfying credit criteria under the LEED® certification. The structures analysed are the opaque building envelope, the transparent envelope and related facilities, whilst activities considered are those regarding commissioning and water management (WM). Once the applicable retrofit measures are identified, these are then cross-referenced with LEED® credits in order to evaluate the improvement in sustainability of the building.

9.1 LEED® Protocol 2009

The leadership in energy and environmental design (LEED) environmental certification is a protocol developed and published by the U.S. green building council (USGBC). This is the most popular protocol in the world (used in over 40 countries), and it is one of the most reliable and widely applied. Furthermore, as [1] underlines, LEED rating systems adopt a holistic evaluation of overall impact of a building on the environment. These are the reasons that led the authors to consider LEED as a reference environmental protocol for the Green Energy Audit.

LEED provides different formulations depending on building type and the phase of the building's life cycle. In this chapter, the 2009 version for New Construction and the 2009 version for Operations and Maintenance are analysed in detail. LEED for New Construction can be used for new buildings and for existing buildings subject to major renovations (for remedial actions that involve significant elements of air conditioning systems, significant remedial actions on the construction itself and renovation of interior spaces) [2, 3]. LEED Operations and Maintenance was designed to certify the sustainability of ongoing operations of existing commercial and institutional buildings.

Elisa Bruni contributed this chapter.

G. Dall'O', *Green Energy Audit of Buildings*, Green Energy and Technology,
DOI: 10.1007/978-1-4471-5064-0_9, © Springer-Verlag London 2013

The rating system is organised into five environmental categories: sustainable site (SS), water efficiency (WE), energy and atmosphere (EA), materials and resources (MR) and indoor environmental quality (IEQ). An additional category known as innovation in operations (IO) for existing buildings or innovation in design (ID) for new construction and renovations addresses innovative practices aimed at sustainability and issues not covered in the five previous categories. The category regional priority (RP) is used to highlight the importance of local conditions in determining best practices.

In 2009, the protocol was updated through a re-weighting of LEED credits. The available points have been re-distributed such that the various credits reflect more accurately their potential to mitigate the negative environmental impacts of the building or to promote its positive impacts. The weighting of the credits is based on scientific methods to guarantee their greater accuracy and transparency. Quantitative methods were introduced to objectively assess the environmental impacts of a building during its entire life cycle.

The 2009 LEED certification for the operation and maintenance of existing buildings and the 2009 LEED certification for new construction and renovations classify buildings according to the following rating scale:

- Base: 40–49 points;
- Silver: 50–59 points;
- Gold: 60–79 points; and
- Platinum: 80 and above.

9.2 Green Energy Auditing and LEED Credits

In the LEED system, as in most systems for the environmental certification of buildings, the designer is not bound by mandatory choices but is offered some flexibility. It is necessary to meet the mandatory criteria, but design choices can be tuned in different ways to favour or exclude certain technologies; what is important is that a certain total score is reached.

Any design choice, or in the case of the Green Energy Audit, any measure, is not uniquely linked to a credit. Design choices and credits are therefore two elements between which there is a correlation that is not bi-univocal, as in the simplified diagram in Fig. 9.1 where in the building's systems, as viewed in the entirety of its design choices are compared with the evaluation system.

A measure in terms of energy and the environment can meet one or more credit criteria from different LEED categories as a function of the choices being made.

Although LEED certification is given based on an assessment of the building as a whole, *the performance of the structures that make up the building and building management strategies can help meet the objectives set by the credits* and thus can contribute to the acquisition of final certification for the building

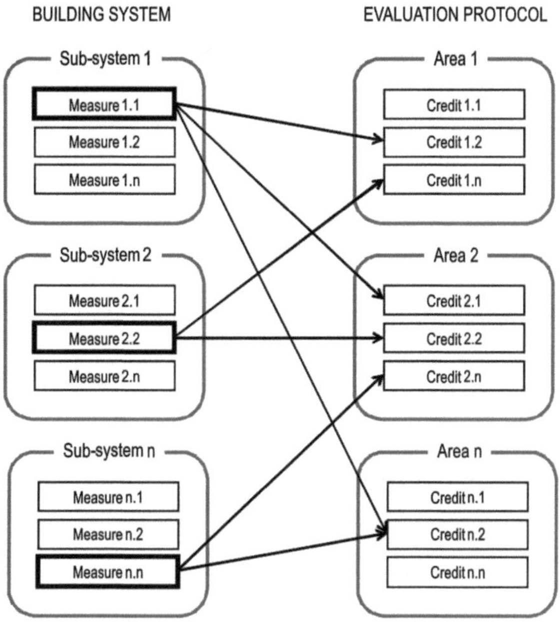

Fig. 9.1 Correlation between the retrofit measures and the credits of the adopted evaluation protocol

The aim of a Green Energy Audit is to evaluate the degree of improvement in sustainability of the building as a whole that can be obtained through the proposed choices; such choices do not necessarily generate an advantage in terms of energy, but they can generate many advantages with respect to sustainability. If the standard of comparison is the LEED Protocol, then the problem is in understanding how application of a certain remedial action can help to meet the credits [4].

With the aim of highlighting the contribution they can make to satisfying credits, different structures and activities are analysed, such as the following: opaque building envelope; transparent building envelope (fenestration); heating, ventilation and air conditioning (HVAC) systems; the commissioning process and water management.

Credits identified as involving structures and activities are then cross-referenced with various remedial actions to clearly identify the relevance of each action to LEED credits, as outlined in Fig. 9.2.

Amongst building systems, the transparent building envelope is a structure of greater complexity from this point of view. According to its technical characteristics and performance, the transparent envelope of a building can help to fulfil certain specific credits for the categories EA, MR and IEQ. In turn, the credit optimising energy performance in the category EA is relevant to window replacement measures since, in the case of building renovation, replacing windows with models of superior properties can improve the energy performance of a building. Figure 9.2 illustrates LEED credits for new construction related to the transparent building envelope, and points are used to indicate the relevance of certain transparent building envelope remedial actions to the LEED credits.

LEED New Construction		Green Energy Audit				
		Area: Transparent envelope				
Area LEED	LEED Credit	Replacement of the windows	Replacing of glass panes	Application of low-emissivity films	Sealing of air infiltration	Installation of a sunspace
EA	Minimum Energy Performance	●	●	●	●	●
EA	Optimize Energy Performance	●	●	●	●	●
MR	Recycled Content	●	●			●
MR	Regional Materials	●				●
MR	Certified Wood	●				●
IEQ	Increased Ventilation					
IEQ	Low-Emitting Materials	●			●	
IEQ	Controllability of Systems-Thermal Comfort					
IEQ	Daylight and Views - Daylight	●	●			
IEQ	Daylight and Views - Views					
		LEED – GEA Appliances				

Fig. 9.2 Transparent building envelope intervention with relevance to LEED credit

Another example concerns the water management of a building. Managing water by stimulating the flow of rainwater and reducing the consumption of potable water may contribute to the building various credits in the categories SS, WE and EA. In the SS category, there is the "Stormwater Design: control of the quantity" credit. This credit in turn is relevant to the remedial action "Rainwater harvesting" as one of the strategies that can reduce the amount of stormwater flowing into water bodies (e.g., water courses).

9.3 Opaque Building Envelope

The opaque envelope, depending on its technical characteristics and performance, can contribute to the satisfaction of some specific credits as defined by LEED. The categories that involve the opaque envelope are those relating to SS, EA, MR and IEQ (see Table 9.1)

The SS category is dedicated to environmental issues related to the relationship between the building and the site in which it is located. It deals in particular with the phenomenon known as the "heat island effect". Incident solar radiation is partly reflected and partly absorbed by a building's surfaces and then radiated as heat, contributing to an increase in the surrounding ambient temperature. Each surface has a own ability to reflect solar radiation (represented by a parameter called:

Table 9.1 Credits of LEED version 2009 USA for new buildings and for existing buildings related opaque envelope [2, 3]. It also shows the points available in the respective credit

Category	Credit	Available points NC 2009	Available points EB 2009
Sustainable sites	Credit 7.2: heat island effect—roof	1	1
Energy and atmosphere	Prerequisite 2: minimum energy performance	Mandatory	Mandatory
	Credit 1: optimize energy performance	1–19	1–18
Materials and resources	Credit 1.1: building reuse—maintain existing walls, floors and roof	1–3	–
	Credit 1.2: building reuse—maintain 50 % of interior non-structural elements	1–2	–
	Credit 2: construction waste management	1–2	–
	Credit 3: material reuse	1–2	–
	Credit 4: recycled content	1–2	–
	Credit 5: regional materials	1–2	–
	Credit 6: rapidly renewable materials	1	–
	Credit 7: certified wood	1	–
	Credit 3: sustainable purchasing—facility alterations and additions	–	1
	Credit 9: solid waste management—facility alterations and additions	–	1
Indoor Environmental Quality	Credit 4.1: low-emitting materials—adhesives and sealants	1	–
	Credit 4.2: low-emitting materials—paints and coatings	1	–
	Credit 4.3: low-emitting materials—flooring systems	1	–
	Credit 4.4: low-emitting materials—composite wood and agrifiber products	1	–

solar reflectance index), concrete surfaces (dark in colour or damaged) have a low ability to reflect solar radiation unlike the surfaces covered with vegetation. These characteristics contribute to the occurrence of a thermal gradient between urban and green areas. The raising of the ambient temperature (heat island effect) which occurs in cities, as well as having a negative impact on the local ecosystem, sensitive to variations in temperature, causes an increase in energy consumption for ventilation and summer cooling of buildings.

The credit "Heat Island Effect-Roof" aims to reduce this effect and suggests several options for coating the surface of the roof. It is possible to use roofing materials with a high solar reflectance, or install a green roof system. The remedial actions of "extrados insulation" with green roof or roof replacement with roofing materials characterised by high solar reflectance index contribute to the satisfaction of this credit. This strategy has many benefits, in addition to reducing the heat island effect, including the decrease in demand for summer cooling and the increase of the thermal insulation of the roof which contribute to the "Minimum Energy Performance" and "Optimize Energy Performance" credits of the environmental category EA.

The energy performance of a building depends in fact also on the performance of its opaque envelope. An envelope made of opaque material with a high insulation can help to increase the energy performance of the building. All measures to insulate roofs, the basement and the vertical walls are able to reduce losses owing to transmission by the opaque building envelope and could then contribute to the acquisition of LEED certification for the building.

The credits of the category MR are intended to reduce the impact on the environment and human health of construction materials throughout the life cycle. In the case of renovation of an existing building, the reuse of both existing building structure and envelope (such as roof, structural floor and exterior skin) is encouraged, along with that of interior non-structural elements (such as floor coverings, ceilings systems, interior walls and doors).

The respect of credits "Building Reuse. Maintain Existing Walls, Floors, and Roof" and "Building Reuse". Maintain 50 % of "Interior Non-Structural Elements" in addition to the environmental benefits of reducing the production of waste and conserving resources has the economic benefits of reducing the cost of demolition, transport and disposal of waste and acquired construction materials.

The credit "Materials Reuse" assesses the percentage of use of materials salvaged, refurbished or reused. The material can be reused at a pre-existing site: in this case, it should no longer be able to fulfil its original function and is required to be installed for a different use, or cannot be pre-existing materials on site, for example bricks purchased as material recovery.

The credit "Recycled Content" aims to reduce the use of virgin material and the production of waste by encouraging the use of building materials with recycled content. To meet the credit there is the requirement to use materials with a post-consumer or pre-consumer recycled content. The content of post-consumer recycled material is defined as the percentage of material of a product that has previously been refused for the consumer, while the content of recycled material pre-consumer is defined as the percentage of material of a product that is derived from manufacturing refuse.

The increased demand of materials extracted, processed and manufactured regionally with a consequent reduction in air pollution generated by transport is instead the objective pursued by the credit "Regional Materials". The distance from the site within which the material is to be extracted, processed and manufactured can increase if transport with low environmental impact, such as transport by sea or by rail, is utilised. In the case of new buildings or buildings subject to major renovation, the use of materials involving certified wood or rapidly renewable items contributes to satisfy "Rapidly Renewable Materials" and "Certified Wood" credits.

For existing buildings the "Sustainable Purchasing-Facility Alterations and Additions" credit encourages the adoption of a programme of sustainable purchasing that involves the materials used in the process of renovation and additions to the building. During the performance period of the building it is necessary to prove the purchase of materials with certain percentages of recycled content, certified wood, rapidly renewable materials or of local material with low emissions of volatile organic compounds.

The "Solid Waste Management-Facility Alterations and Additions" credit involves assigning a score if it is shown that part of the waste produced during renovation activities is reused or recycled.

The category IEQ aims to improve the quality of the indoor environment through a variety of strategies including the use of materials with low emissions of harmful substances.

For new constructions, credits are defined relative to low-emission materials with the purpose of limiting the emissions of contaminants which have a negative impact on indoor and outdoor air quality and therefore cause adverse effects on human health. "Low-Emitting Materials" defines limits of emission of volatile organic compounds for different types of products: adhesives and sealants, paints and coatings and flooring systems.

Finally, the settlement of the credit "Low-Emitting Materials-Composite Wood and Agrifiber Products" assumes that each wood product installed inside the building does not contain added urea-formaldehyde resins. This group of credits under the category IEQ is primarily related to materials used inside the buildings under consideration. The use of materials with low emissions of VOCs also on the outside of the building is not currently provided for in the LEED despite the benefits for the environment. Strategies aimed at this goal can, however, be computable within the innovation of the design.

9.4 Transparent Building Envelope

Transparent envelopes, according to their technical characteristics and their performance, can contribute to the fulfillment of the criteria of some specific credits. The categories that involve mainly the transparent envelope are those relating to EA, MR and IEQ (see Table 9.2).

Although the acquisition of a credit has environmental benefits, the impact of transparent envelopes in meeting the credit is not in itself indicative of sustainability but merely expresses the contribution they can give to obtaining LEED certification. For a correct assessment of the impacts on the environment and on human health, it is necessary to define the impacts during all stages of the life cycle of the product.

In the acquisition of LEED certification, optimisation of energy performance assumes a significant weight since it plays a role in reducing the environmental burdens associated with the production and consumption of energy.

The EA environmental category establishes a prerequisite regarding the "Minimum energy performance" and a credit on "Optimizing energy performance".

The methods identified by LEED to evaluate the energy performance of a building depend upon many factors, including the characteristics of the transparent

Table 9.2 Credits of LEED version 2009 USA for new buildings and for existing buildings related transparent envelope [2, 3]. It also shows the points available in the respective credit

Category	Credit	Available points NC 2009	Available points EB 2009
Energy and atmosphere	Prerequisite 2: minimum energy performance	Mandatory	Mandatory
	Credit 1: optimize energy performance	1–19	1–18
Materials and resources	Credit 4: recycled content	1–2	–
	Credit 5: regional materials	1–2	–
	Credit 7: certified wood	1	–
	Credit 3: sustainable purchasing—facility alterations and additions	–	1
	Credit 9: solid waste management—facility alterations and additions	–	1
Indoor environmental quality	Credit 2 and 1.3: increased ventilation	1	1
	Credit 4.1: low-emitting materials—adhesives and sealants	1	–
	Credit 6.2: controllability of systems—thermal comfort	1	–
	Credit 8.1: daylight and views—daylight	1	–
	Credit 8.2: daylight and views—views	1	–
	Credit 2.4: daylight and views	–	1

envelope. The windows can help to reduce leakage through the envelope and to bring a correct use of solar gains. The potential strategies to improve the energy performance of a building can therefore include remedial actions to reduce losses derived from transmission and ventilation of the transparent envelope, to limit the summer external heat gain contributions by sun protection strategies and take advantage of the winter solar gains through solar greenhouses. It should be noted that, since the scores for the energy performance are calculated by comparing the projected consumption of the building design with a reference building and the transparent envelope of the reference building cannot be more than 40 % of the total area, it is almost impossible to attain a high score with a building designed with only a few opaque elements.

The category MR has the objective of reducing the impact on the environment and human health, of building materials throughout their life cycle. The use of materials with recycled content can contribute to reducing the use of virgin material and the production of waste. The materials extracted, processed and produced at short distance from the building can contribute to the reduction of air pollution generated by transport. The materials in certified wood (e.g., the Forest Stewardship Council) help to encourage responsible forest management. In the case of buildings subject to major renovation, the use of this type of material between the elements permanently installed in the building, such as fenestration, can contribute to the attainment of credits: “Recycled Content”, “Regional Materials” and “Certified Wood”. For existing buildings undergoing renovation, the "Sustainable Purchasing-Facility Alterations and Additions" credit encourages the adoption of a programme of sustainable procurement involving materials with recycled content, certified wood, rapidly renewable materials, composed of local material or with low emissions of volatile organic compounds.

In the evaluation of the type of window to be installed, from an environmental point of view it is desirable to acquire information about the raw materials (wood, glass, aluminium or PVC) of which the window frame is composed and the respective percentage of recycled content, their source and any certification acquired. The credit on "Solid waste management: improvements and extensions" will award a score when it is demonstrated that a portion of the volume of waste products (including windows) during the restructuring activities will be reused or recycled. This would eliminate the incineration of waste, or the transfer of waste to landfill with all its environmental and economic implications.

The credits related to the category IEQ are addressed at improving the quality of the indoor environment through proper ventilation of the rooms, use of materials with low emissions of harmful substances, the possibility for the user to adjust the temperature of spaces and take advantage of natural lighting. Natural ventilation by opening windows contributes to the improvement of the internal comfort through control of the temperature in the room and the maintenance of high values of air quality. One of the possibilities to improve the quality of indoor air is to remove the pollutants present within the indoor environments by increasing the ventilation. In the case of naturally ventilated spaces, the windows may contribute to the acquisition of credit scores for "Increased Ventilation". The credits on low-emission materials define the emission limits for volatile organic compounds for each product type of adhesives, sealants and paint that may be present on surfaces in contact with the interior of the building.

In addition, there is a credit dedicated to "Controllability of Systems-Thermal Comfort", which has the objective of promoting adequate control and regulation of the facilities to ensure the thermal comfort of building occupants. It has indeed been shown that individual temperature control increases occupant comfort and productivity of workers. Amongst the strategies used for the control of the temperature are included opening windows, subject to certain conditions.

Credits related to "Natural light and vision" are intended to ensure the natural lighting of interior spaces and specific connection between indoor spaces and the outdoors.

Whilst the window surfaces allow the entry of natural light for the occupants, contact with the outside and reduction in the need for artificial lighting, there is on the other hand, a necessity to evaluate solar gains, thermal losses and glare phenomena.

The methods used for the assessment of natural light consider, amongst other information, the surface and the visible light transmittance. Glass surfaces with certain characteristics may contribute to the attainment of credits "Natural light and vision" (see Tables 9.3, 9.4, and 9.5).

Table 9.3 Correlation between retrofit measures on transparent envelope and LEED 2009 for new buildings credits related to the transparent envelope

Correlation measures/credits		Transparent envelope							
Category	LEED credits NC 2009	FE01 Replacement of the windows	FE02 Installation of storm windows	FE03 Replacement of glass panes on the existing frame	FE04 Application of low-emissivity films	FE05 Sealing of air infiltration	FE06 Insertion of gaskets to improve air-tightness	FE07 Transparent enclosure of verandas	FE08 Installation of a sunspace
EA	Prerequisite 2: minimum energy performance	v	v	v	v	v	v	v	v
EA	Credit 1: optimise energy performance	v	v	v	v	v	v	v	v
MR	Credit 4: recycled content	v	v	v				v	v
MR	Credit 5: regional materials	v	v					v	v
MR	Credit 7: certified wood	v	v					v	v
IEQ	Credit 2: increased ventilation								
IEQ	Credit 4.1: low-emitting materials—adhesives and sealants	v	v			v			
IEQ	Credit 6.2: controllability of systems—thermal comfort								
IEQ	Credit 8.1: daylight and views—daylight	v	v	v					
IEQ	Credit 8.2: daylight and views—views								

Table 9.4 Correlation between retrofit measures on transparent envelope and LEED 2009 for existing buildings credits related to the transparent envelope

Correlation measures/credits		Transparent envelope							
Category	LEED Credits EB 2009	FE01 Replacement of the windows	FE02 Installation of storm windows	FE03 Replacement of glass panes on the existing frame	FE04 Application of low-emissivity films	FE05 Sealing of air infiltration	FE06 Insertion of gaskets to improve air-tightness	FE07 Transparent enclosure of verandas	FE08 Installation of a sunspace
EA	Prerequisite 2: minimum energy performance	v	v	v	v	v	v	v	v
EA	Credit 1: optimise energy performance	v	v	v	v	v	v	v	v
MR	Credit 3: sustainable purchasing—facility alterations and additions	v	v					v	v
MR	Credit 9: solid waste management—facility alterations and additions	v	v						
IEQ	Credit 1.3: increased ventilation								
IEQ	Credit 2.4: daylight and views	v	v	v					

Table 9.5 Correlation between retrofit measures related to sun protection and daylighting with LEED 2009 for new buildings and major renovation credits related to the transparent envelope. It must be noted that the intervention of shading with vegetation outside can make a direct contribution to the credit heat island effect of the sustainable sites category

Correlation measures/credits		Sun protection systems				Illumination using Daylight		
Category	LEED credits NC 2009	SP 01 Installation of external solar shading systems	SP 02 Application of solar control films	SP 03 Application of internal solar shading systems	SP 04 Shading with vegetation outside	DL01 Painting of internal walls with light colours	DL02 Installation of light pipes	DL 03 Installation of light shelves
EA	Prerequisite 2: minimum energy performance	v	v	v	v		v	v
EA	Credit 1: optimise energy performance	v	v	v	v		v	v
MR	Credit 4: recycled content	v						
MR	Credit 5: regional materials	v		v				
MR	Credit 7: certified wood	v		v				
IEQ	Credit 2: increased ventilation							
IEQ	Credit 4.1: low-emitting materials—adhesives and sealants					v		
IEQ	Credit 6.2: controllability of systems—thermal comfort							
IEQ	Credit 8.1: daylight and views—daylight	v		v				v
IEQ	Credit 8.2: daylight and views—views							

9.5 Building Facilities (HVAC, Renewable Energy Systems and Lighting Systems)

The categories that involve mainly the heating, ventilation and air conditioning systems (HVAC or Heating, Ventilation and Air Conditioning) and the production of energy from renewable sources are those relating to EA and IEQ. If there are any cooling towers or evaporative condensers present, then an involvement of credits concerning WE are to be found: "Water Use Reduction" and "Cooling Tower Water Management" (Table 9.6).

Within the EA criteria in which plants are directly involved, there are credits available for: commissioning, energy performance, refrigerants management, renewable energy and performance measurement.

9.5.1 Energy Performance

The energy performance of buildings is linked to a mandatory prerequisite "Minimum energy performance" and a optional credit "Optimizing energy performance". From the point of view of points acquired, the highest levels are: 19 points for new construction and 18 points for existing buildings. Points can be acquired dependant upon the percentage improvement of the energy performance of the building when compared to that of a reference building. The reference for the evaluation of the energy performance is the ASHRAE 90.1-2007, which defines performance criteria of the various components of the building including the building envelope, the heating, ventilation, air conditioning, water heating, energy electricity and lighting. In addition to the energy performance of the envelope and facilities, the process energy shall be considered. By process, energy use meant the energy consumption of the activities conducted inside the building and related to its intended uses. All remedial actions and works that can improve the energy performance of the various components of the building including facilities can contribute to this particular credit.

9.5.2 Renewable Energy

LEED certification aims to reduce the environmental loads associated with the consumption of energy from fossil fuels both through the improvement of the energy performance and by stimulating the production of energy on site from renewable sources and/or the purchase of green energy.

The installation of power generation such as photovoltaic, wind, solar thermal and power generation based on bio-fuels, geothermal heating and low-impact hydroelectric generation that exploit the waves and tides can contribute to the

Table 9.6 Credits of LEED version 2009 USA for new buildings and for existing buildings related building facilities. It also shows the points available in the respective credit

Category	Credit	Available points NC 2009	Available points EB 2009
Water efficiency	Prerequisite 1: water use reduction	Mandatory	–
	Credit 3: water use reduction	2–4	–
	Credit 4: cooling tower water management	–	1–2
Energy and atmosphere	Prerequisite 1: fundamental commissioning of building energy systems	Mandatory	–
	Prerequisite 1: energy efficiency best management practices	–	Mandatory
	Prerequisite 2: minimum energy performance	Mandatory	Mandatory
	Prerequisite 3: fundamental refrigerant management	Mandatory	Mandatory
	Credit 1: optimise energy performance	1–19	1–18
	Credit 2: on-site renewable energy	1–7	–
	Credit 2.1/2.2/2.3: existing building commissioning	–	2/2/2
	Credit 3: enhanced commissioning	2	–
	Credit 3.1/3.2: performance measurement	–	1/1–2
	Credit 4: enhanced refrigerant management	2	–
	Credit 4: on-site and off-site renewable energy	–	1–6
	Credit 5: measurement and verification	3	–
	Credit 5: enhanced refrigerant management	–	1
	Credit 6: green power	2	–
	Credit 6: emissions reduction reporting	–	1
Indoor environmental quality	Prerequisite 1: minimum indoor air quality performance	Mandatory	Mandatory
	Prerequisite 2: ETS control	Mandatory	Mandatory
	Credit 1: outdoor air delivery monitoring	1	–
	Credit 1.1/1.2/1.3/1.4/1.5: indoor air quality best management practices	–	1/5
	Credit 2: increased ventilation	1	–
	Credit 2.1/2.3: occupant comfort	–	1/1
	Credit 2.2: controllability of systems—lighting	–	1
	Credit 3.1/3.2: construction IAQ management plan	1/1	–
	Credit 5: indoor chemical and pollutant source control	1	–
	Credit 6.1: controllability of systems—lighting	1	–
	Credit 6.2: controllability of systems—thermal comfort	1	–
	Credit 7.1/7.2: thermal comfort	1/1	–

fulfillment of the "On-Site Renewable Energy" credit (unless there are specific restrictions). The "Green Power" credit expects the stipulation of a 2-year contract for the supply of electricity generated from renewable sources to meet 35 % of the electricity needs of the building.

So far as existing buildings are concerned, the "On-site and Off-site Renewable Energy" credit considers both the production and the purchase of energy from renewable sources (see Table 9.7).

9.5.3 Refrigerant Management

The management of refrigerants is addressed in prerequisite "Fundamental Refrigerant Management" and credits "Enhanced Refrigerant Management". The aim is to preserve the ozone layer which is degraded by the presence of compounds in the stratosphere based upon chlorofluorocarbons and chloro-fluoro-hydrocarbon refrigerants. The prerequisite states that neither CFC nor HCFC refrigerants are to be used in air conditioning/refrigeration plant and that halons be removed from fire protection systems. The credits "Advanced management of refrigerants" concerning the impact of refrigerants on the greenhouse effect provide two options: the elimination of refrigerants or the choice of refrigerants with low global warming potential and the adoption of strategies to minimise losses of refrigerant from equipment having a refrigerant charge and to have efficient equipment with a long life cycle.

9.5.4 Performance Measurement

The credit "Measurement and Verification" aims to measure the energy consumption of the building during operation through the implementation of a programme of measurements and audits. The adoption of an automated building management system, able to monitor and control at least the heating, cooling, ventilation and lighting is encouraged in the "Performance Measurement-Building Automation System" credit.

"Performance Measurement-System-Level Metering" requires the adoption of a system of performance measurement that can first provide information about the energy consumption of the building and then identify appropriate strategies for improvement.

Remedial actions and works that can contribute to the fulfillment of these credits are those related to the activation of monitoring procedures and energy accounting and the installation of a home automation system.

Table 9.7 Correlation between retrofit measures related to renewable energy with LEED 2009 for new buildings and major renovation credits related to energy performance and renewable energy systems

Correlation measures/credits		Renewable energy sources				
		RS01	RS02	RS03	RS04	RS05
Category	Credits LEED NC 2009	Installation of a thermal solar system for low temperature uses	Installation of a solar heating and cooling system	Installation of a photovoltaic (PS) solar system	Installation of a biomass boiler	Purchase of certified energy from renewable sources
EA	Prerequisite 2: minimum energy performance	v	v	v	v	
EA	Credit 1: optimise energy performance	v	v	v	v	
EA	Credit 2: on-site renewable energy	v	v	v	v	
EA	Credit 6: green power					v

9.5.5 Indoor Air Quality

The IEQ category involves plants in numerous credits. The systems most affected are those of ventilation and in fact many credits are aimed at improving indoor air quality, which is obtainable through appropriate ventilation.

The prerequisite "Minimum Indoor Air Quality Performance" requires compliance with what is established in ASHRAE Standard 62.1-2007. In the case of existing buildings for which the existing mechanical ventilation system represents an impediment, a minimum ventilation flow rate is still defined.

If the building is equipped with a mechanical ventilation system it is necessary to coordinate the strategies aimed at improving energy performance and thermal comfort of the occupants with those aimed at improving indoor air quality. This is because improving indoor air quality alone could in fact lead to an increase of energy consumption in ventilation systems.

On the ventilation side two points can be acquired, one point for the "Outdoor Air Delivery Monitoring" credit and the other for the "Increased Ventilation" credit. "Outdoor Air Delivery Monitoring" provides for the installation of permanent systems for monitoring the concentration of CO_2 inside occupied spaces and in the case of mechanically ventilated spaces it is necessary to provide a system with certain characteristics for measuring the flow of outside air. "Increased Ventilation" encourages the provision of a spare fan for additional air to further improve the quality of indoor air and consequently the comfort and well-being of the occupants.

Construction and renovation activities can cause the introduction of pollutants into indoor environments and result in problems related to indoor air quality for both the construction workers and the subsequent building occupants. "Construction Indoor Air Quality Management Plan-During Construction" provides for the implementation of an indoor air quality management plan, for the construction and pre-occupancy phases of the building, which defines specific control measures. These are the protection of HVAC systems to limit their contamination, the control of the sources for materials that contain VOCs, interruption of contamination pathways to prevent contamination of occupied spaces or those where works are carried out and, finally, the cleaning of surfaces. At the end of the construction phase there is also the requirement to replace all filtration media. Prior to occupancy of the building, and after the installation of interior finishes, the "Construction Indoor Air Quality Management Plan-Before Occupancy" credit gives 2 options: perform a building flush-out or conduct IAQ testing. The flush-out is performed via the building HVAC systems or other temporary facilities by supplying outdoor air to remove contaminants. IAQ testing is to demonstrate that the contaminant concentration levels measured are below maximum concentration levels.

Indoor air quality also depends on the daily activities within the building, such as the use of printers and of chemicals for cleaning which can increase the presence of pollutants. Similarly occupants entering rooms with shoes and clothes in fact cause the entry of pollutants into buildings. The objective of "Indoor

Chemical and Pollutant Source Control" is to minimise building occupant exposure to chemical pollutants and particulates by means of entryway systems, isolation of the areas where contaminants and chemicals are present and installation of high-level filtration systems.

Five are the credits related to Indoor Air Quality Best Management Practices described in the manual dedicated to existing buildings. "Indoor Air Quality Management Program" refers to the development of an IAQ management programme based on the EPA Indoor Air Quality Building Education and Assessment Model. "Outdoor Air Delivery Monitoring" requires installation of permanent systems for monitoring the CO_2 concentration in the occupied spaces. "Increased Ventilation" encourages increased outdoor air ventilation rates to improve indoor air quality, and consequently the comfort and well-being of the occupants. "Practices-Reduce Particulates in Air Distribution" provides for the development of a protocol for the management of sources of pollutants. Finally, "Indoor Air Quality Management for Facility Alterations and Additions" refers to the implementation of an IAQ management plan for the construction and occupancy phases. In those cases where smoking rooms are allowed inside the building, separate ventilation systems are required to isolate the contaminated air, as defined by the prerequisites of "Environmental Tobacco Smoke (ETS) Control".

9.5.6 Lighting Systems

In the New Construction manual, lighting systems are affected by "Light Pollution Reduction" in the SS category and by "Controllability of Systems-Lighting" in the IEQ category (see Table 9.8).

To minimise light "trespass" from the building during after-hours periods specific requirements for interior lighting systems are defined, such as installing automatic control of interior lighting fixtures. To minimise the impact from exterior lighting on the nocturnal surroundings, and the environment in general, it is possible to shield exterior fixtures so as to prevent light from being emitted into the night sky.

The fulfillment of the requirements for both internal and external lighting systems permits, at an environmental level, a reduction in any negatives effects on the ecosystem and limits nocturnal light pollution. From an economic point of view it also makes a reduction in energy costs possible. Another feature requested by lighting systems is defined by "Controllability of Systems-Lighting" which aims to provided a high level of lighting system control by individual occupants.

All measures capable of improving the energy performance of the lighting system can contribute to credits dedicated to energy performance. Any measures which ensure that occupants can control lighting systems, such as the presence of table lamps in the case of offices, can contribute to credit on the controllability of systems.

Table 9.8 Correlation between retrofit measures related to lighting systems with LEED 2009 for existing buildings

Correlation measures/credits		Lighting systems						
		LI 01	LI 02	LI 03	LI 04	LI 05	LI 06	LI 07
Category	Credits LEED EB 2009	Replacement of lamps for interior lighting	Replacement of lamps for outdoor lighting	Installation of presence detectors	Installation of twilight switches	Installation of daylight sensors	Installation of high frequency electronic ballasts	Installation of dimmers
EA	Prerequisite 2: minimum energy performance	v	v	v	v	v	v	v
EA	Credit 1: optimise energy performance	v	v	v	v	v	v	v
IEQ	Credit 2.2: controllability of systems-lighting			v	v	v		

9.5.7 Thermal Comfort

Another aspect to which LEED certification gives particular attention is the thermal comfort of the occupants, since it has been shown that an indoor environment which is comfortable from a thermal point of view promotes the well-being and productivity of occupants.

For the acquisition of the "Controllability of Systems-Thermal Comfort" credit individual comfort controls for at least 50 % of the building occupants must be provided. This individual control is intended to be an operable window or a regulation system including the individual thermostats and individual radiant panels. In the case of spaces shared by multiple occupants, it is necessary to ensure the presence of at least one accessible comfort adjustment device.

"Thermal Comfort-Design" requires compliance with defined standards both in the case of active systems (HVAC) and in the case of passive systems (natural ventilation). "Thermal Comfort-Verification" includes, in addition to a survey on thermal comfort itself, a system of continuous monitoring of internal environmental variables. The presence in the occupied spaces of a monitoring system for air temperature and relative humidity can contribute also to the acquisition of the "Comfort occupant thermal comfort monitoring" credit.

9.6 Commissioning Process

The category that concerns the commissioning activities of the energy systems is related to EA (see Table 9.9). Commissioning and energy audit are processes that have the common goal of improving the energy performance of the building. LEED certification requires the carrying out of a commissioning process for both new buildings (in which case it is denominated "commissioning") and for existing buildings (in which case it is called "retro-commissioning" or "re-commissioning").

Table 9.9 Credits of LEED version 2009 USA for new buildings and for existing buildings related building commissioning. It also shows the points available in the respective credit

Category	Credit	Available points NC 2009	Available points EB 2009
Energy and Atmosphere	Prerequisite 1: fundamental commissioning of building energy systems	Mandatory	–
	Prerequisite 1: energy efficiency best management practices	–	Mandatory
	Credit 2.1/2.2/2.3: existing building commissioning	–	2/2/2
	Credit 3: enhanced commissioning	2	–

More precisely, retro-commissioning is a commissioning process activated on a building that was not commissioned during its construction phase. Whereas, re-commissioning is a commissioning process applied when an existing building has already been subject of commissioning process and is based on considerations emerging from the energy audit.

9.6.1 New Buildings and Major Renovation

In LEED for new construction, the protocol prerequisites and credits that address the commissioning are "Fundamental Commissioning of Building Energy Systems" and "Enhanced Commissioning". The commissioning is a process of quality control that allows one to verify that the building energy systems have been installed and are operating in accordance with the requirements of the owner, as determined in the design phase. For this reason, the main documents for an effective commissioning process are the Owner Project Requirement and the Basis of Design. The Owner Project Requirement describes the objectives and the requirements in terms of performance of the project and expectations about systems, the Basis of Design describes the performance of commissioned systems and corresponding regulations.

There are many advantages including reductions in energy consumption and in operating costs and improvements in the comfort conditions for the occupants.

Commissioning process activities must be completed, as a minimum, for the energy-related systems: heating, ventilating, air conditioning and refrigeration (HVAC and R) systems, lighting controls, domestic hot water systems and renewable energy systems.

The activities are developed at different stages: design, construction and occupancy of the building and provide for the involvement of the various professionals concerned, including the owner, the design team, the contractor, maintenance personnel and users under the guidance of a Commissioning Authority.

For new buildings, the activities associated with the commissioning which develop during the occupation, relate to reviewing the operation of the building with operations and maintenance staff and with the occupants within 10 months after substantial completion in order to identify and resolve inefficiencies. This activity is mandatory for the "Enhanced Commissioning" credit and is optional for the fulfillment of the "Fundamental Commissioning of Building Energy Systems" prerequisite.

The process of commissioning has origins much more remote than the LEED protocol: it is a standardised quality control system that starts from the definition of the requirements of the owner and continues up to the building management. Fundamental Commissioning and Enhanced Commissioning are therefore firstly, a prerequisite that the LEED protocol considered as mandatory, and secondly the natural completion of a process that ideally one can terminate at the time of the occupation of the building.

9.6.2 Existing Buildings

In LEED for existing buildings, the protocol prerequisites and credits that address the commissioning are "Energy Efficiency Best Management Practices" and "Existing Building Commissioning".

So it is mandatory to develop a building operational plan containing the operating mode and maintenance of the systems or alternatively the execution of a level I energy audit, as defined in ASHRAE is planned. Whereas compliance with the "Existing Building Commissioning" credit is optional.

"Existing Building Commissioning-Investigation and Analysis", provides two different options: the development of a plan of retro-commissioning (of re-commissioning or ongoing commissioning) or the execution of an level II energy audit as defined in ASHRAE. An ongoing commissioning process provides for the monitoring and analysis of data on the performance of the building in order to identify and correct any problems of energy systems.

"Existing Building Commissioning-Implementation" requires the implementation of no-cost or low cost operational improvements and the creation of a plan for major retrofits or upgrades with a corresponding cost versus benefit analysis. Adequate preparation and training is also required for all personnel involved in operations management and maintenance of the building.

"Existing Building Commissioning-Ongoing Commissioning" includes the implementation of an ongoing commissioning programme.

Measures relating to management improvement may contribute to credits related to the commissioning process on existing buildings (see Table 9.10). The commissioning process provides for the monitoring and analysis of data on the performance of the building to fix problems with energy systems, the development of adequate maintenance of facilities and the implementation of operational improvements. There is a strong correlation between Green Energy Audits and re-commissioning. The result of a Green Energy Audit is crucial as a starting point to implement the subsequent strategies to ensure the management of the building in accordance with the energy-saving targets set by the owner (re-commissioning process).

9.7 Water Management

Correct water management, resulting in a reduction in water consumption, can contribute to the fulfillment of certain LEED credits. In LEED protocols, there is a category dedicated entirely to water efficiency. Water efficiency measures are directly related to the energy aspects (EA category) and in fact consumption of potable water causes energy consumption to heat and distribute water and to operate the equipment. There are also two credits in the SS category relating to "Storm-water management" (see Table 9.11).

Table 9.10 Correlation between retrofit measures related to management improvement with LEED 2009 for existing buildings related to commissioning process

Correlation measures/credits		Improvement in Management of the Building								
		MI01	MI02	MI03	MI04	MI05	MI06	MI07	MI08	MI09
Category	Credits LEED EB 2009	Reduction in operation times for HVAC systems	Control of indoor environmental conditions	Deactivating standby	Control of DHW temperature	Maintenance of lighting fixtures	Drafting of an instruction manual for the users	Activation of reward strategies	Scheduling of monitoring procedures	Scheduling of energy accounting procedures
EA	Prerequisite 1: energy efficiency best management practices		v			v	v		v	v
EA	Credit 2.1: existing building commissioning—investigation and analysis		v						v	v
EA	Credit 2.2: existing building commissioning—implementation	v		v	v	v				
EA	Credit 2.3: existing building commissioning—ongoing commissioning		v						v	v

Table 9.11 Credits of LEED version 2009 USA for new buildings and for existing buildings related water management. It also shows the points available in the respective credit

Category	Credit	Available points NC 2009	Available points EB 2009
Sustainable sites	Credit 6.1: stormwater design—quantity control	1	1
	Credit 6.2: stormwater design—quality control	1	–
Water efficiency	Prerequisite 1: water use reduction	Mandatory	–
	Credit 1: water efficient landscaping	2–4	–
	Credit 2: innovative wastewater technologies	2	–
	Credit 3: water use reduction	2–4	–
	Prerequisite 1: minimum indoor plumbing fixture and fitting efficiency	–	Mandatory
	Credit 1: water performance measurement	–	1–2
	Credit 2: additional indoor plumbing fixture and fitting efficiency	–	1–5
	Credit 3: water efficient landscaping	–	1–5
	Credit 4: cooling tower water management	–	1–2
Energy and Atmosphere	Prerequisite 1: fundamental commissioning of building energy systems	Mandatory	–
	Prerequisite 1: energy efficiency best management practices	–	Mandatory
	Prerequisite 2: minimum energy performance	Mandatory	Mandatory
	Credit 1: optimise energy performance	1–19	1–18
	Credit 2.1/2.2/2.3: existing building commissioning	–	2/2/2
	Credit 3: enhanced commissioning	2	–

The topic of water is treated with a dual purpose: improvement in storm-water management to limit disruption of natural hydrology and reduction of the consumption of potable water. The SS category of new buildings is dedicated to environmental issues related to the relationship between the building and the site in which it is located and reserves two credits "Stormwater Design-Quantity Control" and "Stormwater Design-Quality Control" for the storm-water management.

The amount of water that precipitates and reaches the water bodies is proportional to the impermeable surface of the site, so the greater is the extent of the impermeable surface on the site, the greater will be the volume of storm-water that does not infiltrate into the ground and then flows to the water bodies. Increased storm-water runoff can cause erosion of rivers and damage the aquatic ecosystem. To control the amount of storm-water runoff one can implement a storm-water management plan which identifies the best combination between the strategies of reducing the amount of impervious surfaces and that of rainwater collection. Among these options are: the use of pervious paving, the presence of bumps of infiltration and retention ponds and the installation of vegetated roofs and storm-water reuse for non-potable applications.

Another aspect to be addressed because it creates a negative impact on water quality, concerns the quality of runoff containing sediments and contaminants. Measures of the quality control of storm-water provide for the collection and treatment of rainwater with techniques capable of removing suspended solids. In the version of the manual for existing buildings, a single credit is available on monitoring the amount of stormwater: “Stormwater quantity control”.

The reduction in potable water consumption involves the water used indoors in buildings and that consumed outside for irrigation purposes. In order to demonstrate the reduction of potable water consumption indoors it is first necessary to calculate the reference value (baseline). Specifically, the water consumption by occupants is estimated for each fixture, fitting and appliance in the reference case and is compared with the design value, then deriving the percentage of reduction.

It is mandatory to use 20 % less water than the calculated baseline water use (“Water Use Reduction” prerequisite) and points are acquired on the basis of higher rates achieved (“Water Use Reduction” credit). Among the suggested methods for the reduction of water consumption are the installation of low-flow fixtures, dual-flush toilets, waterless urinals and composting toilet systems.

As well as in the design phase for new buildings, even in the management phase of existing buildings LEED requires a reduction in potable water consumption compared to a baseline. A “Minimum Indoor Plumbing Fixture and Fitting Efficiency” prerequisite and an “Additional Indoor Plumbing Fixture and Fitting Efficiency” credit have been defined for this purpose.

Regarding the reduction in the amount of potable water used for irrigation purposes (“Efficient management of water for irrigation purposes” credit), actions that can contribute to this goal are numerous: use of native or adapted plants to reduce the amount of water needed, adoption of methods of efficient irrigation (irrigation systems, humidity sensors) and rainwater collection systems or the use of recycled wastewater.

The “Innovative Wastewater Technologies” indicate two options: reduce the potable water used to convey the sewage produced in the building or treat on-site the wastewater produced. This type of action makes it possible to decrease the load applied to municipal water supply systems and the volume of wastewater which must be treated.

Measuring water consumption (credit “Water Performance Measurement”) is the first step towards the efficient use of an existing building as it allows the identification of wastages and thus the establishment of appropriate strategies for improvement. Finally, the last category of credit WE for existing buildings concerns the “Cooling Tower Water Management”.

Identifying the best strategies to adopt in water management in buildings is necessary in order to assess the corresponding energy consumption. Integrated management of energy and water issues can in fact bring substantial benefits, both from an environmental and from an economic point of view, reducing energy costs. The relevant EA credits are: “Minimum Energy Performance” and “Optimize Energy Performance” (see Tables 9.12 and 9.13).

Table 9.12 Correlation between retrofit measures related to reduction of water consumption with LEED 2009 for new buildings related to water management

Correlation measures/credits		Reduction of water consumption				
Category	Credits LEED NC 2009	WR 01 Installation of water saving devices for terminals	WR 02 Installation of potable water meters	WR 03 Efficient use of water for irrigation purposes	WR 04 Recovery and treatment of grey-water	WR 05 Rainwater harvesting
SS	Credit 6.1: stormwater design—quantity control					v
SS	Credit 6.2: stormwater design—quality control					v
WE	Prerequisite 1: water use reduction	v				
WE	Credit 1: water efficient landscaping			v		v
WE	Credit 2: innovative wastewater technologies				v	
WE	Credit 3: water use reduction	v				
EA	Prerequisite 2: minimum energy performance		v	v	v	v
EA	Credit 1: optimise energy performance		v	v	v	v

Table 9.13 Correlation between retrofit measures related to reduction of water consumption with LEED 2009 for existing buildings related to water management

Correlation measures/credits		Water consumption reduction				
Category	Credits LEED EB 2009	WR 01 Installation of water saving devices for terminals	WR 02 Installation of potable water metres	WR 03 Efficient use of water for irrigation purposes	WR 04 Recovery and treatment of gray-water	WR 05 Rainwater harvesting
SS	Credit 6: storm-water design—quantity control					v
WE	Prerequisite 1: minimum indoor plumbing fixture and fitting efficiency	v				
WE	Credit 1: water performance measurement		v			
WE	Credit 2: additional indoor plumbing fixture and fitting efficiency	v				
WE	Credit 3: water efficient landscaping			v		v
WE	Credit 4: cooling tower water management				v	v
EA	Prerequisite 2: minimum energy performance		v	v	v	v
EA	Credit 1: optimise energy performance		v	v	v	v

References

1. Dakwalea VA, Ralegaonkara RV, Mandavganec S (2011) Improving environmental performance of building through increased energy efficiency: a review, sustainable cities and society
2. U.S. Green Building Council (2009) LEED reference guide for green building design and construction. US Green Building Council, U.S.A
3. U.S. Green Building Council (2009) LEED reference guide for green building operations and maintenance, U.S.A
4. Dall'O' G, Bruni E, Sarto L (2012) An Italian pilot project for zero energy buildings: towards a quality-driven approach, Renewable Energy

Chapter 10
Economic Assessment of the Retrofit Actions

The retrofit measures to adopt within a strategy aimed at enhancing the energy efficiency and the environmental impact of buildings do largely pay for themselves on the basis of energy savings, and consequent savings in expense, obtained thanks to the effects of their implementation. A proper economic evaluation of the outlined remedial actions is thus of great importance, as one of the key elements of the decision-making process among various possible scenarios. A number of schemes have been developed to provide some methods of comparison. This chapter will discuss the basic concepts of economic evaluation of investments in energy conservation and in the use of renewable energy sources.

10.1 Introduction

Possible remedial actions aimed at energy savings can be discerned by cost differences, but also by a different level of effectiveness in relation to the potential energy saving and economic benefits obtainable. To make a choice between the various options (i.e., different scenarios of retrofit measures), and to assess what is actually convenient from the economic point of view, it is necessary to make a detailed cost versus benefit analysis for each of the measures considered.

The economic analysis is not as simple as it might seem. If the objective is to evaluate and compare the convenience or the profitability of an investment, one must consider several economic parameters (such as the life cycle of the investment, the financing costs, the fluctuations of the costs related to energy, risks, taxes, etc.) some of which are not so easy to define and predict.

The objectives of the economic analysis may also vary depending on the interests and the specific objectives of the subject who makes it, as the expectations may differ. For the end-users, the cost reduction achievable whilst obtaining the same service level is of utmost importance, whereas for the energy suppliers (e.g., Energy Utilities, ESCo, etc.) the objective of profit maximisation is achieved through the more profitable business of providing as much energy as possible.

The economic assessment of investment does not rely on a single method but on several methods that are focused on specific aspects of the analysis. An evaluation

G. Dall'O', *Green Energy Audit of Buildings*, Green Energy and Technology,
DOI: 10.1007/978-1-4471-5064-0_10, © Springer-Verlag London 2013

based on more than one method allows the auditor to derive a wider and therefore more complete view of the investment scenario.

Economic-evaluation models can be used in a number of ways to increase the economic efficiency of energy-related decisions. There are models used to obtain the largest possible savings in energy costs for a given energy budget; others can be used to achieve a targeted reduction in energy costs for the lowest possible energy efficiency andrenewable energy investment; yet others determine how much it pays to spend on energy efficiency and renewable energy to reduce total lifetime costs (including both investment costs and energy cost savings). Ruegg and Short [1] state that the first two goal-setting models (i.e., to obtain the largest savings for a fixed budget and to obtain a targeted savings for the lowest budget) have a more limited scope than the third one, which aims to minimise total costs or maximise net savings from expenditure on energy efficiency and renewable retrofit measures.

Later on in this chapter, some of the classic cost versus benefit analysis indicators that contribute to the economic and environmental assessment will be described in a simplified way, giving also indications for a comprehensive picture from more specialised works. [2, 3]. Useful for the economic evaluation for energy systems in buildings is the European standard EN 15489 [4].

10.2 Basic Concepts and Definitions

10.2.1 Economic Parameters

A definition of the parameters and the concepts that significantly affect the economic decision making is useful to understand the economic evaluation methodologies that will be discussed later.

Initial investment (I_0)
The initial investment is the sum of all the costs to be sustained for the complete realisation of a given remedial action, up to the final delivery.

These costs include design, purchase of systems and components, connection to suppliers, installation and the commissioning process. The initial investment costs are the costs presented to the customer.

Costs (C_t)
These are the costs of a remedial action at a time t (months or years): they comprise initial investment costs and annual costs, including running costs, periodic or replacement costs for repair or change of components and systems:

- *running costs* comprise maintenance costs, operational costs (i.e., annual costs for operators), energy costs and added costs;
- *maintenance costs* are annual costs for preserving and/or restoring the desired quality of the installation; they include annual costs for inspection, cleaning, adjustments, repair under preventive maintenance and consumable items;

- *administration and general costs* are annual costs for insurance, other standing charges, taxes (including environmental taxes for energy). Subsidies for renewable energy delivered or produced locally are considered as benefits and are taken into account as negative annual costs or revenue;
- *energy costs* are annual costs for energy and include all the charges listed in the energy bills.

Benefits (B_t)
These are the benefits achieved as a result of a remedial action in a time t (months or years); the most important are the cost savings achieved thanks to a reduction of the energy bills to be paid during the period considered.

Net savings (S_t)
These are the benefits obtained as a result of a remedial action to improve energy efficiency net of any maintenance cost or scheduled repair, therefore

$$S_t = B_t - C_t$$

Calculation period (T)
It is the time span considered for the calculation, and it could be

- the lifespan of the retrofit measure, e.g., the expected lifetime normally expressed in years;
- the lifespan of the complex of the retrofit measures (retrofit scenario);
- the maximum time for which one is willing to support the investment.

Table 13.1 shows average European data, provided by EN 15459 [4] highlighting the lifetime of certain components used for energy efficiency measures in buildings and their maintenance costs in terms of percentage of the initial investment.

Discount rate (r)
This is the parameter indicating the value of money at different times. The discount rate is used to determine the present value of money that will occur in the future, generated by an investment. The technical term for this operation is "actualisation". The value of the discount rate varies, depending upon the project and can be obtained from various sources selected at the discretion of the person making the estimate (Table 10.1).

In the evaluation of energy efficiency projects, the *discount rate* can be defined by the following equation where all the terms are expressed in %:

$$r = i + f - f' \tag{10.1}$$

where:

i is the *inflation rate*, annual depreciation of the currency expressed in %;
f is the real cost of money (capital) gross of the direct taxes (for example, the interest rate charged by a bank for the loan of a sum of money);
f' is the annual rate of change in the energy price (as a basis equal to inflation rate, or available from energy utilities: can be positive or negative)

Table 10.1 Values of useful life of typical components used for energy efficiency measures in buildings and percentage costs, in relation to the initial investment of any maintenance and repair [4]

Component	Lifespan min—max (years)	Annual maintenance cost in % of the initial investment
Air conditioning units	15	4
Boiler—condensing	20	1–2
Control system—central	15–25	4
Control system—room control	15–25	4
Control valves, automatic	15	6
Control valves, manual	30	4
Cooling panels and ceilings	30	2
Duct system for filtered air	30	2
Duct system for non-filtered air	30	6
Electric wiring	50	0.5
Water floor heating	50	2
Evaporators	20	2
Fan coil units	15	4
Fans	20	4
Fans with variable flow	15	6
Heat pumps	15–20	2–4
Heat recovery units, static	20	4
Motors, diesel	10	4
Motors, electric	20	1
Pipes	30	1
Pipes, steel in open system	15	1
Pumps—regulated	10–15	1, 5–2
Solar collector	20	0.5
Tank storage for domestic hot water	20	1
Thermostats for radiators	15	4
Valve—thermostatic	10	1.5
Wiring	30	1

10.2.2 Cash Flow Analysis

In energy audits, it is important to account for the total cash receipts and disbursements associated with an implementation of retrofit measures for each period during the calculation period.

The difference between the total cash receipts (*inflows*) and total cash disbursements (*outflows*) for a given period of time is called a *cash flow* (*CF*). An accurate accounting of all the CFs should be performed over the lifetime of project.

In order to visualise capital flows of investments throughout the economic calculation period, CF diagrams such as the one shown in Fig. 10.1 are often used. The horizontal line is used as a time axis: the choice of the time unit is arbitrary, but the scale is usually graduated in years starting from zero on the left.

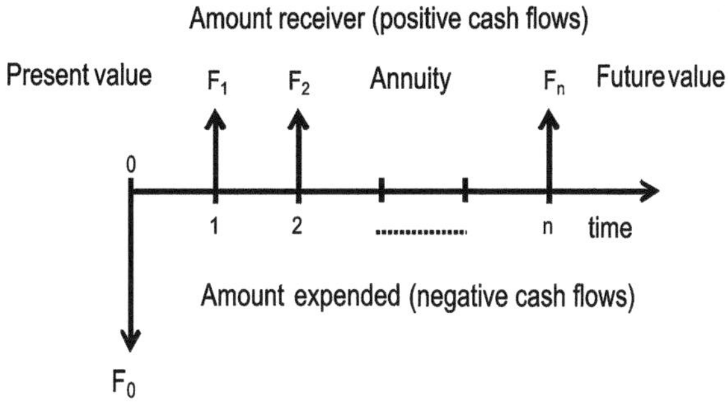

Fig. 10.1 Typical cash flow diagram

When drawing a CF diagram, the following rules are applied:

- each arrow represent the CF at a given time (F_0, F_1, F_2,... F_n);
- arrows always point away from the time axis;
- arrows pointing up represent income (positive CF);
- arrows pointing down represent expenses (negative CF);
- the length of the arrow is proportional to the amount of the capital represented;
- arrows can be summed in the same time period (e.g., year).

10.2.3 Compounding Factors

A correct economic assessment should take into account that the sums of money may vary, over time, as a function of the discount rate r and the number of years n.

The future value F can be calculated, referring to present value P (or present worth) using the equation:

$$F = P \cdot (1 + r)^n \tag{10.2}$$

where $(1 + r)^n$ is known as *compound amount factor*.

The ratio F/P is often called *single payment compound amount factor* and can be expressed as:

$$F/P = (1 + r)^n \tag{10.3}$$

The inverse ratio P/F, called *single payment present value*, allows one to determine the value of the CF P needed to get a given amount of CF F after n years

$$P/F = (1 + r)^{n-1} \tag{10.4}$$

The present value P can be calculated, referring to future value F using the equation

$$P = F \cdot (1 + r)^{n-1} \tag{10.5}$$

The equations described above are used, in the economic assessment of investments, to define the CFs that have a common time base for evaluation.

10.3 Economic Indicators and Methods

10.3.1 Net Present Value

The *net present value* (*NPV*) or *net present worth* (*NPW*) of a time series of CFs, both incoming and outgoing, is defined as the sum of the present values of the individual CFs.

NPV is one of the most often used economic indicators for the evaluation of energy retrofit projects: it can be calculated using the Eq. 10.6.

$$NPV = \sum_{n=0}^{T} \frac{CF_n}{(1+r)^n} = \frac{B_0 - C_0}{(1+r)^0} + \frac{B_1 - C_1}{(1+r)^1} + \ldots + \frac{B_T - C_T}{(1+r)^T} \quad (10.6)$$

If $NPV > 0$, the investment would add value and the project may be accepted, whereas if $NPV < 0$, the investment would subtract value and the project should be rejected. If $NPV = 0$, the investment would neither gain nor lose value; in this case, a decision should be based on other criteria (e.g., strategic positioning or other factors not expressed in the calculation.

Obviously, the higher is the *NPV*, the more economically sound is the project. The NPV method is often called the net savings method since the revenues are often arise from the cost savings from implementing the project.

10.3.2 Profitability Index

The *profitability index* (*PI*), also known as *profit investment ratio* (*PIR*) and *value investment ratio* (*VIR*), is is equal to the ratio between the discounted benefits and costs discounted. *PI* is calculated with the following equation:

$$PI = \frac{\sum_{n=0}^{T} \frac{B_n}{(1+r)^n}}{I_0 + \sum_{n=0}^{T} \frac{C_n}{(1+r)^n}} = \frac{\frac{B_0}{(1+r)^0} + \cdots + \frac{B_T}{(1+r)^T}}{I_0 + \frac{C_0}{(1+r)^0} + \cdots + \frac{C_T}{(1+r)^T}} \quad (10.7)$$

If *PI* is greater than 1, the project could be accepted: otherwise it would have no reason to be. The PI is a very useful indicator for comparison between multiple projects. If there is a need to choose between multiple projects, the most economically advantageous, it would not be sufficient to identify it from having a greater NPV. One must assessed which of the different indices expresses the higher profit.

10.3.3 Internal Rate of Return

The *internal rate of return* (*IRR*), also called *discounted CF rate of return* (*DCFROR*) or *rate of return* (*ROR*) is a ROR used in capital budgeting to measure and compare the profitability of investments. The term *internal* refers to the fact that its calculation does not incorporate environmental factors (e.g., the interest rate or inflation).

As the index of profit, *IRR* has the advantage of being very useful when comparing different projects. Its value is equal to the discount rate r for which *PI* is equal to 1. *IRR* is calculated with the following equation:

$$PI = \frac{\sum_{n=0}^{T} \frac{B_n}{(1+IRR)^n}}{I_0 + \sum_{n=0}^{T} \frac{C_n}{(1+IRR)^n}} = 1 \tag{10.8}$$

To solve accurately this equation, any numerical method can be used. However, an approximate value of r can be obtained by trial and error. This approximate value can be determined by finding the two r-values for which one gives the *NPV* as slightly negative and the other giving a slightly positive *NPV*, and then interpolating linearly between the two values. The *IRR* allows for an immediate comparison with the rates of return offered by banks to deposit the same amount of money as that needed for a specific remedial action. If the IRR is greater than the cost of capital, the project could be accepted, if the IRR is less than the cost of capital then the project should be rejected.

10.3.4 Payback Time

The *payback time* (*PBT*) is the most popular economic indicator, being easier to understand for non-experts. The information returned is the time in which the investment can be paid off, corresponding to the number of years in which the benefits equal the costs of its implementation.

We can define the *PBT* as the smallest value of t (months or years) for which

$$I_0 - \sum_{n=0}^{T} \frac{B_n - C_n}{(1+r)^n} = 0 \tag{10.9}$$

If the *PBT* is less than the lifetime of the project, then the project is economically viable. *PBT* obtained using (10.9) is called also *discount payback time* (*DPT*), since it includes the value of money.

If the time value of money is neglected and the annual savings A are constant, then a simplified method for the economical assessment of the investment can be considered. The *simple payback time* (*SPT*) can be easily calculated as the ratio of the initial investment I_0 to the annual net saving A:

$$SPT = \frac{I_0}{A} \tag{10.10}$$

The values for the SPBT are shorter than for the DPBT since the undiscounted net savings are greater than their discounted counterparts. Therefore, acceptable values for simple payback periods are typically significantly shorter than the lifetime of the project.

10.4 Life-Cycle Cost Method

The life-cycle cost (*LCC*) method is the most commonly accepted method to assess the economic benefits of energy conservation projects over their lifetime. The method is used to evaluate at least two alternatives of a given project (for instance, evaluate two alternatives for the installation of new heating boilers).

The basic procedure of the *LCC* method is relatively simple since it seeks to determine the relative cost effectiveness of the various alternatives.

This method permits estimation through the identification of all costs associated with the remedial action (e.g., initial installed cost, maintenance costs, operating costs including energy, fuel escalation rates, inflation, interest on the investment, salvage value and other lifetime expenses for the equipment). These expenses, over the life of the intervention, are multiplied by their single payment present value factor $P/F = (1 + r)^{n-1}$. The sum of all the present values is called the life cycle cost. Only one alternative will be selected for implementation based on the economic analysis.

As regard the calculation period (T), two factors enter into appraising the life of the system; namely, the expected physical life and the period of obsolescence. The latter factor is the governing time period.

A practical example will clarify the *LCC* approach. Two heating boilers are being evaluated to determine the lowest *LCC*. The two retrofit solutions, A and B, are compared with the baseline (i.e., the old heating boiler).

Alternative A, referred to a new standard boiler, has a first cost of 48,000 €, while alternative B, referred to a new condense boiler, has a first cost of 70,000 €. The calculation period is estimated in 15 years, corresponding to the useful life of the two technologies. The baseline is the existing boiler, an old low efficiency boiler and its first cost, obviously, is zero.

The *LCC* for the three cases are shown in Tables 10.2 (baseline), 10.3 (alternative A) and 10.4 (alternative B). The items of the columns are as follows:

1. reference year;
2. first and replacement costs (F&RC);
3. energy costs (EC);
4. maintenance costs (MC);
5. total annual costs (AC);
6. present value factor (P/F);
7. present value of annual costs (PVA) (column 5 multiplied by column 6);
8. present value cumulative costs (PVC).

Table 10.2 Life-cycle cost (*LCC*) analysis for a heating boiler replacement: baseline

Years	F&RC (€)	EC (€)	MC (€)	AC (€)	P/F	PVA (€)	PVC (€)
0	–	25,000	2,500	27,500	1.00	27,500	27,500
1	–	25,000	2,500	27,500	0.95	26,190	53,690
2	–	25,000	2,500	27,500	0.91	24,943	78,634
3	–	25,000	2,500	27,500	0.86	23,756	102,389
4	–	25,000	2,500	27,500	0.82	22,624	125,014
5	–	25,000	2,500	27,500	0.78	21,547	146,561
6	–	25,000	2,500	27,500	0.75	20,521	167,082
7	–	25,000	2,500	27,500	0.71	19,544	186,625
8	–	25,000	2,500	27,500	0.68	18,613	205,238
9	–	25,000	2,500	27,500	0.64	17,727	222,965
10	25,000	25,000	0	50,000	0.61	30,696	253,661
11	–	25,000	2,500	27,500	0.58	16,079	269,739
12	–	25,000	2,500	27,500	0.56	15,313	285,052
13	–	25,000	2,500	27,500	0.53	14,584	299,636
14	–	25,000	2,500	27,500	0.51	13,889	313,526
15	–	25,000	2,500	27,500	0.48	13,228	326,754
	25,000	400,000	37.500	462,500		326,754	–

Table 10.3 Life-cycle cost (*LCC*) analysis for a heating boiler replacement: alternative A

Years	F&RC (€)	EC (€)	MC (€)	AC (€)	P/F	PVA (€)	PVC (€)
0	48,000	16,500	960	65,460	1.00	65,460	65,460
1	–	16,500	960	17,460	0.95	16,629	82,089
2	–	16,500	960	17,460	0.91	15,837	97,925
3	–	16,500	960	17,460	0.86	15,083	113,008
4	–	16,500	960	17,460	0.82	14,364	127,372
5	–	16,500	960	17,460	0.78	13,680	141,053
6	–	16,500	960	17,460	0.75	13,029	154,082
7	–	16,500	960	17,460	0.71	12,408	166,490
8	–	16,500	960	17,460	0.68	11,818	178,308
9	–	16,500	960	17,460	0.64	11,255	189,563
10	5,000	16,500	0	21,500	0.61	13,199	202,762
11	–	16,500	960	17,460	0.58	10,209	212,970
12	–	16,500	960	17,460	0.56	9,722	222,693
13	–	16,500	960	17,460	0.53	9,259	231,952
14	–	16,500	960	17,460	0.51	8,818	240,770
15	–	16,500	960	17,460	0.48	8,399	249,169
	53,000	264,000	14,400	331,400		249,169	–

The economic evaluation has been prepared considering a discount rate r of 5 %. In order to simplify the calculations, it has been assumed that the annual rate of change of the energy price f' is equal to the inflation rate$i$, so the discount rate r, in this case, corresponds to the real cost of money f.

Table 10.4 Life-cycle cost (*LCC*) analysis for a heating boiler replacement: alternative B

Years	F&RC (€)	EC (€)	MC (€)	AC (€)	P/F	PVA (€)	PVC (€)
0	70,000	12,500	960	83,460	1.00	83,460	83,460
1	–	12,500	960	13,460	0.95	12,819	96,279
2	–	12,500	960	13,460	0.91	12,209	108,488
3	–	12,500	960	13,460	0.86	11,627	120,115
4	–	12,500	960	13,460	0.82	11,074	131,188
5	–	12,500	960	13,460	0.78	10,546	141,735
6	–	12,500	960	13,460	0.75	10,044	151,779
7	–	12,500	960	13,460	0.71	9,566	161,345
8	–	12,500	960	13,460	0.68	9,110	170,455
9	–	12,500	960	13,460	0.64	8,676	179,131
10	5,000	12,500	0	17,500	0.61	10,743	189,875
11	–	12,500	960	13,460	0.58	7,870	197,745
12	–	12,500	960	13,460	0.56	7,495	205,240
13	–	12,500	960	13,460	0.53	7,138	212,378
14	–	12,500	960	13,460	0.51	6,798	219,176
15	–	12,500	960	13,460	0.48	6,474	225,650
	75,000	200,000	14,400	289,400		225,650	–

Examining the tables, it can be observed that:

- at year 10, for alternatives A and B an extraordinary maintenance is scheduled, with a cost of 5,000 €, while for the existing boiler the cost rises to 25,000 € (burner replacement);
- the annual EC decrease with the increase of the performance of the heating boilers (25,000 € for existing boiler, 16,500 € for new standard boiler and 12,500 € for condensing boiler);
- the annual MC decrease (2,500 € for existing boiler, 960 € for new standard boiler and for condensing boiler).

Comparing the two alternatives A and B, the *LCC* analysis estimate alternative B (replacing with a condensing boiler) to be the best one (LCC = 225,650 €), even if the initial investment is higher.

10.5 Uncertainties and Risks Estimation

The methods of economic evaluation of investment are intended to provide criteria to choose amongst several options the best one: the LCC method discussed above is a useful tool to achieve this goal.

If the retrofit actions object of the analysis are particularly complex, for example due to significant elements of uncertainty in the economic evaluation, and the economic investment required is significant, it may be useful to submit the results of the LCCA to verifications in order to assess the degree of uncertainty.

That is to quantify the risks that might be faced when deciding on a particular project alternative rather than another.

Risk assessment on energy audit provides decision makers (auditors) with information about the "risk exposure" inherent in a given decision i.e., the probability that the outcome will be different from the "best-guess" estimate. Risk assessment is also concerned with the risk attitude of the decision maker that describes his/her willingness to take a chance on an investment of uncertain outcome.

Approaches to risk assessment can be grouped into two categories of methods:

- the probabilistic approaches;
- sensitivity approaches.

The need of estimating uncertainty is given by the fact that the entries of the data for the LCCA are largely based on estimates.

If one analyses an investment in improving the energy efficiency of a building or facility, the uncertainties may relate to the internal and external factors.

Internal factors may include, for example, the following:

- the estimated energy savings depend largely upon how the facilities are used: if the initial assumptions are not met, for example the hours of operation are different or the internal temperature is set in a different way compared to the project set-point values, energy consumption can be very different from that initially estimated (the activation of a IPMVP protocol[1] can reduce this type of risk);
- the lifetime of the system components may be less than that estimated, owing to breakdowns requiring replacement which was not provided for (the activation of warranty contracts with suppliers can reduce or eliminate these risks).

External factors are related to the evolution of the economic parameters used for discounting: the cost of money and inflation may change over time, in addition to the rising cost of energy. In order to simplify the approach, it is usually considered that the change in the cost of energy is equal to the change in inflation but this assumption may well not be correct. The uncertainties and risks, of course, are greater, the longer is the calculation period.

The risk assessment techniques range from simple and partial to complex and comprehensive. Though none takes the risk out of making decisions, the techniques, if used correctly, can help the decision maker to make more informed choices in the face of uncertainty. An overview of the probability-based risk assessment techniques is proposed by Ruegg and Short [1].

[1] The International Performance Measurement and Verification Protocol (IPMVP) defines standard terms and suggests best practice for quantifying the results of energy efficiency investments and increase investment in energy and water efficiency, demand management and renewable energy projects. This topic is discussed in Sect. 11.6.

10.5.1 Probabilistic Approach

Probabilistic approach is the most widely used technique for uncertainty analysis. In the probabilistic approach, uncertainties are characterised by the *probabilities* associated with *events*. An event corresponds to any of the possible states a physical system can assume, or any of the possible predictions of a model describing the system.

The *probability* of an event can be interpreted in terms of the frequency of occurrence of that event. When a large number of *samples* or *experiments* are considered, the probability of an event is defined as the ratio of the number of times the event occurs to the total number of samples or experiments. A probability of 0 for an event means that the event will never occur, and a probability of 1 indicates that the event will always occur [5].

10.5.2 Sensitivity Analysis

The approach of the sensitivity of the estimation of the uncertainties is used to determine the probability that, between two alternatives, the one with the lowest value of LCC is effectively such. If this probability is satisfactory, then the results of LCCA are considered conclusive.

The sensitivity approach does not produce a direct measurement of this probability, but is approximated through judgments based on the cost information available, as well as through the results of an analysis of sensitivity and a calculation of the break-even point (break-even point).

The sensitivity analysis is, essentially, a method to determine how a value of a parameter is influenced by changes in the value of a second parameter upon which it depends. These parameters are called *output parameters* and *input parameters*.

The method is particularly useful when the relationship between the two parameters cannot be expressed in a direct way.

It implies

- the assignment of different and reasonable values to the input parameter;
- the calculation of the value of the output parameter that corresponds to each value of the input parameter;
- the analysis of the couples of values (input value and output value) that result from the application of the method; these couples of values may be tabulated, or shown graphically to indicate the relationship between the two parameters.

The method is useful in economic analysis, for example, to determine how the variation of the initial cost of a retrofit measure (in the case of the example of the previous sub-section, the choice of the heat generator) influences the LCC, understood as the output variable.

The cost of life cycles can be calculated for each of the different possible evaluations for the initial cost. The resulting pairs (initial cost and LCC) can then be tabulated or plotted graphically to indicate how the LCC can be influenced by the initial cost. Since this variation is caused by uncertainties of the initial cost, the table or graph derived from it, indicates how the LCC of this alternative is affected by these uncertainties.

Sensitivity analysis can of course be made by considering as input parameters not only the initial cost but also other factors (for example the cost of maintenance, energy saving, the finance cost, etc.). The sensitivity analysis of the output parameter is always the LCC of the alternative at lower cost.

References

1. Ruegg R, Short W (2007) Handbook of energy efficiency and renewable, Chapter 3—economics methods. CRC Press Taylor & Francis, Boca Raton
2. Bhattacharyya SC (2011) Energy economics, concepts, issues, markets and governance. Springer, London
3. Thumann A, Waodroof EA (1997) Handbook of financing energy projects. CRC Press Taylor & Francis, Boca Raton
4. CEN (2007) European Standard EN 15459—Energy performance of buildings. Economic evaluation procedure for energy system in buildings, Brussels
5. Isukapalli SS (1999) Uncertainty analysis of transport-transformation models, New Brunswick

Chapter 11
Contractual Issues

Once the retrofit measures have been identified, the remedial action should be implemented and energy performance must then be maintained. For the implementation of the measures and the management of building and facilities the owner/client can refer to external companies. A careful definition of contractual issues is a key element to promote the sustainable upgrading of a building in a transparent and efficient way. This chapter analyses aspects related to contractual issues within energy retrofit and energy management. The most diffuse contractual schemes are discussed with a particular focus on Third Party Financing schemes which are an interesting opportunity to promote energy efficiency without initial investments for the owner. The issues related to the performance evaluations in accordance with the IPMPV international protocol are also discussed in this chapter.

11.1 Implementation and Managing of the Retrofit Measures

The energy audit (or the green energy audit) permits the definition of retrofit actions which, if performed, allow a reduction in energy consumption and operating costs. The transition from the analysis phase (audit) to the implementation phase (implementation of strategies), however, raises issues that must be dealt with:

- the access to financial resources to implement the retrofit measures;
- the guarantee that the retrofit measures, once implemented, would provide the expected performances;
- the efficient management of the facilities and services over time;
- the management of maintenance activities;
- the energy supply.

The above issues can be outsourced (to, e.g., contractors or providers of technology, management and maintenance companies, energy suppliers, energy

G. Dall'O', *Green Energy Audit of Buildings*, Green Energy and Technology,
DOI: 10.1007/978-1-4471-5064-0_11,

managers, etc.). Any relationship between client and supplier of technologies, energy or services should be managed according to specific contracts, drawn up in order to ensure maximum transparency.

The contractual schemes may be different: the customer, in fact, can manage multiple contracts with multiple providers, but can also outsource the management of everything to a single supplier.

In the simplest contract model, which is widespread, the clients implement all the retrofit measures at their own expense. Subsequently two separate contracts are activated: one for management and maintenance and one for the provision of energy.

This choice may be convenient if the client company has within its organisation a technical competent structure able to effectively manage the external suppliers (e.g., a skilled energy manager or facilities manager). This normally is not the case for private end-users (residential buildings) and even if the client has a financial capacity, facility management may not be his core business.

The client can outsource the facilities management and the energy supply to a single external supplier. The contractual schemes most diffuse in Europe are two: the degree day contract and the energy supply contract.

In the *degree day contract*, typically used for winter heating, the company that provides the service defines an indicator (global cost per degree day[1]) by dividing the estimated total cost (energy supply, management and maintenance) by the number of the standard degree-days of the location. Actual degree days are recorded, day by day, by official meteorological observatories and published at the end of the heating season. With this contract, the cost of heating varies as a function of the seasonal variability, increasing or decreasing depending upon how extreme are the season's low temperatures. All costs of extraordinary maintenance, such as failures, replacements, repair are charged to the client, while the obligations on heating security shall be borne by the supplier company.

In the *energy supply contract* the client pays only the heat supplied into its distribution network and measured with a heat meter installed downstream of the heating boiler. The only interest of the companies that offer this contract is to maintain the efficiency of the heat generator, in order to minimise combustion losses. No action, instead, is guaranteed in the plant downstream of the heat meter (the higher is the consumption, the greater is the gain of the supply company). Even in this case, as in the previous case, the extraordinary maintenance costs are charged to the customer.

Energy performance contract, with third party financing, described in the next section, is the most complete and effective contractual formula: with this type of contract the cost of the implementation of the retrofit measures shall be borne by the company that provides the service, who then has an interest in ensuring that the whole building-plant system operates with the best energy performance.

---

[1] The degree days (DD) are the summation, extended to the whole reference winter season, of the difference between the temperature of balance and the outside temperature for each of the days considered.

11.2 Energy Performance Contracts with Third Party Financing

Third party financing is based on the assumption that the energy savings result in a flow of lower costs and greater efficiency, which, when discounted, is able to pay off the initial investment. The companies who provide this service are called Energy Service Company (acronym: ESCO or ESCo).

An ESCo is a commercial business providing a broad range of integrated energy services including energy audit, design and implementation of energy savings projects, energy conservation, energy infrastructure outsourcing, power generation, energy supply and risk management.

Once the contract, normally a performance guarantee contract is activated, the client (e.g., building occupants) then benefits from the energy savings and pays a fee to the ESCo in return.

Figure 11.1 shows a typical operational framework of an ESCo.

Typically a qualified ESCo should:

- identify and evaluate energy-saving opportunities;
- develop engineering designs and specifications;
- manage the project from design, through to installation, to monitoring;
- arrange for financing;
- guarantee that savings will cover all project costs.

The savings in energy costs are used to pay back the capital investment of the project over a 5- to 20-year period, or reinvested into the building to allow for capital upgrades that may otherwise be unfeasible. If the project does not provide returns on the investment, the ESCo is often responsible for paying the difference.

The economic benefits achieved as a result of the energy retrofit actions can be shared among the interested parties in different ways, the ESCo normally offers three options:

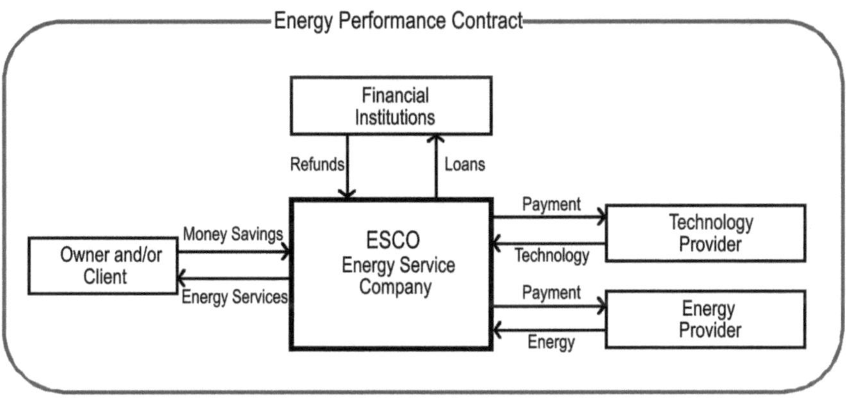

Fig. 11.1 Operational framework of an Energy Service Company (ESCo)

- *first out* (limited global cession), in which 100 % of the savings achieved goes to the ESCo, with a consequent reduction in the duration of the contract;
- *shared savings*, in which the remuneration of the ESCo for the entire duration of the contract is given by a share of the savings achieved, the remaining portion of the savings going to the benefit of the client;
- *guaranteed savings* when the ESCo, through a particular leasing contract, ensures the client that the savings at the end of the contract shall not be less than the amount of the investment, including all fees and expenses.

Figure 11.2 shows a typical scheme of a shared savings TPF contract [1]. In the example, the contract is 5 years. Thanks to the energy retrofit actions made by the ESCO, the annual energy bill is reduced and the savings are shared between the ESCO, which in this way pays the initial costs, and the customer, with percentages that are defined at the time of the contract. At the end of the contract, the customer has achieved a reduction in fuel consumption and can activate a new contract with more favourable terms.

In the energy performance contracts promoted by an ESCo, the energy supply (fuel, heat and electricity) could be included in the contract, or separated from it. In the latter case, the client chooses the energy supplier stipulating a contract with it, while according the ESCo a fee which covers all other costs (initial investment, management and maintenance). The separation of the two payments (energy and fee) should be preferred since the client can check whether the energy retrofit measures have actually generated energy savings and not just a money saving.

Figure 11.3 shows a typical scheme of a guaranteed savings TPF contract. The baseline represents the actual cost that includes energy costs (fuel and electricity) and Operation and Maintenance costs (first column). The implementation of the energy retrofit measures reduce the energy consumption (second column).

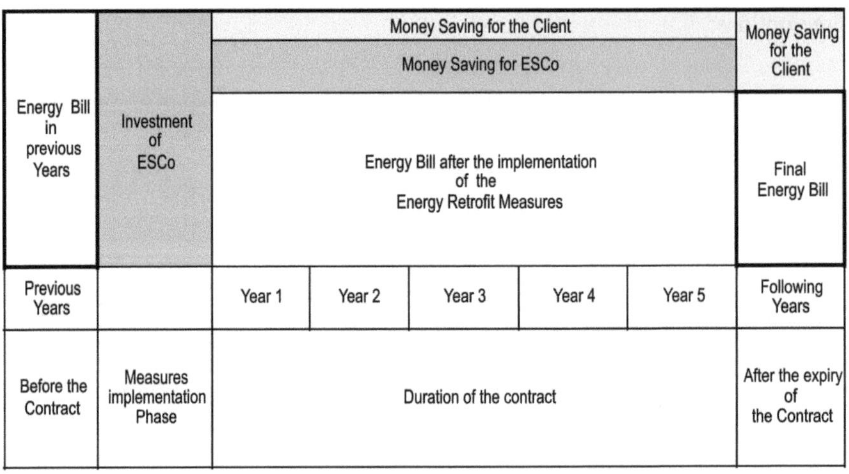

Fig. 11.2 Typical scheme of a shared savings TPF contract

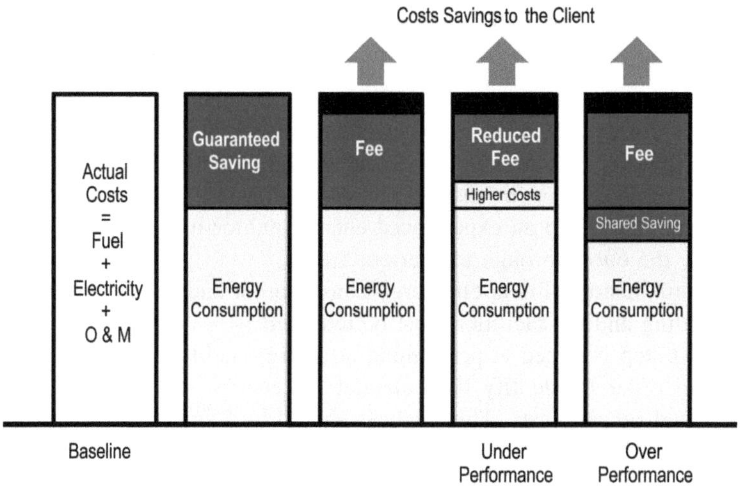

Fig. 11.3 Typical scheme of a guaranteed savings TPF contract

A part of the costs savings is returned to the customer (third column). In case of under performance of the systems, the extra cost of the energy consumption is borne by the ESCo which reduces the fee (that includes O&M costs), so guaranteeing the saving. In case of over performance of the systems, the energy saving is shared between the ESCo and the Client.

11.3 Energy Savings and Money Savings

In defining an Energy Performance Contract with TPF it is important to bear in mind that *energy saving* and *money saving* are two different things, not necessarily directly related.

The big companies, operating on an international market, have the ability to purchase fuel at lower prices and to transfer this discount to the customer, avoiding making investment on retrofit measures. The economic benefit for the customer, in this case, is only apparent, since if the works are not carried out, at the end of the contract the energy efficiency of the building or the facilities is not guaranteed, and the objective of the contract is partially lost.

Energy saving always involves a monetary saving, whereas the opposite is not necessarily true: the TPF formula aims to save energy, which is the premise of a permanent monetary saving.

In the energy performance contracts with TPF it is therefore essential that that share of savings in terms of primary energy which the ESCo undertakes to achieve is clearly highlighted in the contract. Only in this way can an effective cost reduction be guaranteed even after the termination of the contract.

11.4 The Operating Procedure of a TPF Contract

Energy performance contracts with TPF are very convenient for the customer but they are also for the ESCo because they can represent market opportunities. These contracts, however, are quite complex and should be defined carefully.

It is best to proceed in two distinct phases, as shown in the scheme of Fig. 11.4. In a first, guidance step an experienced energy auditor must be chosen who will accompany the client through the various stages.

Afterwards historical data on energy consumption and documentation relating to the building and the facilities must be retrieved.

The next step is aimed at performing an energy audit of the building and the facilities in order to identify the potential for energy savings, possible retrofit measures and related costs. This analysis should be "light", in other words sufficient to determine whether it is convenient to proceed with a request for commercial quotations from different ESCo. Neither in-depth analysis, nor a detailed retrofit project is necessary. The ESCo invited to the tender will carry out their own analysis and their own investigations, on the basis of their own parameters.

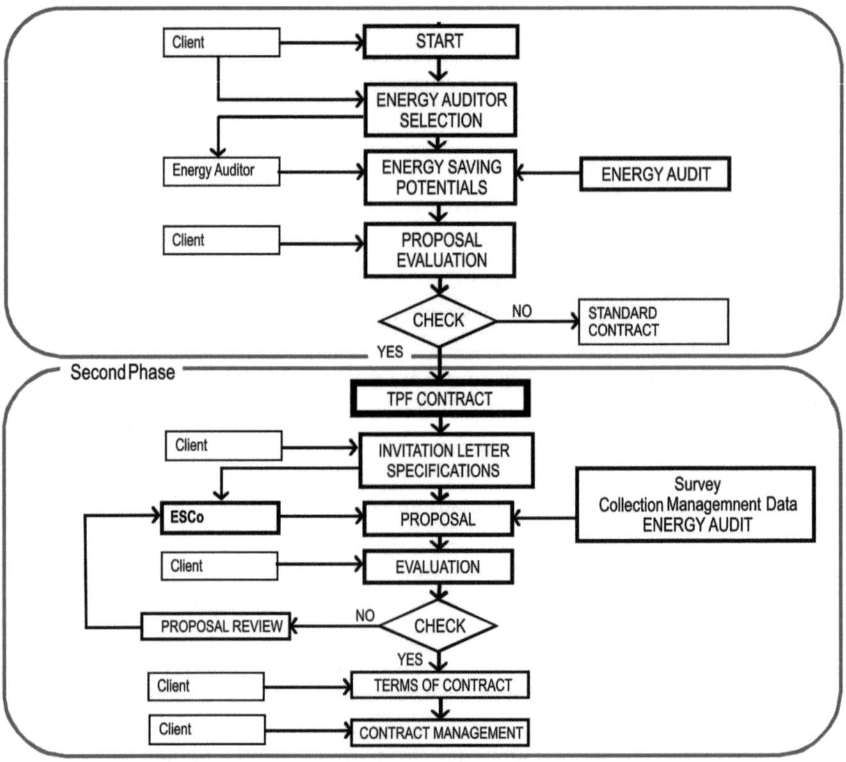

Fig. 11.4 Operative procedure of TPF contract

If the preliminary analysis highlights a good potential for energy savings, it is possible to proceed to the second phase that begins with the TPF contract definition. The objective of the second phase, as shown in Fig. 11.4, is to choose the ESCo to which to entrust the work.

11.5 IPMVP Protocol

When investments for energy saving projects are implemented it is essential to know, from the very beginning, the return on investment and the extent and duration of savings.

Once the measures have been made it is important to assess whether the predictions were really respected and if the remedial action is effectively able to generate the energy savings for which it was created.

The procedures for evaluation of the energy savings achieved have exactly this purpose. They thus become an important tool to support a proper and objective approach to energy management.

But how best to take measurements and make verifications? Through a *measurement* and *verification* protocol (M&V), which is a protocol that allows the application of criteria and rules that are accepted by both the client (or owner) and the supplier of the service as an ESCo.

International Performance Measurement and Verification Protocol (IPMVP) is an international reference protocol set up to promote measurement and evaluation of energy savings that can be applied in various fields (efficient use of energy, use of renewable energy, water conservation, etc.) and in all sectors, from residential to industrial. IPVM has been developed internationally and has been designed as a reference tool to support performance-based contracts and, more generally, to support the activities promoted by the ESCos.

It is a document that describes the standard procedures that, when implemented, allow building owners, ESCos and funding agencies for energy efficiency projects (e.g., banks) to quantify the performance of energy conservation measures and the energy saved.

The application of this protocol ensures transparency in the relationship between customer and supplier and then permits one to:

- increase the certainty, the feasibility and the level of energy savings;
- reduce transaction costs through a methodological approach that meets the industrial market consensus;
- reduce the cost of financing the projects ensuring a standardisation of draft measures and verification (M&V);
- assess the reduction of emissions and the improvement in indoor environmental quality and comfort;
- provide a basis for contract negotiations ensuring that energy efficiency projects do attain their objectives in terms of energy and economic savings.

Table 11.1 Overview of M&V options [3]

M & V options	How savings are calculated	Typical applications
A. Partially measured retrofit isolation Savings are determined by partial field measurement of the energy use of the system(s) to which an ECM was applied, separate from the energy use of the rest of the facility. Measurements may be either short-term or continuous Partial measurement means that some but not all parameter(s) may be stipulated, if the total impact of possible stipulation error(s) is not significant to the resultant savings	Engineering calculations using short term or continuous post-retrofit measurements and stipulations	Lighting retrofit where power draw is measured periodically. Operating hours of the lights are assumed to be one half hour per day longer than store open hours
B. Retrofit isolation Savings are determined by field measurement of the energy use of the systems to which the ECM was applied, separate from the energy use of the rest of the facility. Short-term or continuous measurements are taken throughout the post-retrofit period	Engineering calculations using short term or continuous measurements	Application of controls to vary the load on a constant speed pump using a variable speed drive. Electricity use is measured by a kWh meter installed on the electrical supply to the pump motor. In the base year this meter is in place for a week to verify constant loading. The meter is in place throughout the post-retrofit period to track variations in energy use
C. Whole facility Savings are determined by measuring energy use at the whole facility level. Short-term or continuous measurements are taken throughout the post-retrofit period	Analysis of hole facility utility meter or sub-meter data using techniques from simple comparison to regression analysis	Multifaceted energy management program affecting many systems in a building. Energy use is measured by the gas and electric utility meters for a 12 months base year period and throughout the post-retrofit period
D. Calibrated simulation Savings are determined through simulation of the energy use of components or the whole facility. Simulation routines must be demonstrated to adequately model actual energy performance measured in the facility. This option usually requires considerable skill in calibrated simulation	Energy use simulation, calibrated with hourly or monthly utility billing data and/or endues metering	Multifaceted energy management program affecting many systems in a building but where no base year data are available. Post-retrofit period energy use is measured by the gas and electric utility meters. Base year energy use is determined by simulation using a model calibrated by the post-retrofit period utility data

The IPMVP protocol has been developed in the United States, as of 1994, by the Department of Energy (DOE). Extended and published in 1997, it has been adopted as reference standard for contracts for energy savings in the USA and in other countries. In a few years the protocol has become a landmark of international collaboration: more than 300 experts in 25 countries made it possible to publish a second update in the year 2000. IPMPV refers to ASHRAE (14–2002 Guidelines) recognised as international technical standards [2].

The fields of application of the IPMVP concerns different types of performance contracts, project evaluation, risk sharing between the ESCo and owners/clients. The Protocol provides for the parties to choose, by mutual agreement, between four different measurement options (A, B, C and D) depending upon the size, complexity of the measurement, the cost of measurement and the desired accuracy.

The four options are shown in Table 11.1. The criteria for selection of the various options depend upon several factors including: the cost of M&V in relation to the value of the energy savings generated by energy efficiency measures, the complexity of the measures, the number of measures and their interrelationships, the uncertainty of value of savings and the allocation of risk between the ESCo and the owner of the property.

The flow chart shown in Fig. 11.5 shows the procedure for the establishment of a plan for Measurement & Verification of Energy Savings targets.

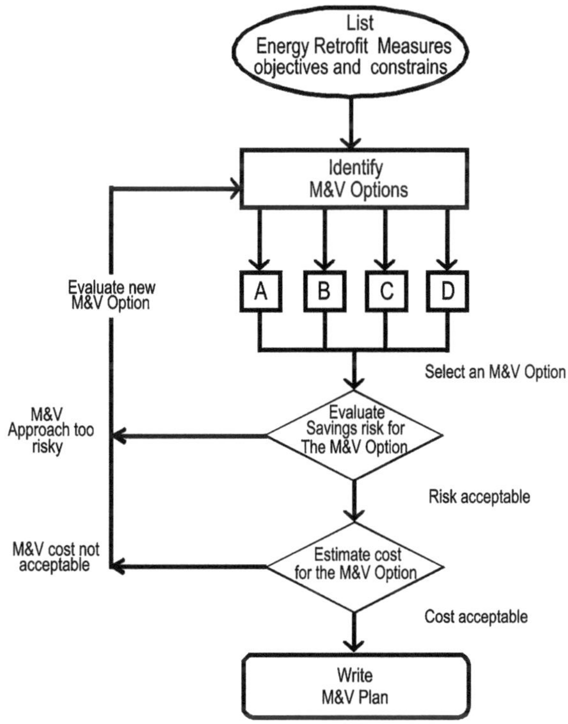

Fig. 11.5 Measure & Verification planning flow chart

It is essential to first define the objectives and constraints of energy savings: based upon the type and complexity of the work: one of the four options for M&V is then chosen. The next step is the evaluation of savings and risk of the option.

If the risk is acceptable one switches to assessment of the cost. If this then also appears to be acceptable, one moves forward to the drafting of the Measurement & Verification Plan which includes the following points:

- description of energy efficiency measures and their objective;
- definition of the impact of these measures on the building in which they are applied;
- a description of the procedures for testing and verification;
- definition of the measurement periods and limitations of the system;
- definition of common conditions (base year) with which to compare the consumption detected;
- choice of the option and the measuring system, the desired precision and the values agreed upon;
- budgeting, responsibilities and reporting procedures.

The IPMVP Protocol can be downloaded from the official website www.ipmvp.org to which reference is made for the necessary details.

References

1. Zabot S (2011) Schemi contrattuali, on green energy audit. Edizioni Ambiente, Milano
2. Haberl JH, Culp C, Claridge DE (2005) Ashrae's guideline 14-2002 for measurement of energy and demand savings: how to determine what was really saved by the retrofit. In: Proceedings of the 5th international conference for enhanced building operations, Pittsburgh, Pennsylvania, 11–13 October 2005
3. DOE/GO (2002) International Performance Measurement & Verification Protocol, International Performance Measurement & OE/GO (2002) International Performance Measurement & Verification Protocol: concepts and options for determining energy and water savings

Chapter 12
Comfort and Well-Being of Occupants

Indoor environmental quality depends on several variables: temperature, lighting, humidity, air quality and acoustics. A good indoor environmental quality requires integration of all these factors and of all ventilation, thermal, acoustic and illumination solutions. This chapter introduces basic aspects of comfort and well-being of the occupants, with the objective of demonstrating that an energy saving approach does not mean a reduction in comfort but, on the contrary, a guarantee of a better relationship between the occupants and their indoor environment.

12.1 Thermal Comfort

The interactions between human beings and the environment are caused by quite complex mechanisms of heat and mass transfer. The concept of thermal comfort cannot be defined in absolute terms, since the reactions of each individual to a particular external thermal stress may be different.

In this section, different approaches to the definition of thermal comfort will be examined: the classical one, which refers to the theory developed by Fanger and which forms the basis of the ASHRAE Standard 55 [1] and 7730 ISO Standard [2], a different interpretation which favours the radiative exchanges and the most recent one that introduces the concept of adaptive comfort, which is addressed by scholars De Dear and Bragher [3] and which lends itself well to set up a new approach to the sustainable design of buildings.

12.1.1 The Classic Approach to Thermal Comfort Evaluation

ASHRAE 55 Standard and 7730 ISO Standard define *thermal comfort* as "that condition of mind which expresses satisfaction with the thermal environment and is assessed by subjective evaluation". The factors that interact are many and tied both to the environment than to those of the individual.

G. Dall'O', *Green Energy Audit of Buildings*, Green Energy and Technology,
DOI: 10.1007/978-1-4471-5064-0_12,

The most important are the following:

- temperature, relative humidity and air velocity;
- temperature of the walls that surround the indoor environment;
- clothing of the individual;
- activity.

Man, like all warm-blooded animals, has the ability to maintain his internal body temperature almost constant, through a mechanism of thermoregulation that ensures a balance between the energy generated by the metabolism and the energy dissipated.

The thermo regulation system, which is based in the hypothalamus, when there is a sensation of heat increases the temperature of the body surface with an increased flow of blood near to its surface (vasodilatation) and perspiration. When instead there is a sensation of cold, the surface temperature is decreased (vaso-constriction) also inducing shivering which is an involuntary muscle activity that requires additional metabolic activity. In the first case, the heat flow from the body to its surroundings (e.g., the indoor environment of a building) increases, in the second case it decreases.

The feeling of well-being of individuals is a direct consequence of the stresses to which is subjected the thermoregulation mechanism.

The complex phenomena of heat (and mass) exchange between the human body and its surrounding environment can be summarised by an equation of the type:

$$S = M - (\pm P \pm E \pm R \pm C) \tag{12.1}$$

where:

S is the change in the internal energy variation of the body per time unit;
M is the thermal power due to metabolic activity;
P is the mechanical power exchanged by the body with the external environment;
E is the thermal power associated with the loss of water by evaporation from the body;
R is the thermal flux exchanged by radiation between the body and environment;
C is the thermal flux exchanged by convection between the body and environment

Since the thermoregulation mechanism of the human body tends to maintain the body temperature constant, it can be assumed that, for a sufficiently long exposure to an environment that is neither too hot and nor too cold, with activity constant, there is no appreciable change in internal energy, and therefore it is permissible to consider $S \cong 0$.

The extent of metabolic activity is closely related to the state of health and the type of nutrition of the person. We define *basal* metabolic rate as that corresponding to the calorie needs strictly necessary for the maintenance of vital functions in conditions of rest or of thermal neutrality. Ambient temperatures of 29 ÷ 31 °C for individuals when naked and 25 ÷ 29 °C for individuals when clothed do not cause variations in body temperature nor evapo-transpiration activity.

With increased activity an individual's metabolic rate increases accordingly, passing, for example, from 43 W/m^2 for the individual who is sleeping to 64 W/m^2 for a typist up to 505 W/m^2 for a wrestler when fighting. To simplify the practical problems of measurement, a unit of metabolic rate has been introduced, the *met*, defined as the energy for metabolic purposes transformed per unit of time and the body surface by an individual of average body (body surface dispersant assumed equal to 1, 8 m^2). For an individual sitting at rest, 1 met = 58 W/m^2.

The term E, the thermal flow exchanged by evaporation is the sum of the three factors.

The first, denoted with E_d in accordance with the annotations of Fanger, represents the thermal power dispersed through perspiration, i.e., by non-sensory evaporation of water in the skin and respiratory system. The adjective non-sensory in this case means "which is independent from the senses" since this phenomenon, slow and continuous, takes place independently of the muscular work and the outside temperature and is not influenced by the regulation system of the human body.

The second, denoted by E_{sw}, takes into account the thermal power lost through transpiration: this is one of the key devices of thermoregulation. When the skin temperature rises above 36 °C, the sweat glands are stimulated and begin to secrete sweat; evaporation allows the rejection of a substantial amount of heat, keeping practically constant the temperature of the skin.

The third, finally, E_{re}, represents the thermal power transferred to the air by evaporation from the mucous membranes of the respiratory system.

The terms E_{sw} and E_{re} are influenced by the value of the air humidity in the environment, the higher the humidity is high, the lower is the value of E_{sw} and of E_{re}.

Under normal conditions of well-being, all the sweat secreted evaporates, whereas when conditions are warm and damp, the skin is covered with a film of sweat that is hard to evaporate, especially if the surrounding air is still.

The thermal flux between the environment and body surface $(R + C)$ is influenced by the type of clothing; to define the overall thermal resistance between the skin surface and the outer surface of the clothes, a dimensionless parameter clo is used: this is given by the relationship between the overall thermal resistance of the dress considered (r_{cl}) and a thermal resistance of reference equal to 0.155 m^{2}K/W:

$$I_{\mathrm{cl}} = \frac{r_{\mathrm{cl}}}{0,155} \tag{12.2}$$

For the naked body $I_{\mathrm{cl}} = 0$ clo; values r_{cl}, and then of I_{cl}, are obtained experimentally, for various types of clothing (the range is between 0 and 4 clo). Experimental studies conducted by Fanger on a large sample of individuals have shown that the surface temperature of the body and thermal flux associated with sweating, when the individual is in a state of well-being, may be related to the specific metabolic activity. The influence of other variables, such as the mean radiant temperature and clothing, is not significant.

In the evaluation of the exchange of sensible heat between the body and the indoor environment $(R + C)$ two elements often influenced by HVAC systems are involved: the *mean radiant temperature* t_{mr} and the *operating temperature* t_{op}.

The mean radiant temperature affects the heat transfer by irradiation of the human body. It can be defined, with good approximation, as the weighted average of the internal surface temperatures of the walls that surround the individual and can be expressed with the equation:

$$t_{mr} = \sum_{i=0}^{n} t \cdot F_{p,i} \tag{12.3}$$

where t is the surface temperature of the walls surrounding the individual and $F_{p,i}$ is the *radiation view factor* that is the fraction of thermal energy leaving the surface of a object 1 (for example the body of the individual) and reaching the surface of object 2 (for example the wall surface), determined entirely from geometrical considerations. Stated in other words, F_{12} is the fraction of object 2 visible from the surface of object 1, and ranges from 0 to 1. This quantity is also known as the *Radiation Shape Factor* and is dimensionless.

The calculation of the view factor can be quite complex, especially if the rooms have irregular geometrical forms. In most situations, the value of the radiation view factor can be calculated as the weighted average of the internal surface temperatures of the walls.

We define *operating temperature* to be the weighted average between the average radiant temperature t_{mr} and the air temperature t_a, each being weighted with the respective heat exchange coefficient. If the air speed is relatively low and the mean radiant temperature is not too high, the mean radiant temperature can be estimated with the following simplified equation:

$$t_{op} = \frac{t_{mr} + t_a}{2} \tag{12.4}$$

The operating temperature defined by (12.4) is a parameter of considerable practical interest, because it takes into account two variables, the mean radiant temperature and the air temperature and may constitute a first evaluation parameter of the indoor environment from the point of view of well-being.

His experiments have allowed Fanger to develop an equation, known as the *equation of well-being*, which is liable to be resolved with respect to any variable one wishes to determine to ensure conditions for thermal comfort in a room.

For example, given a type of activity and a certain clothing, through the equation of well-being, it is possible to evaluate different combinations of indoor environmental conditions (air temperature, mean radiant temperature of the walls, relative humidity, air velocity) that match the conditions of well-being and between them to choose the most appropriate.

The degree of acceptance of indoor environmental conditions can be expressed with the average value of a vote expressed, based on a scale of "thermal sensation"; ASHRAE 55 Standard and 7730 ISO Standard propose the following scale:

+3 = very hot
+2 = hot
+1 = slightly warm
0 = comfortable
−1 = slightly cold
−2 = cold
−3 = very cold

Based on the scale proposed by ASHRAE, Fanger has developed a method for the evaluation of thermal sensation, indicated as (PMV) predicted mean vote. It may sometimes be useful to know a priori what can be the effect on the user of a certain design or management choices of a building or of an HVAC system. In such cases it may be more convenient, or clearer, instead of referring to a vote, to assess directly the percentage of people who may express, if requested, a negative appreciation in a given indoor environmental situation.

On the basis of experiments already mentioned, Fanger has made the relationship between (PMV and PPD) predicted percentage of dissatisfied that defines the percentage of dissatisfied occupiers.

From the graph that shows the relationship between PMV and PPD, shown in Fig. 12.1, it is interesting to observe how PPD does not vanish even when PMV is equal to 0 (maximum comfort), being a function of the fact that conditions capable of satisfying the equation of well-being cannot statistically be of satisfaction for every group of people.

Standard ISO 7730 considers acceptable conditions in which PMV is in the range between −0,5 and +0,5, corresponding to a PPD value of 10 %.

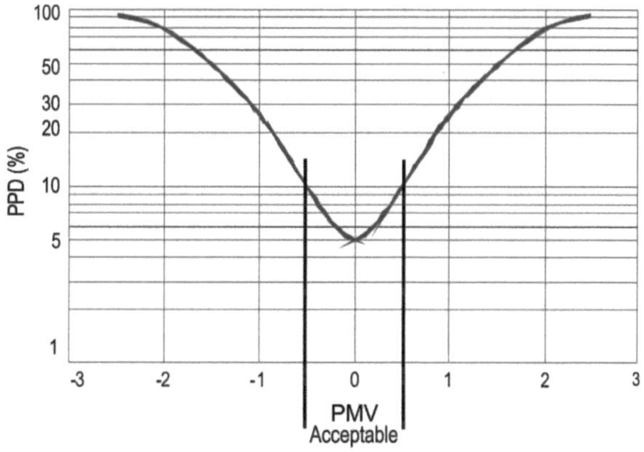

Fig. 12.1 Predicted percentage dissatisfied (PPD) as a function of predicted mean vote (PMV)

12.1.2 Radiant (Ray) Conditioning

In recent years, HVAC systems that make use of radiative heat transfer mechanisms have become more widespread, hence the term *radiant* or *ray conditioning* used to distinguish them from the more classical air conditioning systems that use air and convection as the heat exchange mechanism[1].

Using the air-conditioning technology the mean radiant temperature t_{mr} of the walls is not controlled, the result is that to ensure comfort conditions acceptable in summer (cooling phase) the air temperature inside the conditioned spaces t_{air} must be maintained at a lower value, in practice $t_{mr} > t_{air}$. A consequence of this choice is that the temperature actually perceived, the operating temperature, is 2 ÷ 4 °C higher than expected. In order to ensure conditions of comfort, the solution is to further lower the air temperature, to increase the air speed and to lower the relative humidity in order to permit the body to increase its osmotic evaporation (transpiration). This latter mechanism is of great importance; it occurs in fact across the cell membrane without wetting the skin and not by means of the sweat glands (sweating, through the pores), and for this reason it is comfortable.

The lowering of the relative humidity of the air, however, implies an increase in energy consumption for an HVAC system.

An excessive reduction of the air temperature and the increase of its speed, which moreover increases consumption of energy, also determine situations of physical discomfort.

Our thermal sensations are greatly influenced by the temperature of the surfaces to which the body exchanges by radiation. We irradiate heat in the form of electromagnetic waves. This radiation is not visible to the naked eye but is perceived by the skin as a thermal sensation: if one irradiates too much, one feels cold and if one irradiates too little, one feels warm.

As for all the heat transfer mechanisms, even radiation is influenced by the temperature difference between the warmer body and the cooler body. However, this exchange mechanism does not require a material medium (but only empty "space"). In fact in a vacuum radiation is at its maximum potential, but the air around one, owing to its low density, has a very similar behaviour. So if surfaces that surround individuals have a temperature of only 1–2 °C lower than theirs, only 5–8 W per m^2 of skin, out of the 60–80 W/m^2 which metabolise, can be irradiated, hence the sensation of warmth which those individuals will have.

The ray conditioning acts directly on the surfaces, lowering the temperature (in cooling) and bringing it to 8 ÷ 10 °C below body temperature. As a consequence, the thermal power radiated by the body rises to 20 ÷ 35 W/m^2 of skin and the perceived sensation is of a natural and uniform freshness, at given conditions of temperature, air speed and air humidity. It is clear then that, when irradiation is increased, the air parameters can be moderated. This results in a better condition of well-being and lower energy consumption.

---

[1] Roberto Messana contributed to write this Section.

During the heating phase the parameters are inverted since with the use of ray conditioning the temperature of the surface is greater, and therefore the body radiates less, giving a sensation of greater warmth of the air at the same temperature. This compensates very well, with heavier clothing, the natural decrease in temperature and humidity typical of the season. Also in this case one obtains an improvement in the sensation of comfort and energy savings as a result of the lower temperature of the air and of the lower working temperature of the heat generator.

From what has been discussed above, it is possible to highlight the relationship between the quality of thermal comfort and productivity, an aspect of great importance, whose market value has priority over that derived from simple primary energy savings. Studies and tests carried out [4, 5] have shown a ratio of up to 1:200 between the said two economic values.

Radiant conditioning has the advantage of being able to combine both of these requirements. In fact, it increases the efficiency of the electrochemical energy of the body and that required for system operation. From the technological point of view, the components that form the basis of the ray conditioning application are the following:

- active radiant surfaces whose function is to absorb (in cooling) or supply (in heating) heat energy to infrared radiation (70–100 %) and natural convection (0–30 %) with the masses and passive people; the most powerful is the ceiling of which is usually sufficient for 40–60 % for both functions;
- the control system, whose main function is to control the supply temperature of the panels in relation to the humidity of the air present in the indoor environment, in order to completely avoid the formation of condensation;
- dehumidification system (in cooling) to ensure the right level of humidity in terms of comfort and performance of the panels.

Radiant conditioning is thus based on a different balance between the modes of heat exchange affecting the dissipation of body heat: cooling in greater irradiation, elimination of forced convection, reduction in osmotic evaporation, less radiative heating, greater natural convection and the possibility of a reduction in the amount of clothing.

12.1.3 The Adaptive Comfort

The classical comfort theory discussed in Sect. 12.1.1 demonstrates how even in the presence of air conditioning systems able to perfectly control the indoor environmental factors, it may not be possible to meet fully the comfort criteria of all the occupants.

The awareness of being in an indoor environment in which the control which governs environmental conditions is mechanical, and does not allow the user to effect changes directly, also makes the perception of climatic conditions even more inflexible. In those spaces for which all possible measures have been taken to obtain "natural" environmental conditions (and any "conditioning" is by "natural" methods), of course the same comfort conditions that are obtained in mechanically air-conditioned spaces cannot be reached, yet the perception of the indoor environmental conditions by users in this case is less inflexible.

The possibility of implementing individual strategies to control the comfort (e.g., through the lowering of a blind, the opening of windows, or the switching on of a fan) amplifies the ability to adapt to climatic situations which even respect far less the standard parameters. Monitoring campaigns comparing environmental conditions with the actual perceptions by users, both in buildings with HVAC systems, and in buildings with natural controls, have highlighted that a number of adaptation factors (psychological, physiological and behavioural) contribute to influence the susceptibility of an individual to the perceived climatic condition.

On this basis, these considerations are based theories that have allowed a number of researchers, beginning with De Dear and Brager [3], to develop models to assess what is commonly called *adaptive comfort*.

In surveys conducted by two researchers for the evaluation of the perceived comfort, rather than the assessment scale proposed by ASHRAE (+3 to −3), a simplified and less arbitrary scale was used, which consisted of asking respondents if they preferred hotter, colder or no change to the room temperature. The temperature that the maximum the number of persons accepted, i.e., people who did not consider any change to be necessary was defined to be the *optimum temperature*.

For buildings with natural ventilation, in particular, De Dear and Brager have highlighted that the optimum temperature t_{co} could be related to the average outdoor temperature t_{out}, calculated as the arithmetic average between the average of the daily minimum temperatures and average daily maximum for the month in question, according to the equation:

$$t_{co} = 0.31 \cdot t_{out} + 17.8 \tag{12.5}$$

In Eq. (12.5), the value 17.8 is a constant representing the minimum acceptable temperature, while the value 0.31 defines the correlation between the comfort temperature t_{co} and the outside temperature t_{out} (all temperature values are expressed in °C).

The equation has been transposed by ASHRAE with the following ranges of acceptability (PD is the percentage of dissatisfied):

$$\text{for}\, t_{out} \pm 2.5\,^{\circ}\text{C},\ \text{PD} = 10\,\%$$

$$\text{for}\, t_{out} \pm 3.5\,^{\circ}\text{C},\ \text{PD} = 20\,\%$$

using as a reference value for the outdoor temperature the value of the monthly average temperature.

Equation (12.5) is represented graphically in Fig. 12.2 [1].

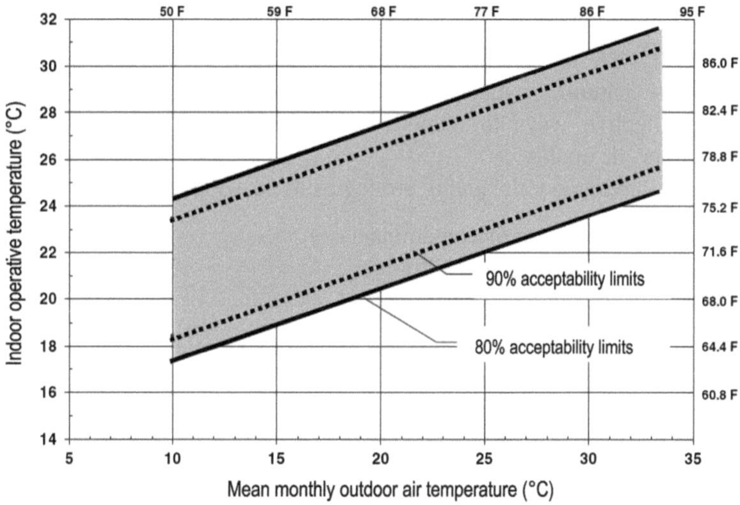

Fig. 12.2 Definition of the limits of the adaptive comfort [1]

The ability to make use of less rigid parameters for the comfort in confined spaces allows one to obtain two advantages: a physiological sensation of indoor climate for the users which is a closer to the natural one and a considerable reduction in the management costs for HVAC systems. This explains the interest raised by this new definition of comfort for adaptive models, i.e., models that take into account the ability of humans to adapt to situations that, while not causing stress, are different from those defined with the classic approach.

12.2 Indoor Air Quality

Edited by Elisa Bruni

Given that people spend most of their time indoors, indoor air quality (IAQ) has a major affect on human health as well as occupant comfort and productivity. Indoor air pollution can be the cause of asthma, allergies and irritation. In addition to issues related to human health, it is necessary to take into consideration the issues related to the productivity of people working in the buildings. It has been shown that good IAQ improves productivity and reduces absenteeism. Thinking about these issues from the point of view of economics, and considering that the cost of the salaries of workers are a major item in the budgets of a company, the importance of maintaining a good IAQ is even more evident.

Some of air contaminants are:

- CO_2;
- environmental tobacco smoke;
- particulates and dust;

- volatile organic compounds (VOCs);
- radon.

Appropriate natural or mechanical ventilation and control of contaminants source are effective ways to reduce the concentrations of contaminants and improve indoor air quality.

In order to have good air quality strategies to be adopted are:

- minimise indoor sources of contaminants;
- use effectively ventilation and filtration;
- minimise entry of outdoor contaminants.

A holistic approach that considers the different strategies at different stages of the building's life cycle, from design, through construction, up to operation and maintenance of the building permits improvement of IAQ and consequently of the comfort and well-being of the occupants [6, 7].

12.2.1 Minimise Contaminants From Indoor Source

It has been shown that the prevention of problems related to IAQ is generally more effective, and therefore less expensive than solving the problems afterwards when these have already manifested themselves. So the first step towards a good IAQ is to minimise contaminants from indoor sources.

Many materials in buildings can contribute to cause health impacts, because they emit volatile organic compounds and urea formaldehyde. Paints, adhesives, sealants, furniture, flooring systems are sources of VOCs in buildings. An appropriate selection of materials with low emissions prevents IAQ problems and reduces exposure to indoor contaminants of both occupants and contractors. Specific limits of VOC emission from buildings products are defined in reference standards and voluntary labelling schemes.

Besides reducing the contaminating compounds at the source, when used it is necessary to use them properly. The agency South Coast Air Quality Management District defined rules regarding the correct use in buildings of adhesives primers, sealants and other primers. For these products, a correct use by skilled and trained personnel can reduce the impact on air quality.

Cleaning products that can contain chemical agents, irritating and harmful for cleaning companies' staff and occupants are sources of contamination. So cleaning operations need to be defined in procedures and protocols that indicate the use of products with minimal emissions along with scheduling methods and timing of cleaning activities.

Indoor air quality also depends upon the daily activities within the building, such as cleaning as well as smoking. Environmental tobacco smoke, in fact, is another contamination source. In the case, smoking rooms being allowed inside the building, indoor designated smoking areas should be defined. A smoking area must be isolated from others area, a negative pressure must be maintained to contain the contaminated air and a dedicated, separate ventilation system must be installed.

Finally, construction and renovation activities can cause production of pollutants in indoor environments with problems for both the construction workers and the building occupants. A management plan during construction activities of the building could define control measures for the protection of HVAC systems and interrupt contamination pathways to prevent contamination by the cleaning of surfaces and replacement of all filtration media at the end of the construction phase. After the installation of interior finishes and before occupancy of the building, a building flush-out can be conducted by supplying outdoor air to remove contaminants.

12.2.2 Effectively Ventilation and Filtration

Both mechanical and natural ventilation systems in buildings allow one to maintain good indoor air quality, providing fresh, clean air inside buildings and carrying exhaust air outside.

The CO_2 concentration level in air is an indicator of indoor air quality. Continuous monitoring of this parameter through the installation of CO_2 monitors is a measure to control air quality and thereby improve it. Another means of measuring IAQ is the fitting of outdoor airflow measurement device.

ASHRAE 62.1-2007: Ventilation for Acceptable Indoor Air Quality [8] specifies minimum ventilation rates for spaces and requirements for ventilation and air-cleaning systems, so that indoor air quality minimises adverse effects on health. An increase in the ventilation rate compared the minimum levels defined by the rule may, however, lead to a further improvement in indoor air quality.

Through ventilation, air contaminants are diluted and in the case of mechanically ventilated spaces, reduced by filtration. The reference standard that defines a test procedure for the evaluation of the performance of air-cleaning devices is ASHRAE 52.2-1999 [9]. This standard evaluates the performance of filtration as a function of the size of particles: Minimum efficiency reporting value (MERV) rating. The higher is the MERV value, the better is the performance in terms of filtration.

If the building is equipped with a mechanical ventilation system, it is necessary to coordinate the strategies aimed at improving its energy performance and the thermal comfort of the occupants with those aimed at improving indoor air quality. Improving indoor air quality using ventilation systems requires energy, so increased ventilation could increase energy consumption. Thus, it is important to have an approach which integrates the different strategies for improving the energy performance and IAQ throughout the life of the building.

IAQ is not just a matter of HVAC plant. Even the most efficient technologies of ventilation are therefore not sufficient to achieve a good IAQ but it is important to manage processes during the different phases of design, construction, operation and maintenance of buildings.

12.2.3 Minimise Entry of Outdoor Contaminants

Outdoor air pollutants (i.e., particles, gases, dust, pesticides, allergens) entering a building through air intakes can have an influence on IAQ.

Outdoor air intakes need to be located at a certain minimum distance at least from external contamination sources to avoid entry of contaminants: prevailing wind conditions must also be taken into account. Another potential pollutant to be kept outside the building is a radioactive gas, called radon present in soil and groundwater. Radon gas can be excluded by maintaining a positive pressure in the room where radon could enter and by sealing potential entryways such as cracks and pores [10].

Finally, occupants entering in buildings with shoes and clothes could cause the entry of pollutants and moisture into buildings. It is possible to minimise their entering by means of entryway barrier systems and effective cleaning strategies for these systems.

To conclude, a building with good indoor air quality is a building in which the pollutants inside and outside have been minimised at source and in which efficient ventilation is provided.

12.3 Lighting Comfort

Edited by Annalisa Galante

12.3.1 The Human Factors on Lighting Design

Visual comfort for a person is that condition of mind which expresses satisfaction with the visual environment (Fig. 12.3). Visual comfort involves two aspects: one is that there must always be enough light, being the quantitative need, where enough light provides the required visibility; the other is the elimination of disturbing effects related to the lighting, i.e., its qualitative side [11].

Visual functions determine the visual comfort, the ability that has the eye to interpret external stimuli, and they are:

- adaptation: the eye encodes sensitivity to light and changes the opening of the pupil;
- beds: movement of the lens depending on the amount of incoming light and the proximity of objects;
- convergence: ability to process a single image despite binocular vision.

Visual capabilities determine, however, the visual performance, i.e., the real eye's ability to perform the visual functions, and these are:

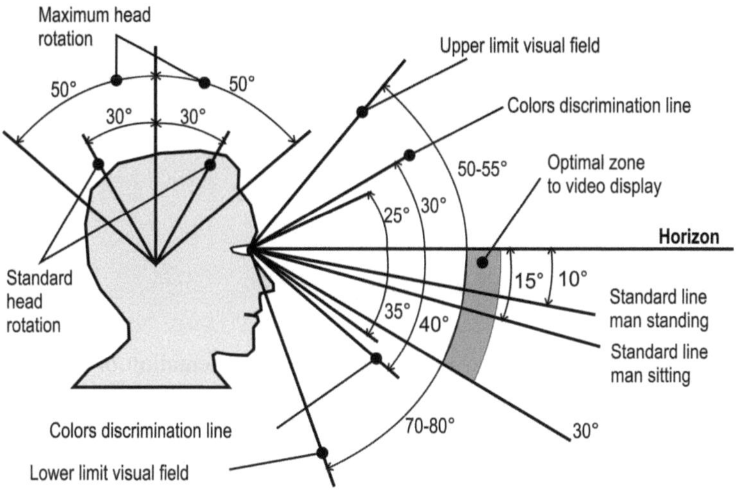

Fig. 12.3 Visual functions and visual capabilities determine visual comfort

- visual acuity: ability to perceive as separate and distinct sensations produced by two nearby objects;
- contrast sensitivity: how the eye is able to perceive differences in (contrasts of) luminance;
- speed of perception: this is the reaction time, i.e., the time which the observer has to perform the visual activity, which increases with increasing illuminance;
- colour perception;
- peripheral vision.

The design of the lighting system must take account of both the comfort of visual capacity and prevention of glare. In fact, different activities require different levels of lighting. In general, the more detailed the task, the greater the light requirement. A process control room should be lit at an illuminance of 300 lux, a corridor or walkway may only require 50 lux, whilst studying an engineering drawing may require 750 lux.

Where individuals are carrying out different activities, they will need control over their local lighting e.g., a control and instrumentation engineer coming into a process control room lit at 300 lux may need a desk with a lamp to study a wiring diagram.

Studies have shown that giving workers in open plan offices local control of lighting can increase job satisfaction (and decrease the experience of stress). Directional sources of light can bounce off reflective surfaces such as display screens and cause glare. Using blinds, correcting the angle of the source of light and using glare filters can help control this, as can use of e.g., up-lighting.

All sources of light have a particular colour. Some of these, such as sodium, can make coloured text and diagrams difficult to read. Sudden contrasts in light levels e.g., coming out of a well-lit area into a dark area or vice versa can be a problem, because it takes the eye several seconds to adapt to new lighting conditions. Changes in lighting levels should be made gradually where possible.

Generally, lighting is designed when the workplace is empty and without consideration of the shadows cast by equipment e.g., lighting of yards where trailers and containers may be parked. Pedestrian walkways in these areas should have specific lighting.

12.3.2 Common Photometric Quantities

The wave theory permits a convenient graphical representation of radiant energy in an orderly arrangement according to its wavelength or frequency. This arrangement is called a spectrum. It is useful in indicating the relationship between various radiant energy wavelength regions. The radiant energy spectrum extends over a range of wavelengths from 10^{-16} to 10^5 m. The Angstrom unit (Å), the nanometer (nm) and the micrometer (μm), which are respectively 10^{-10}, 10^{-9} and 10^{-6} m, are commonly used units of length in the visible spectrum region. The nanometer is the preferred unit of wavelength in the ultraviolet (UV) and visible regions of the spectrum. The micrometer is normally used in the infrared (IR) region. Radiant energy in the visible spectrum lies between 380 and 780 nm [12].

The intensity and spectral properties of a blackbody radiator are dependent solely upon its temperature. A blackbody radiator may be closely approximated by the radiant power emitted from a small aperture in an enclosure, the walls of which are maintained at a uniform temperature.

Data describing blackbody radiation curves were obtained by Lummer and Pringsheim using a specially constructed and uniformly heated tube as the source. Planck, introducing the concept of discrete quanta of energy, developed an equation depicting these curves. It gives the spectral radiance of a blackbody as a function of wavelength and temperature.

The Stefan–Boltzmann Law states that the total radiant power per unit area of a blackbody varies as the fourth power of the absolute temperature. It should be noted that this law applies to the total power, that is, the whole spectrum. It cannot be used to estimate the power in the visible portion of the spectrum alone.

In the period from 1948 to 1979, the luminance of a blackbody operated at the temperature of freezing platinum (2042 K) was used as an international reference standard to define the unit of luminous intensity. Specifically, it has a luminance of 60 cd/m^2. A new definition of the candela was adopted in 1979: the luminous intensity of a 555.016 nm source whose radiant intensity is 1/683 W/sr. The new photometric unit is based on an electrical unit, the Watt, which can be accurately and conveniently measured with an electrically calibrated radiometer. A further advantage of this definition is that the magnitude of the unit is independent of the international temperature scale, which occasionally changes.

Vision can be categorized with reference to the adaptive state of the rod and cone photoreceptors of the retina. The photopic function describes the spectral luminous efficiency function for photopic (cone) vision (that of well-lit conditions), and the scotopic function describes the spectral luminous efficiency for scotopic (rod) vision (that of low light conditions) (Fig. 12.4).

The luminous efficacy of a light source is defined as the ratio of the total luminous flux (in lumens) to the total power input (in Watts). There are 683 lumens/Watt at 555 nm. Since the scotopic luminous efficiency function peaks at a different wavelength (507 nm), it is necessary to establish different scaling factors for the photopic and for the scotopic luminous efficiency functions.

In general, light sources do not emit light homogeneously in all directions, in the International System of Units (SI), the fundamental photometric quantity (Fig. 12.5), Luminous Intensity (I), is expressed in candelas (cd) and it is the amount of light energy emitted in a specific direction.

$$I = \frac{d\Phi}{\mathrm{d}\omega} \tag{12.6}$$

In Eq. (12.5) the $d\Phi$ is light energy emitted and $\mathrm{d}\omega$ is the steradian. The steradian, that is an SI supplementary unit, is defined as the solid angle subtending an area on the surface of a sphere equal to the square of the sphere's radius.

The current definition of the candela is *the luminous intensity, in a given direction, of a source that emits monochromatic radiation of frequency* 540×10^{12} *Hz and that has a radiant intensity in that direction of* 1/683 W/sr. The definition expresses the candela in terms of the Watt and the steradian.

Two important derived units based on the candela are those of Luminous Flux and Illuminance.

The unit of Luminous Flux (Φ) is measured in lumens (lm), a lumen is the luminous flux emitted by a source that radiates in all directions, including within a unit solid angle, a light intensity of the value of 1 cd. The quantity of luminous flux falling on one square unit of the sphere's surface is defined as 1 lm. Note that the unit itself is arbitrary, since the total quantity of flux that will be incident on this area is independent of the size of the sphere.

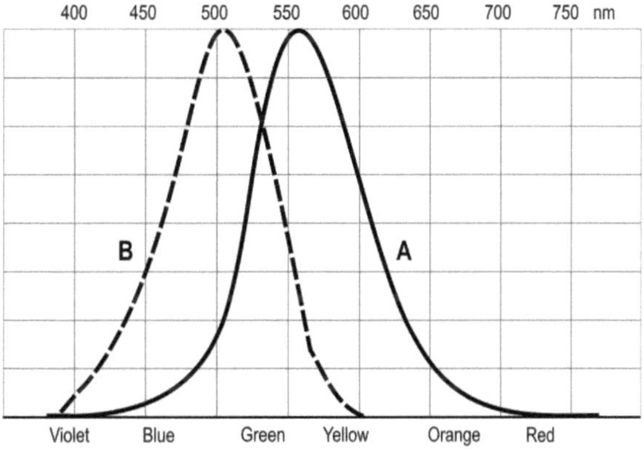

Fig. 12.4 Spectral reference *curve* V (l): the maximum value of relative visibility for radiation of wavelength is equal to 555 nm for photopic vision (**a**) and scotopic (**b**) of 507 nm for that

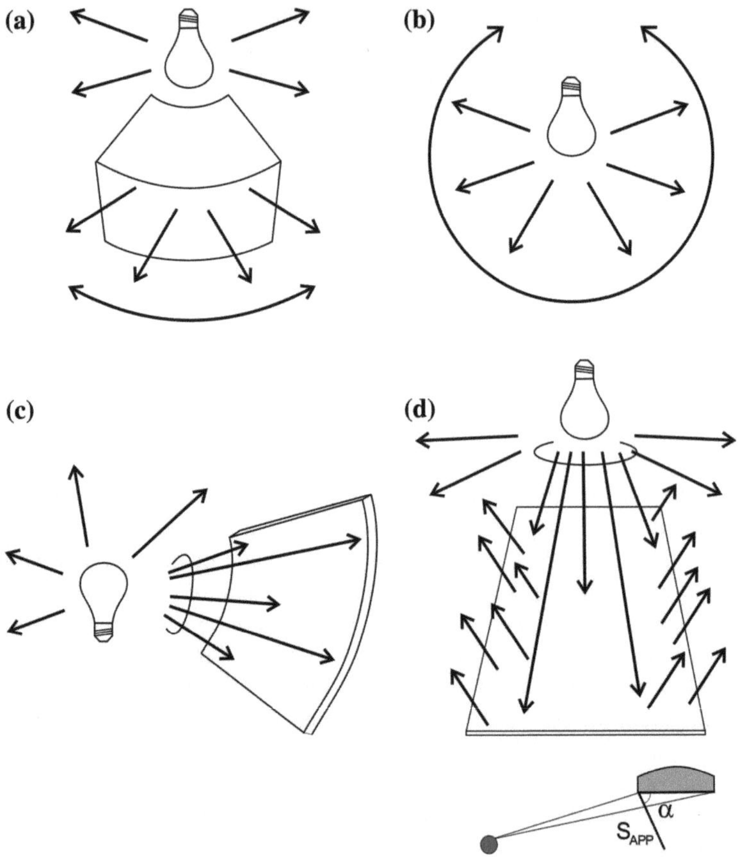

Fig. 12.5 Fotometric quantities: (**a**) Luminous Intensity, (**b**) Luminous Flux, (**c**) Illuminance, (**d**) Luminance

$$\Phi = \frac{I}{d\omega} \tag{12.7}$$

In Eq. 12.7, LI is the luminous intensity (Eq. 12.6) and $d\omega$ is the steradian. Since the visual system is much more sensitive in the green–yellow part of the spectrum than in its red part, the same absolute power of the radiation would provoke a much stronger sensation of green than of the red light.

The concentration of luminous flux falling on a surface, that is, the incident flux per unit area, is called Illuminance (E). To define a unit of illuminance, the sphere must now be given real dimensions, because the flux density diminishes with increasing distance from the source. The Illuminance is measured in lux (lx) and expressed the amount of light that invests a certain surface. This is the ratio between the Luminous Flux ($d\Phi$) impinging on a surface and the area (dS in m^2) of the element taken into consideration (Eq. 12.8).

$$E = \frac{d\Phi}{dS} \tag{12.8}$$

Another important luminous quantity is Luminance (L), which is directly related to perceived brightness, i.e., the visual effect that illumination produces. Luminance depends not only upon the Illuminance on an object and its reflective properties, but also on its projected area on a plane perpendicular to the direction of viewing. There is a direct relationship between the luminance of a viewed object and the Illuminance of the resulting image on the retina of the eye. This is analogous to the exposure requirements in photography. The unit of luminance is the candela per square meter (cd/m^2).

$$L = \frac{dI}{dS_{app}} \tag{12.9}$$

In Eq. 12.9, Luminance (L) is the ration between dLI (difference of Luminous Intensity, Eq. 12.6) of a source in the direction of an observer and the emitting apparent surface as seen by the observer himself (dS_{app})

$$dS_{APP} = dA * cos\,\alpha \tag{12.10}$$

In Eq. 12.10, emitting apparent surface (dS_{APP}) is the emitting surface multiplied by cos where α is the angle between the direction of observation and the axis normal to the surface.

The photopic luminous efficiency function applies to visual stimuli to the fovea and at luminance levels higher than approximately 3 cd/m^2. The scotopic luminous efficiency function applies to visual stimuli in regions outside the fovea and to luminance's below approximately 0.001 cd/m^2. A family of mesopic luminous efficiency functions is required for application to luminous stimuli between approximately 0.001 and 3 cd/m^2. Research in this area is on going [13, 14].

With the exception of special measurements for research purposes, almost all photometric quantities are measured photopically, even at luminance's below 3 cd/m^2 and for peripheral vision.

12.3.3 Towards a Sustainable Lighting

When retrofit measures for lighting are being proposed, one considers the degree of visual comfort that is decided by both daylight and artificial lighting levels. Generally, these two means of lighting can be evaluated separately, since artificial lighting is provided for those situations when there is no or insufficient daylight present. In green buildings, but also in retrofit, however, there is frequently an interaction between these two light sources and their control and regulation. This leads to a soft transition between daytime and evening illumination.

The evaluation of visual comfort in an artificial lighting setting is based, in essence, on the following factors:

- Degree of illuminance, both horizontally and vertically;
- Evenness of illuminance distribution through the room;
- Freedom from glare for both direct and reflex glare settings;
- Direction of light, shading and colour;
- Reproduction and light colour.

Illuminance is defined especially through the beam direction and the beam capacity of the lamps used. The advantages of indirect illumination are a high degree of evenness and a low potential for glare effects. Advantages of direct illumination include low electricity consumption, better contrasts and demand-oriented adjustment.

For indirect illumination, the only way the same illuminance level of 500 lux can be achieved, on the work plane, as for direct lighting is by using twice the amount of electricity. While evenness of room illumination is still achieved, it is monotonous, however, on account of missing shades. With exclusively direct illumination of the room, vertical illuminance is so low that it restricts perception of the room. This does not allow for comfortable communication among the occupants, and further there is uneven illuminance also at the working plane level. It is by the combination of these two means of lighting that both the visual and economic optimum are achieved [15].

Every task requires a different level of illuminance. The minimum limit for tasks requiring a certain amount of concentration is 300 lux. In Fig. 12.6, minimum illuminance requirements as outlined in the European directives are summarised. Office readings show that, with daylight illumination, illuminance of 300 lux is perceived as comfortable. Unfortunately, these settings are not included in the standards for artificial lighting, although they have been demonstrated to apply in practice.

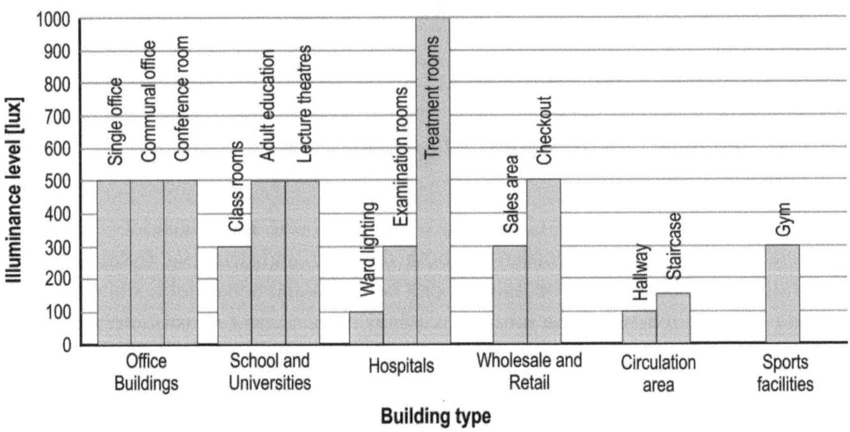

Fig. 12.6 Illuminance levels for different user applications [13]

The prevailing lighting atmosphere inside the room is determined by the reflection characteristics of surface areas, light colour and colour reproduction of the illuminants used. Contemporary quality illuminants are capable of setting light moods for the room that are similar to those in daylight. Available illuminant colours are off-white, neutral white and daylight white. Usually, it is off-white and neutral white light which are perceived as comfortable by office occupants. Daylight white light, at 500 lux, is perceived rather as being cold and uncomfortable. Only at much higher illuminance levels does this particular light colour start to be accepted.

Colour reproduction merits of a lamp stand for its ability to reproduce the colours of people and objects as close to life as possible. For a good level of colour reproduction, the illuminants used should have, at the very least, a colour rendering index of Ra = 80 or, better still, of Ra = 90 and higher.

Evaluation of visual comfort in a daylight setting, independent of artificial lighting used, is much more complex since it is not only the stationary situation that needs to be taken into account but also changes in brightness levels over the course of an entire year. Room shape, obstructions in the immediate vicinity and chosen lighting-technological merits of the façade are all decisive factors for determining daylight quality inside a room. All three factors, however, are linked to architectural and thermal requirements, so that an optimum illumination can be achieved only through an integrated approach. Good daylight quality levels are provided when:

- Indoor brightness, as opposed to outdoor brightness in winter and summer, reaches certain critical values (Daylight Factor and Sunlight Factor);
- Natural lighting inside the room is evenly distributed;
- Indoor brightness changes according to outdoor brightness, so that a day/night rhythm can be felt (this especially applies for rooms not oriented to the North, since they receive different sunlight according to the time of year);
- An outside relationship can be established with concurrent sufficient solar protection;
- Glare, especially as it occurs with work place monitors, can be avoided (near and far field contrasts);
- A large proportion of lighting, during usage hours, stems exclusively from daylight, without the use of electric power or artificial lighting (daylight autonomy).

Correct façade design for maximum use of natural daylight potential present, while also adhering to solar protection considerations and limitation of glare, is one of the most difficult tasks of building design. The reason for this is the high variability factor of sun and sky conditions during the course of the day.

Horizontal illuminance encompasses readings from 0 to 120,000 lux, whereas solar luminance is up to one billion cd/m^2. For rooms with monitor work stations, an illuminance of 300 lux suffices, window surface luminance should not exceed 1,500 cd/m^2. This means that sufficient natural lighting is achieved if a mere 0.3 % of daylight in summer and 6 % in winter can be transported on to the work planes.

The degree of difficulty is mainly due to the fact that simultaneously sky luminance needs to be reduced to between 3 and 13 %, and solar luminance to 0.0002 %.

Daylight and solar factors define daylight quality inside a room. Both values define the relationship of illuminance on the working plane to outdoor brightness. The daylight factor is calculated for an overcast sky, in order to evaluate a given room independently of any solar protective devices or systems. The solar factor, on the other hand, is calculated for a sunny room with solar protection in order to allow evaluation of daylight conditions with the solar protection active.

This distinction is of importance in order to compare façades, with and without systems of illumination using daylight, across the board, for any sky condition. The measurement variable called luminance can be imagined as the level of light perception for the eye. Different luminance levels lead to contrast formation. Contrasts are important for the eye to even identify objects. Yet, if contrasts are too high, they lead to glare effects that are hard on the human organism. In order to attain a comfortable and sufficient visual level at the monitor, contrasts between working field and near field should not exceed 3:1 and between working field and far field should not be greater than 10:1.

The near field runs concentric around the main viewing direction, with a beam angle of 30 °C (See Fig. 12.3). The far field has twice that opening angle. Research shows that higher contrast levels for both near and far field are acceptable to the user. This can be traced back to the fact that, through the psychologically positive effect of daylight on people, higher luminance levels outside the window are not perceived as bothersome. Contemporary monitors are mainly non-reflecting and have their own luminance levels of between 100 and 400 cd/m^2.

Contrary to artificial lighting, a high level of evenness for one-sided daylight illumination is much harder to achieve. Illumination equability is defined as the ratio of minimum illuminance level and medium illuminance level of a given area of the room. For artificial lighting, the ratio should be larger than 0.6. For daylight illumination, however, this value can scarcely be achieved, or only by a decrease in the overall illuminance level. For this reason, equability evaluations for daylight illumination cannot be based on the same criteria as those for artificial lighting. Rather, practically attainable values need to be consulted.

The aim here is to achieve a reading of more than 0.125 for equability. Essential factors of influence in this are downfall size and reflection grades of the materials used indoors.

12.3.4 Visual Comfort and Energy Efficiency

Lighting technology has produced many recent developments in energy-use reduction; many industries have upgraded their lighting systems, and lighting manufacturers have brought more efficient products on to the market. However, in most facilities, the lighting system still presents significant opportunities for the reduction of electricity costs. For example, purchasing the most efficient lighting system available today does not have to cost more than using standard fixtures and standard design. In fact, the project's first cost may even be reduced by using the most efficient available fixtures and designs. To achieve that, three key concepts should be adhered to:

- use only recommended lighting levels;
- use parabolic fixtures with T-lamps and electronic ballasts;
- take advantage of lower A/C size and costs.

The first step in reducing electricity costs related to lighting is to survey the facilities in question to discover whether the lighting equipment in each area is appropriate for the work performed there and whether it is the most energy-efficient type available for the task [16].

Lighting surveys often reveal one or more of the following energy management opportunities:

- *Lights left on in unoccupied areas.* Even the most efficient lights waste energy when they are left on unnecessarily. The best way to ensure that lights are turned off when they are not needed is to develop the occupants' sense of responsibility so that they take care of turning off unneeded lights. The installation of timers, photocells and occupancy sensors or integrating the lighting system into an energy management control system can also be considered. Lights (and other powered equipment, such as fans) left on unnecessarily in refrigerated areas add substantially to the refrigeration load. The same applies to air-conditioning systems.
- *Dirty lamps, lenses and light-reflecting surfaces.* Dust and grease deposits on lighting fixtures can reduce the light that reaches the target area by as much as 30 %. Lighting fixtures should be cleaned at least once every 2 years, and more often when they are installed in greasy, dusty or smoky locations and/or when they are part of a heating, ventilating and air-conditioning (HVAC) system.
- *Overlit areas.* In areas with more lighting than the activities require, remove some lights or install dimming systems. Lighting requirements vary widely within a building, and a reduction in general area lighting combined with an increase in task or workstation lighting often increases the occupants' comfort while decreasing electricity costs. When de-lamping areas that are lit with fluorescent and high-intensity discharge fixtures, ensure that the ballasts are disconnected since they consume electricity even when the bulb is removed. Dimming systems are useful for areas where several types of activity take place. For example, plant production areas can be fully lit during production periods and dimmed when cleaning and security staff are on duty.
- *Obsolete lighting equipment.* Updating lighting system with more energy-efficient equipment is usually cost-effective.

Consider increasing the use of illumination using daylight, where feasible. Cutting energy use for lighting reduces not only the electricity bill but also the load on the air-conditioning system. Note that measures taken to reduce electricity consumption by lighting systems help to reduce emissions from thermo-electricity-generating stations.

For further details of these concepts see Sect. 13.5.2., where all the measures for retrofit lighting efficiency are described.

References

1. ANSI/ASHRAE (2003) Thermal environmental conditions for human occupancy, Inc, Atlanta, 55
2. ISO international standard 7730 (2005), Ergonomics of thermal environment—Analytical determination and interpretation of thermal comfort using calculation of the PMV and PPD indices and local thermal comfort criteria
3. Brager GS, De Dear R (2001) Climate, comfort and natural ventilation: a new adaptive comfort standard for ASHRAE Standard 55, Proceedings: moving thermal comfort standards into 21st Century. Oxford Brookers Univesity, Winsor
4. Messana R (2011) Capire il comfort, elementi di climatizzazione radiante. Tecniche nuove, Milano
5. Wargocki P, Seppänen O, Andersson J, Boerstra A, Clements-Croome D, Fitzner K, Olaf Hanssen S (2008) Clima interno e produttività negli edifici. Translator: Schiavon S. Collana Tecnica AICARR, Milano
6. U.S. green building council (2009) LEED reference guide for green building design and construction, U.S.A
7. U.S. green building council (2009) LEED reference guide for green building operations and maintenance, U.S.A
8. ANSI/ASHRAE 62.1-2007: Ventilation for acceptable indoor air quality, american society of heating, refrigerating and air-conditioning engineers, Inc., Atlanta
9. ANSI/ASHRAE 52.2-1999: Method of testing general ventilation air-cleaning devices for removal efficiency by particle size, american society of heating, refrigerating and air-conditioning engineers, Inc., Atlanta
10. Indoor air quality guide, best practices for design, construction, and commissioning, american society of heating, refrigerating and air-conditioning engineers, Inc., Atlanta
11. Majoros A (2005) Visual comfort by daylighting. Daylight Symposium, Budapest
12. Illuminating engineering society of North America (2000) The IESNA lighting handbook—Reference and application, mark stanley Rea, 9th edn, U.S.A
13. He Y, Bierman A, Rea MS (1998) A system of mesopic photometry. Light Res Tech 30(4):175–181
14. He Y, Rea MS, Bullough J (1997) Evaluating light source efficacies under mesopic conditions using reaction times. J Illum Eng Soc 26(1):125–138
15. Bauer M, Mosle P, Schwarz M (2010) Green building—guidebook for sustainable architecture. Springer, New York
16. Canadian industry program for energy conservation (2002) Energy efficiency planning and management guide, Her Majesty the Queen in Right of Canada, Ottawa

Chapter 13
Retrofit Measure Sheets

The retrofit measures for improving the energy performance of a building and the reduction of its environmental impact are crucial to achieving the goal of an overall improvement in sustainability. In this Chapter, one hundred retrofit measures that can be implemented in existing buildings are presented and discussed. The information contained, however, may in some cases also be used for buildings of new construction. For each of these measures, essential information needed to direct the auditor to the aspects to check and the necessary details to effect the measure correctly are provided.

13.1 Introduction

The definition of the *green energy plan* for the improvement of the sustainability of a building is brought to fruition mainly with the choice of retrofit measures that should be implemented. The 97 measures presented and discussed in this chapter are sufficient to cover most needs for residential, educational and commercial buildings but can also be useful for industrial buildings, excluding obviously the measures aimed at the optimisation of production processes. The proposed measures relate to:

- parts of the building (building envelope and systems) that whilst "operative" and not subject to urgent redevelopment, still cannot guarantee the best performance in terms of energy, comfort and indoor environment;
- parts of the building which are instead obsolete and are subject to urgent redevelopment or replacement;
- improvement of management and maintenance.

The different sections, one for each measure, constitute separate sheets organised in the following manner:

- a table containing the main parameters for a preliminary selection (code of the measure, short description, rating assessment);
- a general description of the measure;

G. Dall'O', *Green Energy Audit of Buildings*, Green Energy and Technology,
DOI: 10.1007/978-1-4471-5064-0_13, © Springer-Verlag London 2013

- tips and warnings to consider before selecting the measures.

Retrofit measures have been combined into the following sections:

- Building envelope (shell)
- HVAC Systems
- Plumbing and DHW Systems
- Electrical Systems
- Renewable energy sources
- Management improvement.

The retrofit measures described below are treated independently: the auditor should take into account that, when provided with multiple measures (scenarios) there may be interactions between them, with positive, but sometimes negative effects.

13.2 Building Envelope (Shell)

13.2.1 Pitched Roofing Insulation

13.2.1.1 Extrados Insulation with Under-Tile Insulation (1.TR.01)

The measure consists of the insulation from the outside of a pitched roof by means of the insertion of a thermal insulation layer under the tiles (Table 13.1).

General Description

Insulation of the roof from the outside can reduce heat loss by transmission of the building and improve the indoor thermal comfort. The application of this measure is convenient if the space below the pitched roof is normally used/occupied, otherwise it is more convenient to thermally insulate the floor slab which borders with the space below (see measures 1.TR.04 and 1.PR.04).

The measure requires remedial work from the outside and the removal of the roof tiles or roof covering, which are normally replaced. It is therefore a retrofit

Table 13.1 Extrados insulation, insulating under the tiles (1.TR.01): main parameters for a preliminary choice

Working	C	Obsolete	A	New	-	O&M	-
Saving potential			■■■■	Payback			■■■
Reliability			■■■	Feasibility			■■■
Environmental effects			EC				

Fig. 13.1 Example of extrados insulation, insulating under the tiles (courtesy of Rockwool Co.)

action which is justified if the building is subject to mayor renovation. When working from outside it is clearly necessary to consider the costs for the related scaffolding and those involved in ensuring the safety of workers. Where the roof has steep slope, the fixing of the elements, including the insulation layer, should be mechanical.

Wherever the roof could need to be capable of withstanding external loads (such as those caused by snow), a high density, preferably fire-resistant. thermal insulation should be chosen (Fig. 13.1).

Tips and Warnings

In this measure it should be considered that the cost of insulating material is low compared to the overall cost. Therefore the choice of a greater thickness of insulation does not increase excessively the cost, improving on the other hand the thermal performance.

Before programming the remedial work it is nonetheless necessary to verify whether the roof structure is really able to withstand the additional load: it is also necessary to make a hygrometer check to avoid the formation of condensate.

13.2.1.2 Extrados Insulation with Ventilated Roof (1.TR.02)

The measure consists of the replacement of the existing roof by an insulated and ventilated roof, installed on the existing roof structure (Table 13.2).

Table 13.2 Extrados insulation with ventilated roof (1.TR.02): main parameters for a preliminary choice

Working C	Obsolete A	New -	O&M -
Saving potential	■■■■	Payback	■■
Reliability	■■■	Feasibility	■■■
Environmental effects	EC		

General Description

For this measure the same comments as those made for the measure 1.TR.01 are valid, with the difference that the new roof is not only thermally insulated but is ventilated. In a ventilated roof, in fact, an intermediate hollow space between the thermal insulation and the outer layer of covering (tiles or metal panels) is provided: in this way a system of natural ventilation is created. There are several advantages of a ventilated roof:

- in winter the natural air flux will keep dry the insulating material, avoiding in this way the creation of condensation and ensuring the durability of the structural elements of the roof;
- in summer the cool air coming from the eaves is heated by the effect of the cavity, becoming lighter and exiting from the roof ridge, removing the accumulated heat in the roofing material. Furthermore the hollow space between the external elements of the roof and the underlying structure avoids the generation of a direct effect on the whole structure by the solar radiation (Fig. 13.2).

With this measure the indoor environmental conditions of the spaces below the roof will significantly improve, particularly in summer.

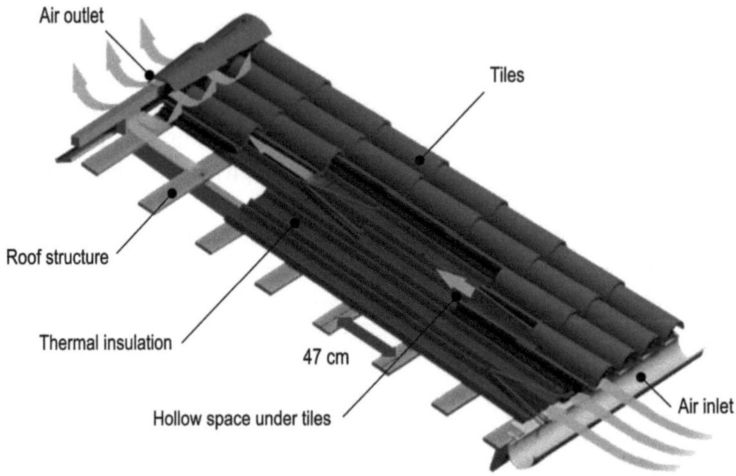

Fig. 13.2 Example of a ventilated roof system (courtesy of Onduline Co.)

Table 13.3 Insulation of the roof from the inside(1.TR.03): main parameters for a preliminary choice

Working B	Obsolete A	New -	O&M -
Saving potential	■■■■	Payback	■■■
Reliability	■■■	Feasibility	■■■
Environmental effects	EC		

Tips and Warnings

The same observations as those made for measure 1.TR.01 are applicable.

13.2.1.3 Insulation of the Roof from the Inside (1.TR.03)

The measure consists of the insulation of the roof from the inside through the application of insulating panels and coating panels (Table 13.3).

General Description

Insulation of the roof from the inside can reduce heat loss by transmission of the building and improve the indoor thermal comfort (Fig. 13.3).

The application of this measure is convenient if the space below the pitched roof is normally used/occupied, otherwise it is more convenient to thermally

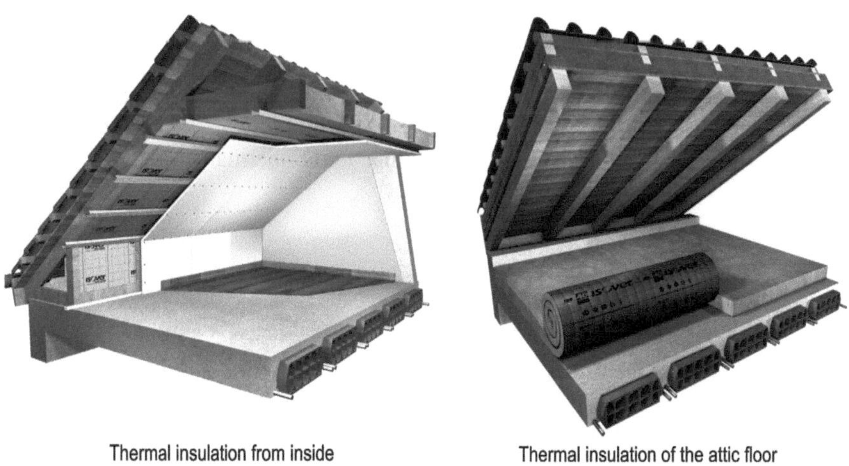

Fig. 13.3 Example of insulation of the roof from inside and of attic floor (courtesy of Isover Co.)

insulate the floor slab which borders with the space below (see measures 1.TR.04 and 1.PR.04).

The internal insulation of a pitched roof significantly increases the thermal performance of the roof with lower costs than those of isolating from the outside (1.TR.01). In this case, in fact, the installation is simpler and more economical and requires no scaffolding. Working from the inside selective remedial action can be programmed, even within individual areas or parts of the building.

The technique of coupling the insulating panels with the coating panels depends on the state of the supporting structure of the roof and therefore must be defined on a case by case basis.

Tips and Warnings

Before programming the remedial work it is necessary to verify that the roof structure is really able to withstand the additional load: is also necessary to make a hygrometer check to avoid the formation of condensate.

Also in this case the cost of the insulating material does not have a significant impact on the total cost which also depends on the chosen finish. Thus it is appropriate to use the highest thickness reasonably possible taking into account, however, that the height of the roof spaces should not be reduced excessively.

13.2.1.4 Insulation of the Attic Floor (1.TR.04)

The measure consists of the insulation of the last floor slab by means of the laying of an insulating layer on the floor of the attic (Table 13.4).

General Description

If the attic is not used at all, the most cost-effective solution is to simply place low density insulation in rolls (e.g., mineral wool or glass fibre), directly on the floor. Obviously the greater the thickness, the better is the thermal insulation, which is advisable.

Table 13.4 Insulation of the attic floor (1.TR.04): main parameters for a preliminary choice

Working	B	Obsolete	A	New	-	O&M	-
Saving potential			□■■■	Payback			□■■■
Reliability			■■■	Feasibility			■■■
Environmental effects			EC				

If the attic is used, then there is the need to walk above the floor or place objects on the floor and so the use of high-density insulation is advisable. In some cases a surface finish (flooring), suitable for walking on may be required.

Tips and Warnings

During the installation of the insulating materials on the floor, continuity of coverage must be ensured. In the case of multiple layers of material it is appropriate to effect an installation of criss-crossed layers.

If there are pipes of water circuits in the attic, the absence of leakage must be verified in order to avoid that the insulation is wetted by liquid leaks, nullifying the thermal performance but especially by increasing its weight.

If considerable layers of high-density insulation are used, and the slab is made of wood, then a structural static test should be performed.

13.2.2 *Flat Roofing Insulation*

13.2.2.1 Extrados Insulation with an Inverted Roof System (1.PR.01)

The measure involves the application of an inverted roof system made with the layer of the insulating material placed over the waterproofing layer (Table 13.5).

General Description

In the inverted roof system insulation is laid over the waterproofing layer and suitably weighted-down to restrain it against flotation and wind uplift and hence to protect it against damage.

Inverted roof constructions can be categorised as heavyweight or lightweight, with reference to the form of building construction involved. The basic roof structure may be of concrete, metal or timber: it must be strong enough to withstand the maximum predicted loads with a suitable safety factor included.

Table 13.5 Extrados insulation with the inverted roof system (1.PR.01): main parameters for a preliminary choice

Working	C	Obsolete	A	New	-	O&M	-
Saving potential			■■■■	Payback			□■■■
Reliability			■■■	Feasibility			■■■
Environmental effects			EC				

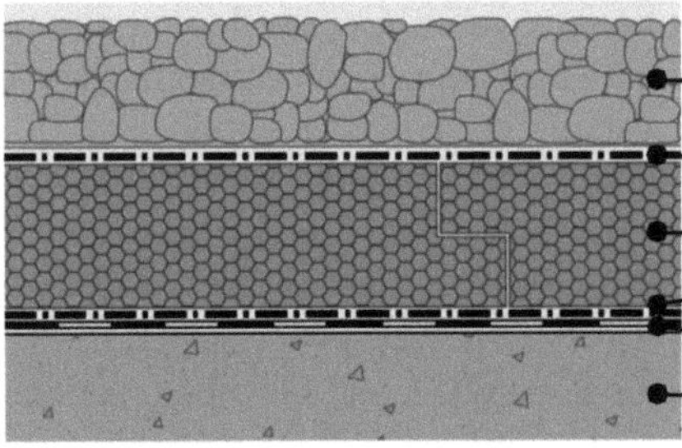

Fig. 13.4 Example of an inverted roof system (courtesy of Dow Corp.)

The application of this technique is convenient for summer operation, when the presence of water below the insulating layer favours the cooling of the underlying layers.

In winter, on the contrary, the presence of water leads to a reduction in the performance of the insulating material. In an inverted roof construction some rainwater will run off beneath the insulation boards and in doing so may draw heat from the deck. To compensate for this intermittent heat loss it is usual to increase the thickness of insulation by 20 % (rainwater cooling penalty).

The inverted roof construction can greatly reduce the risk of condensation in an existing building by keeping the roof structure and the waterproof layer at a temperature above the dewpoint (Fig. 13.4).

Tips and Warnings

In renovation projects the inverted roof concept can be used to upgrade thermal performance of the roof: if the existing waterproof layer is in good condition it can be retained but it may be desirable to overlay it with a new waterproof layer.

The inverted roof concept can be used with a wide range of waterproofing materials, including mastic asphalt and high performance, built-up bituminous felt (whereas bituminous roofing felt with a core of organic fibre is not suitable).

13.2.2.2 Extrados Insulation with the Warm Roof System (1.PR.02)

The measure involves the application of a inverted roof system made with the layer of the insulating material placed under the waterproofing layer (Table 13.6).

General Description

With this type of measure the roof insulation is placed above the roof deck but below the weatherproofing. There should be no insulation below the deck and no

Table 13.6 Extrados insulation with the warm roof system (1.PR.02): main parameters for a preliminary choice

Working C	Obsolete A	New -	O&M -
Saving potential	■■■■	Payback	□■■■
Reliability	■■■	Feasibility	■■■
Environmental effects	**EC**		

ventilation is required. This choice is particularly convenient as it permits insulation of flat roofs with a simple technique of installation.

The positive effects achievable by applying this retrofit measure are not only in winter (heating) but also in summer (cooling). The choice of a reflective external coating can help to significantly reduce the effects of solar radiation with many advantages:

- a reduction of the heat load in summer, thus the reduction of energy consumption for summer cooling;
- an improvement in the thermal comfort for the occupants;
- a contribution to the reduction of the "heat island" effect.

Tips and Warnings

With this solution the layer of insulating material constitutes a barrier to the passage of heat; the outer layer, then, is subject to the effect of solar radiation, to considerable mechanical stresses caused by the variation of temperature and by ultraviolet radiation that can modify the physical (and chemical) and hence the functional characteristics of the layer. The choice of materials must be accurate and installation must be carried out by experienced contractors.

13.2.2.3 Extrados Insulation with a Green Roof System (1.PR.03)

The measure consists of covering the surface of a roof, usually flat, with a layer of vegetation using the technique known as the "green roof system" (Table 13.7).

Table 13.7 Extrados insulation with a green roof system (1.PR.03): main parameters for a preliminary choice

Working C	Obsolete B	New -	O&M -
Saving potential	■■■■	Payback	■■
Reliability	□■■	Feasibility	■■■
Environmental effects	**ECI**		

General Description

A green roof is a roof partially or completely covered with vegetation and a growth medium, planted over a waterproofing membrane. It may also include additional layers such as a root barrier and drainage and irrigation systems. this technique can be applied both on pitched roofs and on flat roofs, however, the latter is decidedly more diffuse thanks to the considerable advantages that can be obtained.

The green roof can be applied, without distinction to any building regardless of the intended use: there are many examples of application of green roofs on residential constructions or office buildings, industrial warehouses, shopping malls, schools and recreational buildings.

There are two types of green roofs:

- intensive roofs, which are thicker and can support a wider variety of plants but are heavier and require more maintenance;
- extensive roofs, which are covered in a light layer of vegetation and are lighter than an intensive green roof.

Green roofs allows to obtain many advantages for a building, such as absorbing rainwater (water is stored by the substrate and then taken up by the plants from where it is returned to the atmosphere through transpiration and evaporation), providing thermal insulation (depending on the used material and the thickness of the layer), reducing summer cooling needs (by evaporative cooling), helping to lower urban air temperatures and mitigate the heat island effect (green roofs can contribute to obtain credits on LEED® protocol), creating a habitat for wildlife (Fig. 13.5).

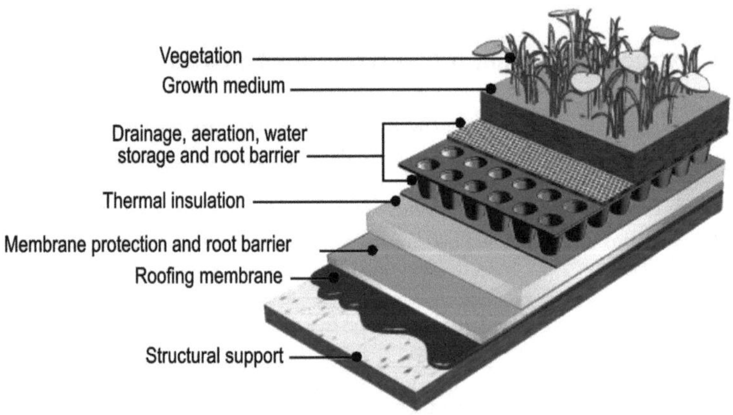

Fig. 13.5 Example of green roof system (courtesy of Heart Pledge Co.)

Tips and Warnings

Applications of green roofs in existing buildings as a retrofit measure, which usually occurs on flat slabs, require a static check of the building structure. Also important is the choice of the most suitable vegetation, which must take account of the climatic zone.

The installation of a green roof offers considerable potential: to the functional advantages one should add the aesthetic aspect, thanks to which it is possible to significantly revalue the architectural and environmental quality of the building. However, one must not forget that it is a complex technique which requires specialist skills both in the design phase and in that of realisation.

13.2.2.4 Intrados Insulation with Insulating Panels (1.PR.04)

The measure consists of the thermal insulation of the upper slab in intrados, by applying a false ceiling with insulating panels (Table 13.8).

General Description

This technique is used when it is difficult or impossible to operate from outside the structure. The installation involves placing a layer of thermally insulating material.

The installation can be performed in two ways:

- using self-supporting insulating panels fixed directly to the slab with coupling systems (for example screw anchors);
- installation of a support structure to which the self-supporting insulating panels are fixed (technique of the false ceiling), thus creating a hollow space.

The technique is simple and cost-effective and can be used in a selective manner, for example by taking the remedial action only in some areas of the building. It is particularly convenient in the case in which the spaces need renovation.

In some cases facilities equipment, e.g., lighting appliances, wiring, pipes and HVAC terminals, are fitted in the original ceiling. These systems must be removed and installed once again: the related additional costs must be considered in the

Table 13.8 Intrados insulation with insulating panels (1.PR.04): main parameters for a preliminary choice

Working	B	Obsolete	A	New	-	O&M	-
Saving potential			□■■■	Payback			■■■■
Reliability			■■■	Feasibility			■■■■
Environmental effects			**EC**				

definition final cost of the measure. There is also the cost for the external finish of the insulating panel, including the painting of the ceiling to be considered.

Tips and Warnings

The application of a layer of insulating material, which may include the thickness of the hollow space, reduces the height of the rooms or spaces below: it is therefore necessary to check whether this is compatible with the local health regulations. The application of this measure leads to an increase in the load that must be borne by the upper slab: a static verification is required, especially if the insole is made with lightweight structure. The cavity may be used for the passage of new plants (e.g., pipes and electrical cables).

If the HVAC system is being renovated, consider the opportunity to install, instead of a normal insulation panels, radiant ceiling panels (refer to 2.RE.04).

13.2.2.5 Height Reduction of Internal Spaces (1.PR.05)

The measure consists in the reduction of the height of the internal spaces by mean of the installation of a false ceiling with insulated panels (Table 13.9).

General Description

The measure consists in lowering the height of the internal spaces through the installation of a insulated countertop-ceiling, deriving two advantages:

- a reduction of the thermal losses by transmission;
- a reducing in thermal losses in ventilation thanks to the reduction of the useful volume (and of the air changes).

The measurement can be conveniently applied to areas characterised by considerable internal height (e.g., historical buildings, industrial buildings, warehouses etc.) for which a change in the intended use or, more generally, renovation

Table 13.9 Height reduction of internal spaces (1.PR.05): main parameters for a preliminary choice

Working	**B**	Obsolete	**A**	New	-	O&M	-
Saving potential			■	Payback			■■■
Reliability			■■■	Feasibility			■■■
Environmental effects			**EC**				

is envisaged. A lowering of the height may also be expected in some other parts of the building (e.g., in corridors, bathrooms or service rooms).

The installation can be performed through the creation of a support structure to which the self-supporting insulating panels are fixed (technique of the false ceiling), thus creating an hollow space that could be used for the passage of electrical systems and HVAC systems, which however must be anchored independently to the slab. If the HVAC system is being renovated, consider the opportunity to install, instead of a normal insulation panels, radiant ceiling panels (refer to 2.RE.04).

Tips and Warnings

For this measure the same remarks as those made for measure 1.PR.04 are valid (Fig. 13.6).

13.2.3 Thermal Insulation of Slab on Basement or Pilotis

13.2.3.1 Extrados Insulation with Insulating Plaster (1.BA.01)

The measure consists of thermal insulation in the extrados of the lower slab through the application of insulating plaster (Table 13.10).

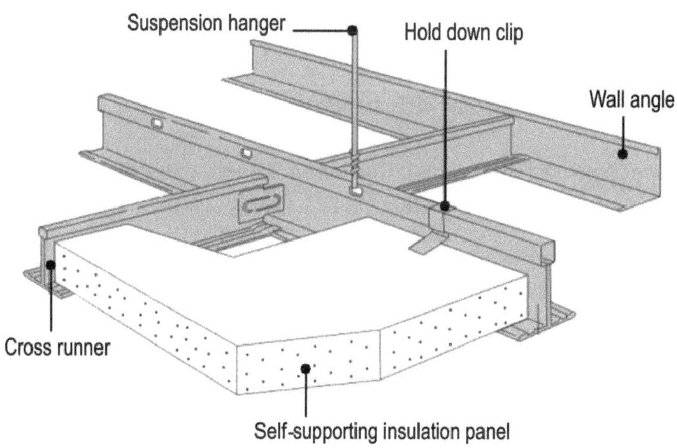

Fig. 13.6 Example of a suspended false ceiling (courtesy of Shakti ceiling Systems Co.)

Table 13.10 Extrados insulation with insulating plaster (1.BA.01): main parameters for a preliminary choice

Working B	Obsolete A	New -	O&M -
Saving potential	■	Payback	■■■
Reliability	■■■	Feasibility	■■■
Environmental effects	EC		

General Description

The thermal performance of the lower slab bordering an unheated space (for example, a cellar or a garage) or an external space (pilotis) can be improved by applying from the outside a layer of insulating plaster.

This technical solution may be chosen as an alternative to that provided by the measure 1.BA.02.: the cost is lower but the performance that can be achieved is more limited, owing to the much lower thermal resistance of the insulating plaster compared to that of true insulation.

Tips and Warnings

The added layer of plaster reduces the height of the space below: one must check whether this reduction is compatible with the building regulations.

The choice of the material must be compatible with the characteristics of the external space (for example wear resistance in the case of external spaces, fire resistance in the case of technical plant rooms subject to safety regulations, such as central boiler rooms).

13.2.3.2 Extrados Insulation with Insulating Panels (1.BA.02)

The measure consists of thermal insulation in the extrados of the lower slab through the application of insulating panels (Table 13.11).

Table 13.11 Extrados insulation with insulating panels (1.BA.02): main parameters for a preliminary choice

Working B	Obsolete A	New -	O&M -
Saving potential	□■■■	Payback	■■■
Reliability	■■■	Feasibility	■■■
Environmental effects	EC		

General Description

The thermal performance of the lower slab bordering an unheated space (for example, a cellar or a garage) or an external space (pilotis) can be improved by applying from the outside thermal insulation panels supported by a structure.

The measure is very similar to measure 1.PR.05:

The installation can be performed through the creation of a support structure to which the self-supporting insulating panels are fixed (technique of the false ceiling), thus creating an hollow space that could be used for the passage of electrical systems and HVAC systems, which however must be anchored independently to the slab.

Tips and Warnings

The added thickness of this measure reduces the height of the space below: one must check whether this reduction is compatible with the building regulations.

The choice of the material must be compatible with the characteristics of the external space (for example wear resistance in the case of external spaces, fire resistance in the case of technical plant rooms subject to safety regulations, such as central boiler rooms).

13.2.3.3 Floor Thermal Insulation (1.BA.03)

The measure consists of the thermal insulation of the floor of the space by means of the application of a layer of insulating material (Table 13.12).

General Description

The thermal insulation of a slab floor (floor) placed in contact with the ground or bordering with a space that cannot be inspected, for example ventilated crawl space, can be applied only from the inside as defined by this measure.

The layer of insulating material can be applied directly on the existing floor. Then above this layer a new flooring, consisting of a screed and a surface finish

Table 13.12 Floor thermal insulation (1.BA.03): main parameters for a preliminary choice

Working	D	Obsolete	C	New	-	O&M	-
Saving potential			□■■■	Payback			□□■
Reliability			■■■	Feasibility			■■■
Environmental effects			EC				

(tiles, carpet, parquet etc.), must be installed. With this measure the effective floor slab undergoes an increase in thickness which may partially be used for the passage of new plant components (e.g., pipes and electrical cables).

Since the operation is performed from the inside, the height of the room is reduced. It is possible to lessen the increase in slab thickness by removing the existing floor but this alternative does increases the final cost of the retrofit measure and produces waste material that must be taken to waste disposal.

Tips and Warnings

For this measure an insulating material with high density should be chosen, which is also resistant to the static load that must be calculated as a function of the upper layers and the intended use of the space.

This measure, as indicated above, reduces the height of the space: one must check whether this reduction is compatible with the building regulations.

The creation of a new floor could be an opportunity to assess the installation of a floor heating system.

13.2.4 External Walls

13.2.4.1 External Insulation with ETICS (1.EW.01)

The measure consists of the thermal insulation of the external walls of the building envelope from outside using the ETICS insulation technique (Table 13.13).

General Description

The External Thermal Insulation Composite System (ETICS) is an optimal solution for the energy retrofit of the external walls of existing buildings (this technology is also widespread as solution for new buildings).

ETICS is a system applied from the outside of the wall, usually including an adhesive, a levelling mortar, an insulation panel, an alkali-resistant reinforcement

Table 13.13 External insulation with ETICS (1.EW.01): main parameters for a preliminary choice

Working	C	Obsolete	A	New	-	O&M	-
Saving potential			■■■■	Payback			■■■
Reliability			■■■■	Feasibility			■■
Environmental effects			**EC**				

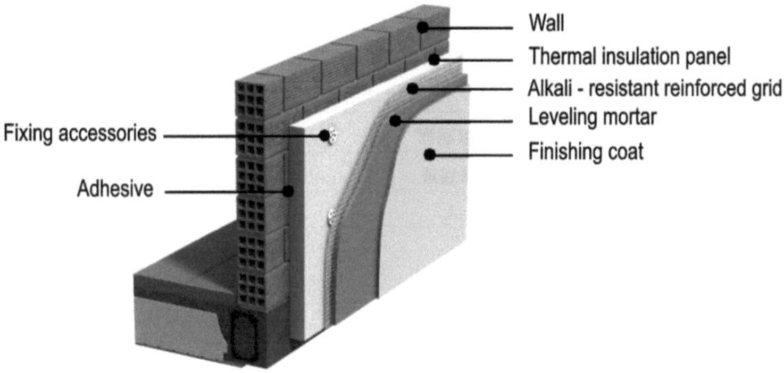

Fig. 13.7 Example of an external thermal insulation composite system (ETICS) (courtesy of Knauf Co.)

grid, a primer and a finishing coat, as well as sealants and accessory materials for the installation.

The main advantage of this measure derives from the continuity of the insulation, which eliminates thermal bridges and the risk of surface and interstitial condensation. The measure increases the thermal inertia and the sound insulation of walls, thus improving both the thermal and the acoustic comfort for users. The operation is carried out relatively quickly and completely from the outside, with very limited works discomfort for users.

Tips and Warnings

ETICS being a multicomponent system, the compatibility between the components is a key factor in obtaining the performance of the overall system as well as its durability: only correct application guarantees optimum results (Fig. 13.7).

The fundamental rules should therefore have already been taken into consideration at the planning stage. These are important prerequisites for a perfect job. The European Association for ETICS has drawn up a Guideline of Quality in cooperation with the responsible national bodies (www.ea-etics.eu).

The cost of this measure depends greatly on the complexity of the façade: this measure encounters difficulties if applied to historical buildings or buildings with a lot of decoration.

13.2.4.2 Thermal Insulation with Ventilated Façade (1.EW.02)

The measure consists of the thermal insulation of a wall applying from the outside of a ventilated façade (Table 13.14).

Table 13.14 Thermal insulation with ventilated façade (1.EW.02): main parameters for a preliminary choice

Working	C	Obsolete	A	New	-	O&M	-
Saving potential			■■■■	Payback			■■
Reliability			■■■■	Feasibility			■■
Environmental effects			EC				

General Description

The system called *ventilated façade* or *ventilated wall* consists, from interior to exterior, of an insulating layer anchored to the supporting structure (e.g., a wall) and by a layer of cladding tied to the building through means of an appropriate fixing system. Between the thermal insulating layer and the cladding an air space is thus created which, due to the "chimney effect", creates effective natural ventilation, with considerable benefits for the entire system.

The advantages of a ventilated façade are present both in summer and winter. A ventilated façade can reduce the amount of heat that buildings absorb in hot weather thanks to partial reflection of solar radiation by the covering and the ventilated air gap and to the application of insulating material, thus achieving considerable reduction in the running costs of the HVAC systems. In winter, these systems guarantee a thermal insulation of the façade, and a consequent reduction in the heating needs (Fig. 13.8).

Compared with cladding applied directly to the wall structure, the advantages of a ventilated façade are:

- energy-saving, thanks to the insulating layer;
- elimination of thermal bridges;
- elimination of surface condensation, thanks to the presence of the air gap;
- reduced risk of cracking and detachment;

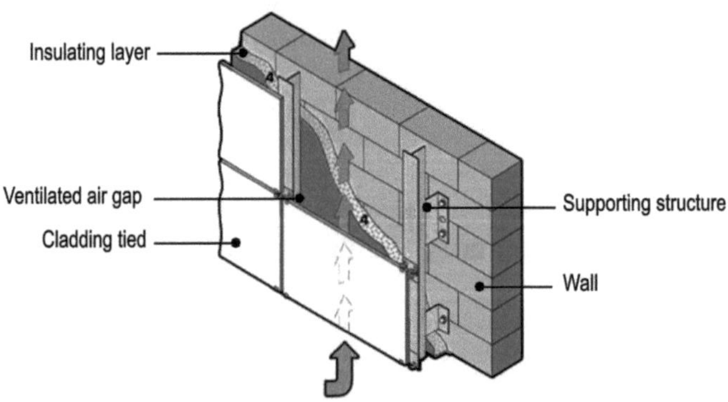

Fig. 13.8 Example of a ventilated façade (courtesy of Grantec Co.)

- easy installation; elements are assembled "dry" on site by means of mechanical anchoring devices;
- maintenance: work may be done separately on each individual slab;
- protection of the wall structure against atmospheric agents;
- acoustic protection of the building.

The chimney effect sets up efficient natural ventilation, aiding heat and moisture removal and ensuring a high level of comfort for users. In addition, ventilated façades tend to increase the reflection of external noise since the particular construction, consisting of those layers of facing, air gap and insulating material, ensures a certain level of acoustic absorption.

Tips and Warnings

Ventilated façades, compared with ETICS (I.EW.01), are more suitable when it is necessary to greatly reduce the summer thermal loads. This retrofit measure can also become an opportunity to significantly enhance the architectural appearance of the building. The complexity of this system requires a careful design, not only in terms of the energy efficiency but also for the static requirements, and specialised, experienced firms for installation. A maintenance programme should be scheduled. One must verify on the basis of building type compliance with the applicable fire safety regulations.

13.2.4.3 External Insulation with Insulating Plaster (1.EW.03)

This measure consists of thermal insulation in the extrados of outside walls by means of the application of insulating plaster (Table 13.15).

General Description

The thermal performance of the external walls can be improved by applying from the outside a layer of insulating plaster.

Table 13.15 External insulation with insulating plaster (1.EW.03): main parameters for a preliminary choice

Working	D	Obsolete	B	New	-	O&M	-
Saving potential			■■	Payback			□■■
Reliability			■■■	Feasibility			■■■
Environmental effects			**EC**				

This technical solution may be chosen as an alternative to that provided by the measures 1.EW.01 or 1.EW.02: its cost is lower but the performance that can be achieved is more limited, owing to the much lower thermal resistance of the insulating plaster compared to that of true insulation.

Tips and Warnings

The measure can be considered if the budget is limited and maintenance action is scheduled. The existing plaster must be removed and the costing should include for scaffolding. Since the energy performances are much lower than those of a ETICS, before choosing this measure one should check on the technical–economic feasibility of applying ETICS.

13.2.4.4 Wall Cavities Insulation (1.EW.04)

The measure consists in filling with insulating material the air gaps between walls (Table 13.16).

General Description

This technique is applicable in the case in which the external walls have been built with layers that create an air gaps. In order to increase the thermal resistance of the wall, the cavities can be filled with insulating materials in bulk. This measure is normally effected by application from inside, with the aim of maintaining the integrity of the external façade.

To insulate the cavity walls, the contractors drills small holes of around 22 mm diameter at intervals of around 1-1.5 m in the inside wall. With specially designed equipment, they then blow insulation into the cavity. Once all the insulation is in place, the contractor fills in the holes.

Cavity wall insulation can be composed of various different types of material: mineral wool, glass wool, beads or granules and insulating foam. The thickness of the cavity should be 5 cm or more.

Table 13.16 Wall cavities insulation (1.EW.04): main parameters for a preliminary choice

Working	B	Obsolete	A	New	-	O&M	-
Saving potential			■■	Payback			■■■
Reliability			■■■	Feasibility			□■■
Environmental effects			**EC**				

Tips and Warnings

In existing cavity walls, often the air gap is not sealed but is subject to infiltration by outside air. For this reason the thermal insulation effect of the air gap is only theoretical and the filling of the cavities with insulating materials is therefore cost-effective.

Before cavity wall insulation is installed, a survey needs to be carried out to ensure that the wall is suitable for such insulation.

After application, an infrared audit is advisable in order to check whether the insulating material is uniformly distributed in the wall cavities. This measure does not solve the problems deriving from thermal bridges.

13.2.4.5 Internal Insulation with Counter-Wall (1.EW.05)

The measure consists of the thermal insulation of external walls by applying an insulated counter-wall from inside (Table 13.17).

General Description

Where neither thermal insulation of the walls from outside nor cavity insulation are possible, the thermal resistance of the wall can be increased through internal insulation using a counter-wall.

Figure 13.9 shows three different solutions. The simplest (left) consists in applying the insulation layer directly onto the wall. Using a support structure (centre) to maintain the insulating layer detached from the wall, so creating an air gap that may be used for plant components (pipes, electrical wiring etc.). The third solution (right) consists of constructing a brick-built counter-wall.

Tips and Warnings

With this measure the layer of insulating material is positioned inside, therefore the thermal inertia of the wall is reduced: its implementation goes better in spaces used in a discontinuous manner.

Table 13.17 Internal insulation with counter-wall (1.EW.05): main parameters for a preliminary choice

Working	B	Obsolete	A	New	-	O&M	-
Saving potential			□■■	Payback			■■■
Reliability			■■■■	Feasibility			■■■
Environmental effects			EC				

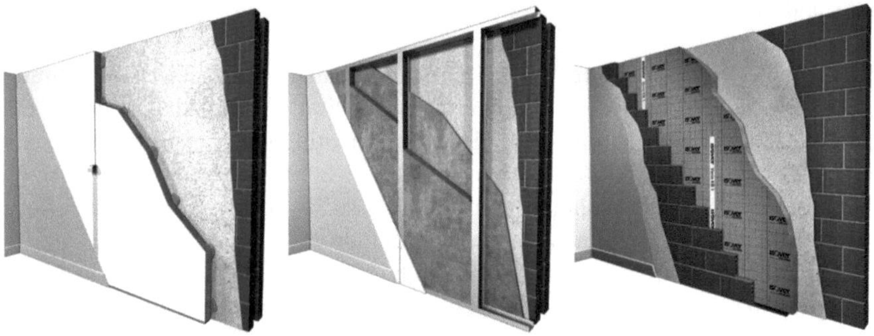

Fig. 13.9 Examples of internal insulation with counter-wall (courtesy of Isover Co.)

Since the application of the system from the inside reduces the useful space, it is advantageous to use materials with very low thermal conductivity values (e.g., micro-porous silica) in order to reduce thickness necessary.

13.2.4.6 Internal Insulation with Insulating Plaster (1.EW.06)

The measure consists of the thermal insulation of walls by applying an insulating plaster from inside (Table 13.18).

General Description

The thermal insulation of a wall of the building envelope can be increased by applying (from indoors) insulating plaster directly on to inner surfaces.

This technical solution may be chosen as an alternative to that provided by the measures 1.EW.05: the cost is lower but the performance that can be achieved is more limited, owing to the much lower thermal resistance of the insulating plaster compared to that of true insulation.

Table 13.18 Internal insulation with insulating plaster (1.EW.06): main parameters for a preliminary choice

Working **B**	Obsolete **A**	New -	O&M -
Saving potential	■■	Payback	□■■
Reliability	■■■	Feasibility	■■■■
Environmental effects	**EC**		

Tips and Warnings

The measure can be considered if the budget is limited and maintenance action is scheduled. The existing plaster must be removed and the costs of this operation should be taken into account. Since the energy performances are much lower than those of the internal insulation with counter-wall (1.EW.05), before choosing this measure verify the technical–economic feasibility of a counter-wall application.

13.2.4.7 Insulation of Subwindows (1.EW.07)

The measure consists of the thermal isolation of the (embedded) subwindows of the outside walls by applying an insulating panel (Table 13.19).

General Description

In existing buildings, in some cases the thickness of the outer wall below the embedded windows (subwindows) is lower than that of the remaining external wall. The purpose of such a reduced localised thickness was to permit the installation of the radiator limiting its overall dimensions. However, this solution is not energetically efficient since the reduction of the wall thickness decreases its thermal resistance and a part of the heat emitted from the radiator is dissipated to the outside.

In order to drastically reduce these thermal losses, it is possible to apply an insulating panel from inside of the subwindow. This measure involves the following steps:

- removal of the radiator (if it exists);
- application of an insulating panel of appropriate thickness using an adhesive material;
- application of an aluminium film having the function of vapour barrier;
- application of a plasterboard panel;
- painting or tile covering of the surface;
- re-installation of the radiator.

Table 13.19 Insulation of subwindows (1.EW.07): main parameters for a preliminary choice

Working	C	Obsolete	B	New	-	O&M	-
Saving potential			■	Payback			■■■■
Reliability			□■■	Feasibility			□□■■
Environmental effects		**EC**					

If the space between the radiator and the existing wall does not permit insertion of this insulating system, it is necessary to replace the existing fixtures with new ones that make it possible to retrieve additional space. In this case there will also be a need to modify and update the hydraulic connections and piping.

Tips and Warnings

This retrofit measure is useful when it is not possible to apply other solutions from outside (see measures 1.EW.01, 1.EW.02). The quality of the measure can be verified with a thermographic analysis from outside of the wall, whilst keeping the heating system activated. A cheaper, but less effective and less aesthetically satisfactory measure, consists of placing behind the radiator, glued on the wall, a reflective insulated panel.

13.2.4.8 Insulation of the Roller Shutter Boxes (1.EW.08)

The measure consists on the thermal insulation of the roller blind containment-boxes by applying insulating panels (Table 13.20).

General Description

The measure consists in applying insulation panels within the roller blind containment-boxes which, in the winter season, represent a critical point of energy dispersion and wastage.

The thickness of the insulating panel must be defined as a function of the available space, considering the maximum dimensions of the roller blind containment-box (up position). The thermal insulation of the box also includes sealing of any cracks in order to limit any unwanted air infiltrations.

Tips and Warnings

This measure is relatively simple and can also be implemented directly by the user.

Table 13.20 Insulation of the roller shutter boxes (1.EW.08): main parameters for a preliminary choice

Working	**C**	Obsolete	**B**	New	-	O&M	-
Saving potential			□■■	Payback			■■
Reliability			■■■■	Feasibility			□■■
Environmental effects			**EC**				

13.2.4.9 Installation of a Green Façade (1.EW.09)

The measure consists of the installation of a green façade (Table 13.21).

General Description

Among the green technologies for the built environment green façades, or green walls, are interesting solutions that can be used for energy and/or environmental improvement retrofit of existing buildings.

The greening systems for vertical walls can be divided into two categories: green façades and living walls (vertical gardens). Green façades are made of green climbing plants that grow directly on the façade or in front of the façade if supported by structures such as grids and cables. Living walls systems (LWS) are constituted by pre-vegetated panels or integrated systems fixed to a structural wall or to a frame, with hydroponics for the growth of vegetation.

The installation of a green wall provides several advantages.

in summer there is a reduction of the solar load on the wall surface protected estimated around 30–40 % shading generated by the vegetation reduces the maximum surface temperature by about 10 °C on days with higher radiation. The phase shift of the peak of heat is around 8 h.

In winter, the presence of the vegetation on the façade reduces convective phenomena, increasing the overall thermal resistance of the wall.

One of the additional benefits of the vegetation regards the capture and retention of fine dust particles on the surface of the leaves. Particles smaller than 10 μm, present in particular in the urban areas, may cause damage to health because they can be inhaled into the human respiratory system.

Green façades may have a positive impact on both physical and mental health and well-being. Green views and access to green spaces in cities help to relieve the everyday pressures of crowds and noise.

Tips and Warnings

The installation of a green façade requires a careful design of the system, a thorough understanding of the climatic conditions and the choice of the most

Table 13.21 Installation of a green façade (1.EW.09): main parameters for a preliminary choice

Working	-	Obsolete	-	New	C	O&M	-
Saving potential			■■	Payback			■
Reliability			■■	Feasibility			■
Environmental effects			ECI				

suitable type of vegetation which must be compatible with the climate and orientation. The necessary maintenance of the system, including the costs of irrigation must be considered in the economic evaluation.

13.2.5 Transparent Envelope

13.2.5.1 Replacement of the Windows (1.FE.01)

The measure consists of the replacement of the old windows with new, high energy-performance windows (Table 13.22).

General Description

In the energy balance of a building, windows are critical elements. In fact within this single component many needs must be satisfied: thermal insulation, solar control, environmental comfort, visual comfort, acoustic comfort and ventilation.

Where energy performance or structural performance cannot be guaranteed, one should verify the state of the windows and, possibly, plan a global replacement.

The energy performance of a new window must be assessed by considering the average U value, being a weighted average between the U value of the frame and the U value of the glass. In a correct choice of the energy performance of the new window, the performance of the frame and the performance of the glass should be balanced: a high performance glass pane would be useless if combined with, on the other hand, a low performance frame.

So far as the energy performance of the glass is concerned, by increasing the number of panes (and the number of the air interspaces) the U value is reduced, thus increasing the thermal performance: on the other hand transparency to solar radiation decreases. Since in the energy balance of a window not only the reduction of heat transmission but also the solar gains must be considered, the solution to increase the number of the glass panes (for example from double glazing to triple glazing) is not convenient for all locations and for orientations.

In order to reduce the U value of the glazing, two strategies can be adopted: a surface coating treatment of the glass panes (e.g., low-emissivity, selective

Table 13.22 Replacement of the windows (1.FE.01): main parameters for a preliminary choice

Working	C	Obsolete	A	New	-	O&M	-
Saving potential			□■■■	Payback			■■
Reliability			□■■■	Feasibility			■■■
Environmental effects			EC				

coatings etc.) and the filling of the interspaces between the panes with gases with low thermal conductibility (e.g., Argon, Xenon, Krypton) instead of air.

A window must guarantee the energy performances not only in winter but also in summer. Whereas in the winter the effects of solar radiation are positive (solar gains), in the summer solar radiation is the main source responsible for the thermal loads. On designing the new windows, the control of solar radiation must be considered in order to reduce the cooling demand of HVAC systems and increase the thermal comfort of the users (see measures 1.SP.01, 1.SP.02, 1.SP.03, 1.SP.04).

The complexity of the replacement of windows requires a careful evaluation of the performance requirements that must take into account the environmental conditions and the conditions of usage.

Tips and Warnings

Replacement of windows in a building is a measure which is not always cost-effective if the assessment only considers energy savings: in fact the payback period can be more than 20 years. However this measure offers a large number of benefits: improvement of the thermal comfort, elimination air infiltration owing to poor air-tightness between the frame and the shutters, improvement of acoustic comfort, and of course aesthetic improvement of the façade and the indoor spaces.

A new window also gives a higher commercial value to the building and contributes to achieving better energy performance of the building as a whole. Within a global renovation of the building envelope this measure should be included in the planning.

The installation of a new window is a critical phase. The new frame, in fact, must be fixed properly and the interstices between the frame and the wall must be eliminated by means of proper sealing. If the old window frame was equipped with a roller blind, the state of the roller blind containment-box (see measure 1.EW.08) must be verified.

In historic buildings, replacing windows usually requires a careful study to check whether there are architectural constraints.

The proper installation of a new window ensures air-tightness avoiding unwanted infiltration. The air-tightness of the new windows should not affect the ventilation of internal spaces: users should be advised to periodically open the windows in order to guarantee a manual air change. Some models of window have a release mechanism which allows the air-tightness to ensure natural ventilation for the selected period. New models of windows equipped with mechanical ventilation system with heat recovery integrated in the frame are available on the market The installation of a mechanical ventilation system guarantees a proper air change (see measures 2.VE.01).

In the evaluation of the total costs of this measure the cost of removing the old window, those for transporting the old window to the disposal, masonry and painting must also be included.

Table 13.23 Installation of storm windows (1.FE.02): main parameters for a preliminary choice

Working	B	Obsolete	A	New	-	O&M	-
Saving potential			■■■	Payback			■■■
Reliability			□■■	Feasibility			□■■
Environmental effects			EC				

13.2.5.2 Installation of Storm Windows (1.FE.02)

The measure consists on the installation of storm windows in order to increase the overall energy performance of the existing windows (Table 13.23).

General Description

In some cases it is either not possible or not cost-effective to replace the old windows with new ones: a practical solution for increasing the energy performance consists in the upgrading of the existing windows by installing storm windows. Storm windows are windows which are mounted outside or inside of the main glass windows of a building.

Storm windows can be mounted externally or internally; can be made of glass, rigid plastic panels, or flexible plastic sheets; and may be permanently or temporarily mounted. They function similarly to insulated glazing.

Wood, aluminium, and vinyl are the most common storm window frame materials. There are advantages and disadvantages to all types of frame materials. Although very strong, light, and almost maintenance free, aluminium frames conduct heat very well.

Although storm windows add little to the insulating performance of single-glazed windows (that are in good condition), field studies have found that they can help reduce air movement into and out of existing windows. Therefore, they help reduce heating and cooling costs.[1] The installation of storm windows, in addition to improving the energy performance contributes to improving the sound quality of the existing windows by reducing the noise coming from the outside.

Tips and Warnings

Interior storm windows offer greater convenience than exterior storm windows. They are easier both to install and remove and require less maintenance, not being exposed to the elements. Since they seal tightly to the primary window, they are more effective in reducing air infiltration.

[1] source: http://energy.gov/energysaver/articles/storm-windows.

13.2.5.3 Replacement of Glass Panes on the Existing Frame (1.FE.03)

The measure consists of the replacement of glass panes on the existing frame of the windows in order to increase the overall energy performance (Table 13.24).

General Description

If the existing glass of a window is a single pane, it may be convenient to evaluate the possibility of replacing it with a high efficiency double glazing (e.g., low emission coating), so maintaining the original frame. This solution is cheaper than the global replacement of the existing windows (see measure 1.FE.01), also if the overall energy performances (U value and tightness) that can be achieved are lower.

The modification of the frame consists in creating greater depth of emplacement for the new glass (normally double glazing) and the adaptation of the fixing system for the new glass (glass beading).

The techniques of installation depend on the type of frame: wooden frames can be modified more easily than metal (steel or aluminium) frames.

If the original glass is a double pane, this could be substituted with a high efficiency double pane glass (e.g., low emission). In this case the thickness of the new glass should not be different or slightly greater from the existing one and there should be no mechanical problems.

Tips and Warnings

The replacement of the existing glazing requires an assessment of the conservation status of the existing frame of the window that will carry a greater load (two plates instead of one). Important is also the verification of the hinges (which may possibly have to be replaced) and the closure fittings.

13.2.5.4 Application of Low-Emissivity Films (1.FE.04)

The measure consists in the application of a low-emissivity film over existing glazing (Table 13.25).

Table 13.24 Replacement of glass panes on the existing frame (1.FE.03): main parameters for a preliminary choice

Working	D	Obsolete	B	New	-	O&M	-
Saving potential			□□■■	Payback			■■■
Reliability			■■■	Feasibility			□■■
Environmental effects			EC				

Table 13.25 Application of low-emissivity films (1.FE.04): main parameters for a preliminary choice

Working	B	Obsolete	-	New	-	O&M	-
Saving potential			■	Payback			□■■■
Reliability			■	Feasibility			■■
Environmental effects		EC					

General Description

Low-emissivity films, or low-e films, are plastic films that contain metal or metal oxide. The metal coating reflects the infrared radiation towards the inside by reducing heat loss and maintaining thermal comfort conditions.

Depending upon which side of the windows these films are placed, manufacturers claim they reflect between 70 and 80 % of solar heat gain in the summer, or conserve over 50 % of interior heat in the winter.

Before purchasing and installing low-e window film, one must consider the local climate and the orientation of the windows. The labels of the packaging of the window film should be examined to identify both the solar heat gain coefficient (SHGC) and the visible transmittance rating (VT). For cold climate a high SHGC rating is advisable, while lower ratings are better for warm climate. Products with higher VT ratings admit more natural light.

The application of the film is convenient if the existing glass is double glazing, the window frame is in good condition and the window maintains a good air tightness.

Tips and Warnings

Also if end users can buy low-emissivity films and install by himself, it is recommended to delegate their application to trained personnel.

13.2.5.5 Sealing of Air Infiltration (1.FE.05)

The measure consists on the sealing of the junction between the window opening and the frame in order to reduce the of unwanted air infiltration (Table 13.26).

General Description

The infiltration of unwanted air into the building envelope often occur at the joints between the window opening and the frame and between the glass (single) and the frame.

Table 13.26 Sealing of air infiltration (1.FE.05): main parameters for a preliminary choice

Working B	Obsolete A	New -	O&M -
Saving potential	□■■	Payback	□■■■
Reliability	■■■■	Feasibility	■■■
Environmental effects	EC		

The sealing of these junctions can reduce this infiltration and reducing energy consumption and improving the summer and winter comfort.

The sealing material may be silicone, polyurethane, acrylic or butyl. The choice of material to use depends on several factors: the type of joint to be sealed, the application substrate (steel, plastic, wood etc.), the most common however, is that silicone-based.

Tips and Warnings

The cost-effectiveness of the intervention increases in windy areas, higher buildings and where the pressure differences are high.

An alternative to this measure, more expensive but are able to ensure the best energy performance, is represented by the installation of a storm window (1.FE.02).

13.2.5.6 Insertion of Gaskets to Improve Air-Tightness (1.FE.06)

The measure consists of the air-tight sealing of the junction between the fixed and movable frames of windows and doors in order to improve air-tightness (Table 13.27).

General Description

Windows and doors in existing buildings are not always able to maintain high performance air-tightness over time. Poor air-tightness is in most cases the main cause of unwanted air infiltration and energy wastage.

Table 13.27 Insertion of gaskets to improve air-tightness (1.FE.06): main parameters for a preliminary choice

Working B	Obsolete B	New -	O&M -
Saving potential	□■■	Payback	□■■■
Reliability	■■■	Feasibility	■■■
Environmental effects	EC		

The reasons for this may be of different origin: in wooden frames, for example, the original shape could alter over time and that may affect their closure. In the case of frames equipped with gaskets, the material of these could gradually deteriorate and become ineffective.

This measure provides for a check on the closure of windows and doors and the correction of the air-tightness through the first-time installation of new gaskets or the replacement of the old ones.

The gaskets are preformed, flexible elements which, by changing their volume and their shape, are able to seal the cracks that can occur between the different parts of the frames. The gaskets can be made of elastomeric material, such as neoprene or silicone, with thermoplastic rubbers or with plastomers.

Elastomeric seals are characterised by excellent resistance to aging, to ultra-violet radiation, to temperature changes and by having good mechanical performance.

Tips and Warnings

The cost-effectiveness of this remedial action increases in windy areas, higher buildings and where the pressure differences are high.

An alternative to this measure, more expensive but are able to ensure the best energy performance, is represented by the installation of a storm window (1.FE.02).

13.2.5.7 Transparent Enclosure of Verandas (1.FE.07)

The measure consists on the a transparent enclosure of verandas (Table 13.28).

General Description

The enclosure of a veranda with a transparent partition permits the creation of a buffer zone which limits heat losses from the building. The improvement of energy performance is also guaranteed by the greenhouse effect created by the transparent

Table 13.28 Transparent enclosure of verandas (1.FE.07): main parameters for a preliminary choice

Working	-	Obsolete	-	New	C	O&M	-
Saving potential			□□■	Payback			■
Reliability			□■■	Feasibility			■
Environmental effects			EC				

partition through which the visible component of solar radiation can pass but which has limited losses in the infrared frequency band.

This measure also allows the recovery of a useful closed space that can be put to good use by building occupiers particularly during the winter season.

The evaluation of the increase in energy performance is quite difficult, since the energy balance is complex, depending upon several factors (geometry of the closed space, orientation of the wall, characteristics of the glass pane, characteristics of the frame of the transparent partition, climatic characteristics of the location etc.).

A precise evaluation of the energy balance of this measure requires a simulation programme. In most cases, however, the reasons that encourage users to enclose a veranda do not derive from energy saving requirements but on the need to create extra living space.

Tips and Warnings

Before planning this measure, which changes the external appearance of the building, one must ensure that this is permitted by the applicable building codes and hygiene regulations. Sometimes it is necessary to submit the project design to the local public planning authorities and wait for approval in order to proceed.

The transparent enclosure of veranda can lead to benefits in the winter season, reducing the consumption of energy for heating. During the summer, however, the transparent closure may generate an increase in the thermal load owing to its greenhouse effect. The best solution is to design an enclosure of which all the transparent parts can be opened or removed during the summer season.

Another aspect to consider is that the enclosure of a space limits the natural ventilation since the opening of the windows facing on to the buffer space provides no regular air change. In this case the solution is to install a mechanical ventilation system.

13.2.5.8 Installation of a Sunspace (1.FE.08)

The measure consists on the installation of a sunspace attached to the building (Table 13.29).

Table 13.29 Installation of a solar greenhouse (1.FE.08): main parameters for a preliminary choice

Working	-	Obsolete	-	New	B	O&M	-
Saving potential			□□□■	Payback			■■
Reliability			□■■	Feasibility			■
Environmental effects		EC					

General Description

The sunspace is the most common component in bioclimatic architecture, however the use of this technology can also be considered for existing buildings when major renovation works are planned.

A sunspace is a solar greenhouse attached to the building: it consists of a direct gain room normally placed on the equatorial side of the building, separated from the main living or working area by a wall with openings that can be closed, such as doorways or windows.

Its operating principle make use of the greenhouse effect created by the transparent enclosure through which the visible component of solar radiation can pass but which has limited losses in the infrared frequency band. The thermal mass of the building components placed behind the transparent closure (e.g., floor slabs, walls) contribute to the operation of the sunspace, providing the necessary thermal inertia.

During the day, warm air is transferred from the sunspace to the rest of the building, primarily by convection of warm air through the connecting doors and windows, which are left open during the day and closed at night The temperature in the sunspace will drop somewhat below room temperature at night but will remain well-above outside temperatures.

The sunspace contributes to reducing the heating requirements and costs. The evaluation of the increase in energy performance is quite difficult, since the energy balance is complex, depending upon several factors (geometry of the closed space, orientation of the wall, characteristics of the glass pane, characteristics of the frame of the transparent partition, climatic characteristics of the location etc.).

A precise evaluation of the energy balance of this measure requires a simulation programme.

The main advantage of the **sunspace, however,** is its provision of an additional area within the building which can be used in many ways (greenhouse, passage-way, air-lock entry etc.) (Fig. 13.10).

Fig. 13.10 Example of sunspace integrated with the building structure

Tips and Warnings

Before planning this measure, which changes the external appearance of the building, one must ensure that this is permitted by the applicable building codes and hygiene regulations. Sometimes it is necessary to submit the design to the local public planning authorities and wait for approval in order to proceed.

Before planning this measure, which changes the external appearance of the building, one must ensure that this is permitted by the applicable building codes and hygiene regulations. Sometimes it is necessary to submit the project design to the local public planning authorities and wait for approval in order to proceed.

The technical warnings are similar to those of measure 1.FE.07.

When planning a sunspace one must be careful about summer operation during which the greenhouse effect can greatly increase the indoor air temperature making the space uncomfortable: openings for natural ventilation, shading systems and, possibly, means of partial removal of the transparent panels must be provided.13.2.6

13.2.6 Sun Protection Systems

13.2.6.1 Installation of External Solar Shading Systems (1.SP.01)

The measure consists of the installation of external solar shading systems in order to control the effects of solar radiation (Table 13.30).

General Description

An important requirement of the sustainable design process is to ensure that the building is shaded from the sun for as many hours as possible throughout the whole day, during the course of the summer season.

The aim of external solar shading systems is to control the effects of solar radiation on the buildings and the internal spaces, thus contributing to maintaining optimal thermal and comfort conditions. The main effects of the external solar shading systems are summarised below:

Table 13.30 Installation of external solar shading systems (1.SP.01): main parameters for a preliminary choice

Working	C	Obsolete	A	New	A	O&M	-
Saving potential			□□■■	Payback			■■
Reliability			□■■	Feasibility			□■■
Environmental effects			EC				

- control of the incoming solar radiation in summer and reduction of the thermal load for space cooling;
- control of the incoming solar radiation in winter and contribution to managing the solar gains in order to reduce the heating demand;
- control of natural lighting while avoiding glare effects.

For a predominately equator facing façade, effective solar shading can be achieved using a fixed horizontal solar shading system. During the day in both summer and spring/autumn, a fixed horizontal system projecting out from the window can be designed to shade the building. The difference in solar altitude between winter and summer gives rise to different effects of these types of shading devices.

With a predominantly East or West facing façade, a fixed system will not perform well throughout the whole day since the altitude of the sun is much lower. For these orientations, horizontal (or vertical) motorised shading systems should be preferred. The design of a system of external shielding is complex since it must take account of several factors, namely:

- the orientation of the façade;
- the shadows of other buildings;
- the technical and architectural features of the façade;
- any architectural constraints (Fig. 13.11).

Issues that must be addressed on designing these systems do not only relate to energy aspects (an energy simulation is however recommended) but also include structural aspects (how to fix the systems to the façade).

Tips and Warnings

The external solar shading systems guarantee high energy performance because the solar radiation absorbed by the opaque components generates heat that is

Fig. 13.11 Example of solar shading systems

dissipated outside. To ensure the functionality of these systems over time one must provide for a maintenance plan.

The shading systems can integrate solar PV cells into the shading elements: in this way the systems have a double function: protect the façade from the sun and generate renewable energy for the building.

13.2.6.2 Application of Solar Control Films (1.SP.02)

The measure consists of the application of a solar control film over existing glazing (Table 13.31).

General Description

The application of a solar control film can reduce the effects of solar radiation incident on the transparent building envelope, improving thermal and visual comfort and reducing the effects of glare. The practical effects of this measure, that is cost-effective if the conditions of the windows are good, are related to the reduction of the summer loads and the energy costs of summer cooling.

Tips and Warnings

Some solar control films are designed for outdoor installation (in this case it may be necessary to provide scaffolding) while others can be installed from inside. Installation is simple and takes little time, however to get good results, and durability, experts applicators are strongly recommended. Maintenance (mainly cleaning) must be made using procedures and products compatible with the characteristics of the films applied.

13.2.6.3 Application of Internal Solar Shading Systems (1.SP.03)

The measure consists of the installation of internal solar shading systems in order to control the effects of solar radiation (Table 13.32).

Table 13.31 Application of solar control films (1.SP.02): main parameters for a preliminary choice

Working	B	Obsolete	-	New	-	O&M	-
Saving potential			■■■	Payback			■■■
Reliability			□■■	Feasibility			■■
Environmental effects			EC				

Table 13.32 Application of internal solar shading systems (1.SP.03): main parameters for a preliminary choice

Working C	Obsolete B	New B	O&M -
Saving potential	□■■	Payback	■■■
Reliability	■■■	Feasibility	■■■
Environmental effects	EC		

General Description

The internal solar shading systems are typically applied in cases where it is neither possible, nor economically convenient, to operate from the outside (see measure 1.SP.01). Their function is similar to that of the external solar shading systems, although their energy and lighting performance are lower.

The technologies used are *fixed* and *mobile textile blinds*, *Venetian blinds* and *slat blinds*. *Venetian blinds* are basic *slatted blinds* made of metal or plastic which are suspended by strips of cloth called tapes, or by cords, by which all slats can be rotated in unison through nearly 180°. The slats can be rotated such that they overlap with one side facing inwards and then in the opposite direction such that they overlap with the other side facing outwards. *Slat blinds* consist of many horizontal slats, usually of metal or vinyl, connected with string in a way that they can be rotated to allow light to pass between the slats, rotated up to about 170° to hide the light, or pulled up so that the entire window is clear. These systems can have manual or automated operation.

Tips and Warnings

This shading system is cost-effective, independent from the windows type and easy to install. Their installation does not require highly skilled personnel.

13.2.6.4 Shading with Vegetation Outside (1.SP.04)

The measure consists of the installation of a shading system with vegetation outside, in order to control the effects of solar radiation (Table 13.33).

Table 13.33 Shading with vegetation outside (1.SP.04): main parameters for a preliminary choice

Working -	Obsolete -	New B	O&M -
Saving potential	■	Payback	■
Reliability	■■	Feasibility	■■
Environmental effects	ECI		

General Description

The use of vegetation for sun protection of façades or windows is a natural solution that can contribute to an increase in the sustainability of the building. The objective is to provide an independent green structure that could be installed adjacent to the walls to be protected or positioned at a certain distance from it.

The possibility of using vegetation with deciduous leaves allows the use of the energy benefits of solar radiation in winter (solar gains) but at the same time protects the walls or the windows in summer.

These systems use the same technology as green façades: for further details see the measure 1.EW.09.

Tips and Warnings

The installation of a vegetation system for shading purpose requires a careful design of the system, a thorough understanding of the climatic conditions and the choice of the most suitable type of vegetation which must be compatible with the climate and orientation. In the economic evaluation one must take an account of the necessary maintenance of the system, including the costs of irrigation.

13.2.6.5 Painting of External Walls with a Low Absorption Finish (1.SP.05)

The measure consists of painting the external walls with a finishing coat/finish with low absorption coefficients in order to reduce the negative effects of solar radiation in summer (Table 13.34).

General Description

The painting of the exterior walls of the buildings with finishes which are characterised by low values of solar absorption coefficient α (and high values of

Table 13.34 Painting of external walls with a finishing coat of low absorption (1.SP.05): main parameters for a preliminary choice

Working	C	Obsolete	A	New	-	O&M	-
Saving potential			■	Payback			□□□■
Reliability			□■■	Feasibility			■■■■
Environmental effects			ECI				

low temperature emissivity ε) can help to reduce the negative effects of solar radiation in summer, thus reducing the summer thermal loads.

The surface treatment of walls, which in some cases can also be applied to flat roofs, contributes to reduce the "heat island" effect. This measure is particularly suitable in those locations with a warm climate and high levels of solar radiation in summer.

Tips and Warnings

Normally light colours better reflect solar radiation, however, one should not influenced by normal perception of light and dark. Since the human eye can see only the "visible" component of the radiation, the evaluation of the material most appropriate must be made on the basis of the solar absorption coefficients provided by manufacturers and obtained through experimental tests (different materials or pigments could have the same colour but different solar absorption coefficient α).

13.2.7 Illumination Using Daylight

13.2.7.1 Painting of Internal Walls with Light Colours (1.DL.01)

The measure consists of painting the internal walls with light colours in order to increase visual comfort reducing energy demand for lighting (Table 13.35).

General Description

The colouring of the inner walls of a room have a great influence on the visual comfort of the users. In fact the surface finish influences the uniformity of the distribution of natural light.

This measure is cost-effective, easy to apply and has significant advantages. Visual comfort is improved not only in the periods in which the lighting is natural but also when artificial lighting is switched on.

Table 13.35 Painting of internal walls with light colours (1.DL.01): main parameters for a preliminary choice

Working	C	Obsolete	A	New	-	O&M	-
Saving potential			■	Payback			□■■■
Reliability			■■■	Feasibility			■■■■
Environmental effects			**EC**				

The energy savings are derived from a reduced use of artificial lighting. The benefits from the improvement of visual comfort, although difficult to quantify, are significant.

Tips and Warnings

This measure contribute to diffusing natural light inside the room, for this reason is compatible with the other measures using daylight for illumination (see 1.DL.02, 1.DL.03).

13.2.7.2 Installation of Light Pipes (1.DL.02)

The measure consists of installing light pipes in order to light naturally poorly lit areas of the building (Table 13.36).

General Description

Inside a building it may happen that not all spaces have natural direct lighting. Using devices called *light pipes* which capture, transport and diffuse daylight it is possible to illuminate, in a natural way, the interior spaces of the building that cannot be served by side lighting through ordinary windows.

The efficiency of fixed light pipes is affected by the absorption that occurs when light is reflected from the walls of the pipe. Unless the sun is lined up with the axis of the pipe, the light is reflected repeatedly as it travels through the pipe. Even if the surfaces of the pipe have high reflectance, say 90 %, a large fraction of light entering is lost with a few reflections.

In some models, the intercepting element remains fixed whilst in others it moves, driven by a motor and by a control system, so as to modify its position according to the position of the sun, thereby minimising reflection losses. A light pipe with this feature is called *sun tracker*) (Fig. 13.12).

Tips and Warnings

The light loss is proportional to the length-to-width ratio of the pipe. Therefore efficiency is sacrificed if the pipe is long in relation to its width. The installation of

Table 13.36 Installation of light pipes (1.DL.02): main parameters for a preliminary choice

Working	C	Obsolete	A	New	-	O&M	-
Saving potential			■	Payback			□■■■
Reliability			■■■	Feasibility			■■■■
Environmental effects			EC				

Fig. 13.12 Example of a light pipe for residential application (courtesy of Solatube Co.)

light tubes necessarily requires works to be effected on the roof with the cutting of a hole of diameter equal to that of the tube: the sealing of the tube is a critical element.

The design, the configuration and implementation of the system are quite complex operations, the choice of a specialised supplier is therefore recommended.

In order to maintain high system performance, the collecting element must be cleaned regularly.

13.2.7.3 Installation of Light Shelves (1.DL.03)

The measure consists of installing light shelves in order to increase the diffuse lighting using natural light (Table 13.37).

General Description

This is a device designed to diffuse natural light in the internal spaces, avoiding phenomena of glare and thermal overheating. It consist of a white or reflective metallic light shelf outside the window on the equator-facing side of a structure.

Usually the window will be protected from direct summer season sun by a projecting eave. The light shelf projects beyond the shadow created by the eave

Table 13.37 Installation of light shelves (1.DL.03): main parameters for a preliminary choice

Working -	Obsolete A	New A	O&M -
Saving potential	■	Payback	□■■■
Reliability	■■■	Feasibility	■■■■
Environmental effects	EC		

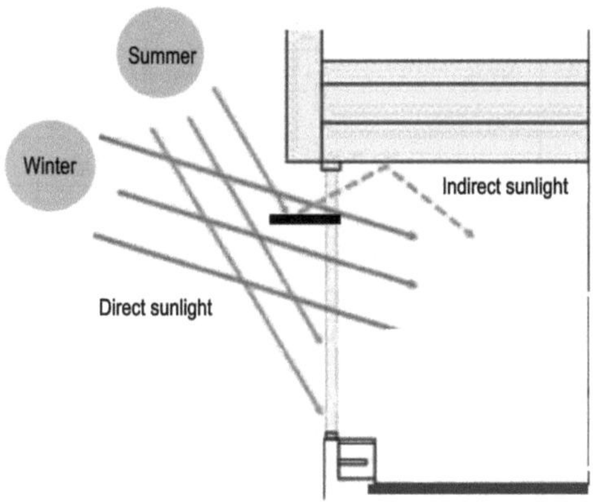

Fig. 13.13 Example of an application of light shelves

and reflects sunlight upward to illuminate the ceiling (that should be painted with white colour). This reflected light can contain little heat content and the reflective illumination from the ceiling will typically reduce deep shadows, reducing the need for general illumination.

Light shelves avoid the glare problem that limits the use of diffusers to exploit the use of daylight through windows.

Tips and Warnings

The inclusion of life shelves in existing buildings is possible where existing windows are sufficiently high and may be modified by positioning the opening part at the bottom, independently of the upper part which serves for the input of the direct and indirect light. For historic buildings, before this measure is planned verify that there are no architectural constraints (Fig. 13.13).

A critical aspect of light shelves is that only the portion of windows above head height is usable for illumination using daylight. Light shelves require periodic cleaning in order to maintain their reflecting characteristics.

13.3 HVAC Systems

13.3.1 Heat Generation

13.3.1.1 Replacement of Boilers (2.HG.01)

The measure consists on the replacement of the existing boiler with a new high performance boiler in order to increase the energy performance of the plant (Table 13.38).

Table 13.38 Replacement of boilers (2.HG.01): main parameters for a preliminary choice

Working	C	Obsolete	A	New	-	O&M	-
Saving potential			□■■	Payback			□■■■
Reliability			■■■	Feasibility			■■■■
Environmental effects			**EC**				

General Description

The replacement of an existing boiler with a new one of higher performance (e.g., condensing boiler) helps to increase the heat generation efficiency of the system, hence reducing operative costs for winter heating and DHW.

Boiler replacement can also be convenient if the existing boiler is operative but with an age above 15 years (this indeed is considered the average period of useful life of a boiler).

Amongst the new technologies, that which offers the best energy performance is the technology of condensation. In condensing boilers a high efficiency (typically up to 108 % if referred to lower calorific value) is achieved by using the waste heat in the flue gases to preheat the cold water entering the boiler. The water vapour produced during combustion is condensed into water, which leaves the system via a drain. In many countries their use is compulsory or encouraged with financial incentives.

Condensing boilers, in order to ensure the best performance need to operate with a fluid temperature lower than that the fluid temperature of conventional heating systems (maximum 40–45 °C instead of 70–80 °C). Given that traditional heating systems operate at high temperatures, theoretically speaking it does not seem convenient to replace a conventional boiler with a condensing boiler. However this reasoning is not entirely correct.

Boiler output temperature can be controlled by an external temperature sensor, and the lower is the heating load, the lower can be the temperature of water supplied to the heating terminals (Fig. 13.14).

By modifying the water circulation system, through the replacement of the circulation pump, it is possible to increase the temperature difference between supply and return of the hydronic circuit of the heating system: in this way the return water temperature can be lower, reaching values compatible with optimum operation of the condensing boiler.

Condensing boilers can therefore be perfectly compatible also with high temperature heating systems, using radiators.

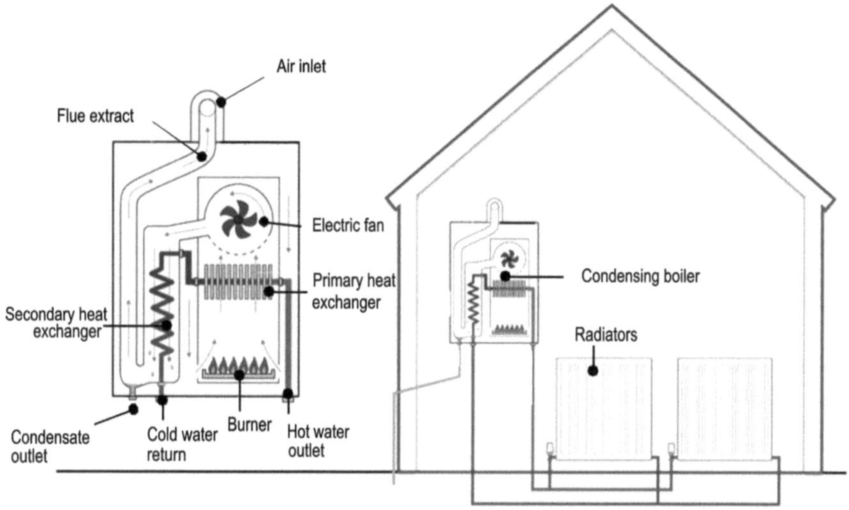

Fig. 13.14 Scheme of a condensing boiler and of a heating system (source: http://www.evoenergy.co.uk)

Tips and Warnings

The replacement of an existing old boiler with a high performance boiler, for example a condensing boiler, is one of the most cost-effective retrofit measures. In order to obtain the best overall energy performance it is advisable to adhere to some rules.

It is advisable to perform a global calculation of the energy needs of the building, so that the new boiler is chosen with a corrected heating capacity. Many existing buildings, indeed, have been realised before laws or rules on energy efficiency were issued. Since the size of the boiler was not required, the boilers were normally oversized with a heat capacity in most cases double if compared with the required capacity.

A check on the hydronic circuit of the heating system is useful for controlling the quality of the water, avoiding that impurities can be deposited or can generate corrosion in the new boiler; it is also appropriate to verify the pumping system, eventually planning the replacement of the pumps (see measure 2.HD.02).

The control and regulating system of the heating plant should be checked over (see measures 2.RE.01, 2.RE.02, 2.RE.03).

The condensate expelled from a condensing boiler is acidic, with a pH between 4 and 5. Condensing boilers require a drainpipe for the condensate produced during operation.

An existing boiler could be replaced with another type of heat generator, for example a compression cycle heat pump (see measure 2.HG.03) or gas fired or an absorption cycle heat pump (see measure 2.HG.04).

Table 13.39 Installing of independent DHW heaters (2.HG.02): main parameters for a preliminary choice

Working C	Obsolete A	New A	O&M -
Saving potential	□■■	Payback	□■■
Reliability	■■■	Feasibility	■■
Environmental effects	**EC**		

13.3.1.2 Installing of Independent DHW Heaters (2.HG.02)

The measure consists of the installation of an independent heater for domestic hot water (DHW) production (Table 13.39).

General Description

In plants with DHW using combined boilers (space heating and DHW) in the summer season, when the heat demand for heating does not exist, the heating capacity of the heat generator can be excessive when compared with the reduced needs and the efficiency of the generator could be low. This measure consists in the installation of an independent DHW boiler, chose and sized on the basis of the DHW demand.

A condensing boiler should be preferred in order to increase the energy performance. Since the DHW system needs to be modified, the auditor should consider also other possibilities such as the installation of a heat pump for DHW (see measure 2.HS.03), or the installation of a solar thermal system (see measures 4.RS.01 or 4.RS.02). Verify if it is possible to deactivate the recirculation circuit (see measure 2.HS.02).

Tips and Warnings

The best performance of a DHW system can be obtained starting from a comprehensive audit. In fact the installation of an independent boiler should be considered an opportunity to check other aspects, implementing all the measures aimed at reducing the water consumption (see measures 2.WR.01, 2.WR.02).

13.3.1.3 Installation of a Compression Cycle Heat Pump (2.HG.03)

The measure consists of the installation of a compression cycle heat pump as replacement of an existing boiler (Table 13.40).

Table 13.40 Installation of a compression cycle heat pump (2.HG.03): main parameters for a preliminary choice

Working	C	Obsolete	A	New	-	O&M	-
Saving potential			□■■■	Payback			□■■
Reliability			□■■	Feasibility			□■■
Environmental effects			EC				

General Description

A heat pump is a device used to transfer thermal energy from a heat source (e.g., outdoor air, water of a river or a lake, or ground) at lower temperature to a heat sink at higher temperature. In building applications the heat generated by a heat pump is normally used for the requirements of space heating and DHW production: a heat pump can then replace a classical boiler.

Heat pumps are able to move thermal energy in a direction opposite to that of spontaneous heat flow: this can be done using a thermodynamic cycle (e.g., compression cycle or absorption cycle) consuming energy (electricity, fuel, high temperature heat depending upon the heat pump type).

In a compression cycle heat pump the following components can be distinguished:

- the *evaporator*, which absorbs the heat from the cold source evaporating the refrigerant fluid;
- the *compressor*, which compresses the refrigerant gas by increasing the temperature and pressure of the refrigerant;
- the *condenser* which condenses the refrigerant, providing heat to the hot sink;
- the *lamination valve* which lowers the pressure and the temperature of the refrigerant by closing the cycle.

In electrically powered heat pumps, the heat transferred can be three or four times larger than the electrical power consumed, giving the system a Coefficient of Performance (COP) of 3 or 4, as opposed to a COP of 1 of a conventional electrical resistance heater, in which all heat is produced from input electrical energy. For this reason, heat pumps are very convenient both energetically and economically speaking (Fig. 13.15).

Reversible heat pumps are designed to work in either thermal direction, in order to provide heating or cooling to the indoor spaces: in this way a single device (reversible heat pump) can meet the demands of heating or cooling of a HVAC system.

Air-source heat pumps are the most widespread because of their simplicity of installation and lower cost. However in order to increase the COP, and to reduce energy consumption, other types of heat pumps are becoming common in the HVAC market: geothermal heat pumps and ground-water heat pumps.

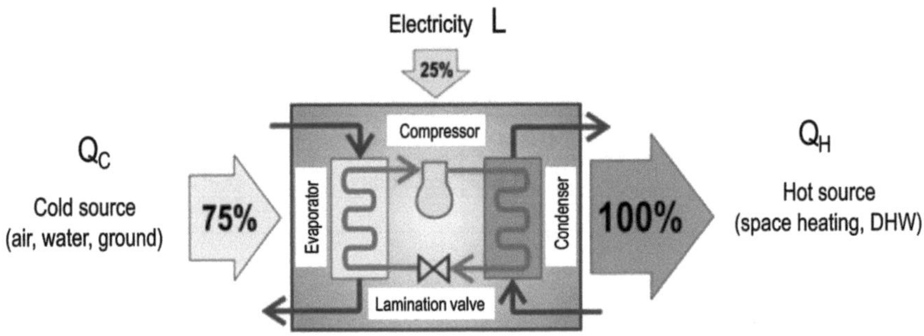

Fig. 13.15 A simple stylised diagram of a heat pump's vapour compression refrigeration cycle

Geothermal heat pumps, best known as *Ground Source Heat Pumps* (GSHP), exploit the ground as the cold source. In order to extract heat from the ground, two types of heat exchangers are used: horizontal heat exchangers, constituted by slinky loops of pipes (generally polyethylene is used) placed in the ground at a depth of 1–2 m, and vertical heat exchangers i.e., vertical closed loop fields composed of pipes that run vertically. For the latter a hole is bored in the ground, typically at 25–150 m depth, then pairs of pipes in the hole are joined with a U-shaped cross connector at the bottom of the hole. The borehole is commonly filled with a bentonite grout surrounding the pipe to provide a thermal connection to the surrounding soil or rock in order to improve the heat transfer.

In groundwater heat pumps, the secondary loop pumps natural water from a well or body of water into a heat exchanger inside the heat pump. Heat is either extracted or added by the primary refrigerant loop, and the water is returned to a separate injection well, irrigation trench, tile field or body of water.

Tips and Warnings

Heat pumps have a much higher cost than that of condensing boilers. Before programming their installation it is important to make an analytical calculation of the heat output required to avoid any unnecessary oversizing.

The efficiency of a heat pump is affected by the temperature of the two sources. It is therefore convenient to lower the operating temperature of the heating system in order to increase the COP value.

For warm climates it is convenient to install an air-source heat pump: a reversible heat pump could be used both for heating and for cooling. For continental climates, a geothermal heat pump (GSHP) or a groundwater heat pump is the better solution. As the cost is higher, it is not cost-effective to size the heat capacity for the maximum required: 60 % of the heat capacity can cover about 90 % of the energy requirements for the whole winter season. A convenient technical solution is to install in parallel to the heat pump a condensing boiler sized for 100 % of the heat capacity (the additional cost is relatively limited) which can then be used to cover the peaks of heat demand on the coldest days and as a back-up.

The installation of an electric heat pump is interesting because it is possible to use electricity supplied by a solar PV system (see measure 4.RS.04).

13.3.1.4 Installing a Gas Fired Heat Pump (2.HG.04)

The measure consists of the installation of a gas fired heat pump to replace an existing boiler (Table 13.41).

General Description

As an alternative to an electric heat pump it is possible to install a gas-driven heat pump. Using gas (normally methane) one can use heat pumps which operate according to a thermodynamic absorption cycle or alternatively heat pumps with a compression cycle in which the compressor is driven by a combustion engine.

The efficiency of a gas-fired heat pump is measured by the gas utilisation efficiency (GUE) that is the ratio between the energy supplied (heat transferred to the medium to be heated) and the primary energy consumed. The GUE depends upon the characteristics of the heat pump and the operating conditions. An average value is between 1.4 and 1.5, therefore for a unit of energy contained in the fuel it is possible to obtain 1.4–1.5 units of useful thermal energy.

Tips and Warnings

The efficiency of a gas-fired heat pump is higher than the efficiency of a condensing boiler, but on the other hand the cost is much higher. Hence it is not cost-effective to size the heat capacity for the maximum required: 60 % of the heat capacity can cover about 90 % of the energy requirements for the whole winter season. A convenient technical solution is to install in parallel to the heat pump a condensing boiler sized for 100 % of the heat capacity (the additional cost is relatively limited) which can then be used to cover the peaks of heat demand on the coldest days and as a back-up.

Table 13.41 Installation of a gas fired heat pump (2.HG.04): main parameters for a preliminary choice

Working	C	Obsolete	A	New	-	O&M	-
Saving potential			□■■■	Payback			□■■
Reliability			□■■	Feasibility			■■■
Environmental effects			EC				

It is not possible to compare the performance of an electric compression cycle heat pump, defined by the COP, with the performance of a gas-fired heat pump defined by GUE. Whereas the GUE compares the thermal energy supplied by the heat pump with the primary energy of the fuel, the COP compares the thermal energy supplied by the heat pump with the electrical power consumption. A correct comparison must refer in both cases to the primary energy: if for example a heat pump has a COP equal to 4 and the electrical efficiency of the network is equal to 2.3, the efficiency of the heat pump reported to primary energy equates to $4/2.3 = 1.74$: this value is comparable with the GUE.

13.3.1.5 Installation of a CHP system (2.HG.05)

The measure consists of the installation of a cogeneration (combined heat and power) system for combined production of thermal energy and electricity (Table 13.42).

General Description

The cogeneration systems (CHP is the acronym of combined heat and power) allow to self-produce electricity using different fuels supplied to endothermic engines or turbines and, at the same time, to recover thermal energy from the transformation process. In the conventional production of electric energy, on the basis of plant type, only 35–55 % of the primary energy supplied to the system is converted into electrical energy, while the rest is dispersed into the environment with serious economical and environmental consequences. Through cogeneration, instead, only 10 % of primary energy is wasted; on average the 37 % is converted into electrical energy and the 53 % recovered and converted into thermal energy as hot water, superheated water or steam.

The continuous evolution of engine technologies as well as control systems, coupled with the costs and conditions of fuel supply made favourable by energy conservation laws, strengthens the economic benefits derived from the adoption of a CHP systems. Further cost benefits derive from the possibility of transferring to the electricity supply network the excess electricity produced, receiving a payment.

Table 13.42 Installation of a CHP system (2.HG.05): main parameters for the choice

Working	C	Obsolete	B	New	-	O&M	-
Saving potential			□■■■	Payback			□■■
Reliability			□■■	Feasibility			■■
Environmental effects			**EC**				

The coupling of an engine, a generator and the heat recovery system create the cogeneration group. In the thermodynamic conversion of primary energy into electrical energy, the thermal energy developed is a by-product of transformation as the main product is the mechanical energy. The thermal energy developed by engine, working to produce mechanical energy, is recovered and transferred to a fluid to be used for heating, industrial processes, domestic hot water etc. There are several applications of cogeneration:

- industrial plants;
- hospitals;
- shopping centres;
- hotels;
- buildings and sites in general where the total use of thermal energy made available by the unit is guaranteed.

A further development of cogeneration system is trigeneration, which, in addition to producing electricity, uses the waste heat to produce chilled water for air-conditioning or industrial processes.

The transformation of thermal energy into cooling energy is made possible by the use of the absorption refrigeration cycle whose operation is based on the phase change of a cooling fluid (e.g., water) in combination with the substance (e.g., lithium bromide) used as absorbent.

A properly designed cogeneration system can be used to supply electricity to the site even in case of public network failure. This prevents outages and overcomes any blackouts.

Tips and Warnings

A cogeneration plant represents an interesting opportunity for a fuller use of the primary energy contained in the fuel which, with the sole production of electricity, would be wasted. This interesting solution, on the other hand, constitutes a considerable financial commitment that must be carefully evaluated on the basis of a feasibility study which takes account of all aspects, both technical and economic. The proper sizing of the CHP/cogeneration plant will be of fundamental importance, so that it operates for the greatest possible number of hours, producing electricity when rates are higher and, at the same time, utilising the totality of thermal energy that is made available from the heat recovery circuits. Amongst the advantages of a cogeneration plant are to be noted:

- high efficiency even at partial load.;
- wide range of fuels can be used: liquid (diesel fuel, vegetable oil) or gaseous (natural gas, LPG, blast furnace gas, biogas, sewage gas, landfill gas, coal gas);
- construction modularity which facilitates maintenance, reducing the risks of a complete out of service situation, while ensuring flexibility of installation and of exercise;

- broad size range, electric power starts from less than 15 kWe up to 10 MWe.
- Amongst the disadvantages:
- lack of capacity to meet thermal loads requiring heat at high temperature (over 140 °C);
- significant emissions of NO_X that must be reduced with special devices in accordance with the regulations in force;
- increased incidence of maintenance costs (particularly for the lubrication oil).

13.3.2 Cold Generation

13.3.2.1 Replacement of Compression Cycle Chillers (2.CG.01)

This measure provides for the replacement of the existing compression cycle chiller with a new, more efficient and more environmentally sustainable chiller (Table 13.43).

General Description

The measure provides for a full check-up of existing chillers and the replacement of those using any refrigerant fluid containing chlorofluorocarbons (CFCs) and chlorofluorohydrocarbons (HCFCs), substances responsible for the impoverishment of the atmospheric ozone layer.

An existing chiller may be replaced by another compression-cycle chiller, but using refrigerants that do not have any impact on the atmosphere (i.e., impact on the ozone layer or impact on global warming), or by a chiller operating on an absorption cycle (e.g., lithium bromide or ammonia).

A new chiller will have also a better energy efficiency, leading to a reduction in electricity consumption. According to the LEED protocols, credits can be achieved in two ways: either by not using, or by choosing refrigerants and/or HVAC systems that minimise or eliminate the emission of compounds that contribute to global warming.

Table 13.43 Replacement of compression cycle chillers (2.CG.01): main parameters for the choice

Working	C	Obsolete	A	New	-	O&M	-
Saving potential			□■	Payback			□■■
Reliability			■■■	Feasibility			■■■■
Environmental effects			ECI				

Tips and Warnings

Before choosing the model of chiller to be replaced, the auditor should analyse carefully the existing HVAC system, assessing the compatibility of the new chiller with the actual operating conditions.

It is advisable to perform analytical calculations of the cooling demand of the building in order to define the cooling capacity of the new chiller according to actual needs.

The heat transferred from the condenser of the new chiller could be used for thermal applications compatible with the values of the operating temperatures (for example DHW heating or pre-heating).

The cooling capacity of the new chiller can be significantly reduced if a cold thermal storage system is installed: this choice, when associated with a suitable application and proper sizing, permits the reduction of the peak cooling power and the use of the chiller in the hours in which the cost of electricity is lower.

13.3.2.2 Installation of Absorption Cycle Chillers (2.CG.02)

This measure provides for the replacement of compression chillers with absorption chillers fed with gas fuels or with thermal fluid at high temperature (Table 13.44).

General Description

As an alternative to measure 2.CG.01, this measure covers the replacement of existing inefficient and environmentally impacting compression chillers with absorption chillers fed with gas fuels or with high temperature thermal vectors.

The absorption cycle uses a heat-driven concentration difference to move refrigerant vapours (usually water) from the evaporator to the condenser. The high concentration side of the cycle absorbs refrigerant vapours (which, of course, dilutes that material). Heat is then used to drive off these refrigerant vapours thereby increasing the concentration once again.

Lithium bromide is the most common absorbent used in commercial cooling equipment, with water used as the refrigerant. The market offers single and

Table 13.44 Installation of absorption cycle chiller (2.CG.02): main parameters for the choice

Working	C	Obsolete	B	New	-	O&M	-
Saving potential			□□■	Payback			□□■■
Reliability			■■	Feasibility			■■■
Environmental effects			EC				

double-effect absorption chillers. Smaller absorption chillers sometimes use water as the absorbent and ammonia as the refrigerant.

The absorption chiller, depending on the model, can be fed with hot water at temperatures above 75–80 °C, high pressure hot water, steam, or directly fired by a flame of natural gas, LPG, or diesel oil.

Tips and Warnings

Absorption chillers are particularly convenient when hot fluids are available for recovery from industrial processes. Their use combined with CHP systems is very interesting: the heat recovered from the cooling of the engine, in the summer season is used to produce, by mean of the absorption chiller, chilled water for summer cooling. This type of plant configuration is called *trigeneration* or CCHP (combined cooling, heat and power).

The absorption chiller can be supplied with hot water produced by solar thermal systems: this is known as *solar cooling* (see measure 4.RS.03).

Absorption chillers have a much higher cost than electrical compression chillers and their sizing must be made with special care in order to meet the real needs and the constraints of the plant. Sometimes it is appropriate not to size the chiller for all the cooling capacity required but rather only a part of it, by installing in parallel a compression chiller that can be used to cover the cooling peaks or as back-up/reserve capacity.

It is not possible to compare the performance of an electric compression cycle chiller, defined by the EER (Energy Efficiency Ratio), with the performance of a gas-fired chiller defined by GUE. Whereas the GUE compares the net cooling capacity with the primary energy of the fuel, the EER compares the net cooling capacity with the electrical power consumption. A correct comparison must refer in both cases the primary energy.

13.3.3 Hydraulic Distribution

13.3.3.1 Thermal Insulation of the Pipes of Distribution Circuits (2.HD.01)

This measure concerns checking the state of the thermal insulation of distribution circuits (of heating and cooling fluids) and the reinsulation of the pipes for which the existing insulation is inadequate (Table 13.45).

Table 13.45 Thermal insulation of pipes of distribution circuits (2.HD.01): main parameters for the choice

Working C	Obsolete A	New -	O&M -
Saving potential	□□■■	Payback	□■
Reliability	■■■■	Feasibility	■■■
Environmental effects	**EC**		

General Description

The heat losses along the distribution circuits of media can be significantly reduced through an effective insulation of their piping.

This measure provides for a check of the distribution circuits in order to verify the quality of the thermal insulation. The poorly insulated pipe sections must be restored through improved thermal insulation. For these pipes the existing insulation, in many cases, must be removed and replaced by new insulation.

The distribution circuits that should be isolated are as follows:

- distribution circuits of heating and cooling systems;
- DHW distribution circuits;
- distribution circuits of cold water plumbing (at least for the section crossing occupied areas in order to avoid condensation on piping surfaces).

The analysis of the state of the insulation of the pipes can be quickly performed by mean of an infrared audit.

Tips and Warnings

It is not easy to check the quality of the thermal insulation of pipes for those sections that are not visible (e.g., inside walls structures, inside the ground or inside non-accessible spaces): an infrared audit can in any case assist the inspection phase.

Sometimes in the plants of old buildings asbestos was used as insulating material and remedial action in these situations must be undertaken with great care.

13.3.3.2 Installation of High Efficiency Pumps (2.HD.02)

This measure concerns the replacement of old or inefficient pumps and hydronic circuit boosters with more efficient models (Table 13.46).

Table 13.46 Installation of high efficiency pumps (2.HD.02): main parameters for the choice

Working C	Obsolete A	New -	O&M -
Saving potential	□□■	Payback	□□■■
Reliability	■■	Feasibility	■■■
Environmental effects	**EC**		

General Description

The energy performance of distribution circuits can be improved by replacing the existing pumps and booster pumps with appliances that use less electricity.

Controlling the speed of an electric motor by means of a VFD/frequency inverter is the most effective way to adjust the energy performance of pumps and booster pumps that must operate at variable speed.

The replacement of the electric pumps and of circulators installed in existing hydronic circuits with appliances of high efficiency is a cost-effective measure since the electrical energy consumed can be reduced by up to 80 %.

Tips and Warnings

The European Union Directive 2005/32/EC established a framework for the setting of eco-design requirements for energy-using products. Pumps and boosters are amongst the equipment considered in this Directive: for this reason the products available on the E.U. market must comply with the requirements of high efficiency.

However it is clearly appropriate to promote the replacement of existing equipment long before any break-down occurs.

13.3.4 Aeraulic Distribution

13.3.4.1 Thermal Insulation of Air Ducts (2.AD.01)

This measure provides for checking the state of the thermal insulation of air ducts and the reinsulation of those ducts for which the existing insulation is inadequate (Table 13.47).

General Description

The heat losses along the air distribution circuits can be significantly reduced through an effective insulation of air ducts.

Table 13.47 Thermal insulation of air ducts (2.AD.01): main parameters for the choice

Working D	Obsolete A	New -	O&M -
Saving potential	□□■■	Payback	□□■
Reliability	■■■■	Feasibility	■■■
Environmental effects	EC		

This measure involves a check of the air distribution circuits in order to verify the quality of the thermal insulation. The poorly insulated duct sections must be restored to proper effectiveness through improved thermal insulation. For these ducts the existing insulation, in many cases, must be removed and replaced by new insulation.

The analysis of the state of the insulation of the air ducts can be quickly made by mean of an infrared audit.

Tips and Warnings

For insulated air ducts, by removing the existing insulating material it is possible to check the state of the duct and if necessary to replace the damaged parts (see measure 2.AD.02).

The removal and disposal in disposal of the existing insulation must be included in the evaluation of the total costs.

13.3.4.2 Revision of Air Distribution Ducts (2.AD.02)

The measure involves a check of the functionality of the air distribution system and eventually any necessary repairs and/or replacement (Table 13.48).

General Description

For existing air distribution systems installed in old existing buildings, it is advisable to check in order to prevent inefficiencies and energy wastage. The critical aspects are as follows:

Table 13.48 Revision of air distribution ducts (2.AD.02): main parameters for the choice

Working -	Obsolete B	New -	O&M -
Saving potential	■	Payback	□■■
Reliability	■■■	Feasibility	■■■
Environmental effects	EC		

- presence of air leakage at the joints or where there are zones of degradation;
- presence of dust or of polluting materials within the ducts;
- cross-sections of the channels which are inadequate to meet the real needs;
- lack or inadequacy of thermal insulation (see measure 2.AD.01).

The verification of the air velocity inside the channel can be made using an anemometer: the air flow rate can be calculated on the basis of the air speed and the cross-sectional area of the duct.

Air leakage can be checked using the infrared audit or tracer gases. Parts of air ducts which have deteriorated or appear inadequate should be replaced.

Tips and Warnings

The check of the proper operation of the air distribution should be done every time changes from the original design are made.

13.3.4.3 Installation of VSD/Inverter-Controlled Fans (2.AD.03)

The measure provides for the replacement of the fan motors with variable speed motors (Table 13.49).

General Description

Frequently in HVAC air systems the fans are made to run at full speed when the air flow requirements could in fact be lower. This results in a greater consumption of electric power for the fans.

The installation of a VSD/inverter systems allows an adjustment of the speed of the motors, and hence of the fans, helping to reduce the consumption of absorbed electrical energy.

In order to save electricity the strategies are as follows:

- control the electric motor with a VSD/inverter instead of a simple contactor;
- remove the existing limitation device (for example the main airflow regulating damper);

Table 13.49 Revision of air distribution ducts (2.AD.03): main parameters for the choice

Working	B	Obsolete	A	New	-	O&M	-
Saving potential			□■■	Payback			■■■
Reliability			■■■	Feasibility			■■■
Environmental effects			EC				

- install a device which detects the pressure downstream of the fan and provides the information to the control system to obtain an optimised setting.

In fans driven by a VSD/inverter, energy saving can be up to 50 %, this measure is therefore cost-effective. For a discussion of high-efficiency motors, see measure 3.EL.02.

Tips and Warnings

The European Union Directive 2005/32/EC established a framework for the setting of eco-design requirements for energy-using products. Fans of HVAC systems are amongst the equipment considered in this Directive: for this reason the products available on the E.U. market must comply with the requirements of high efficiency.

It is in any case appropriate to promote the upgrading of existing equipment long before any break-down occurs.

A variable speed air handler also runs much quieter because the blower slowly ramps up to the preset speed so there's no sudden blast of air when it starts up.

13.3.4.4 Installation of Air Destratification Devices (2.AD.04)

The measure provides for the installation of air destratification devices able to ensure a better distribution of the air temperature in the spaces of considerable height (Table 13.50).

General Description

In spaces of considerable height (e.g., industrial buildings, warehouses, galleries, shopping malls/centres, airports etc.) the air heated from the heating system in the winter season, owing to natural convective motions, accumulates in the upper part of the space, close to the roof. This heat, unused, is eventually dissipated to the outside air.

In order to ensure comfort conditions near floor level for the occupants, the best solution is to install air destratification devices which are able to move the air continuously from the upper part of the space towards the bottom.

Table 13.50 Installation of air destratifiers (2.AD.04): main parameters for the choice

Working	-	Obsolete	-	New	A	O&M	-
Saving potential			□□■	Payback			□■
Reliability			□□■	Feasibility			□□■
Environmental effects			EC				

The market offers various solutions according to the characteristics of the space (size, intended use, height): the simplest one is made up of one or more fans with a vertical axis positioned near the upper part of the room. An alternative solution is to construct an air duct, equipped with a fan, which draws air from above and blows it downwards.

Tips and Warnings

The application of this measure permits two benefits to be obtained: energy savings and a significant improvement in comfort conditions for the occupants. The low cost of this measure makes it a cost-effective solution.

In spaces used for production activities, the auditor should investigate the possibility of replacing the air heating system with a radiant strips heaters, with which it is possible to heat the space in a selective manner (i.e., only where there are occupants).

13.3.5 Ventilation and Air Handling Systems

13.3.5.1 Installation of Mechanical Ventilation Systems (2.VE.01)

The measure provides for the installation of a mechanical ventilation system in order to control the ventilation and improve energy efficiency (Table 13.51).

General Description

Mechanical ventilation systems have the function of ensuring a correct control of the air changes in relation to the activities that are carried out within a space.

Typically one can identify two types of systems: single flow systems and double flow systems.

Figure 13.16 shows the differences between the two solutions. In single flow systems the exhaust air is extracted by an exhaust fan installed on the top of the exhaust/discharge duct, while the fresh air is drawn through the air vents positioned in correspondence to the windows or the outer wall. In some systems it is possible to

Table 13.51 Installation of mechanical ventilation systems (2.VE.01): main parameters for the choice

Working	D	Obsolete	B	New	-	O&M	-
Saving potential			■	Payback			■■■■
Reliability			■■■	Feasibility			■■■
Environmental effects			EC				

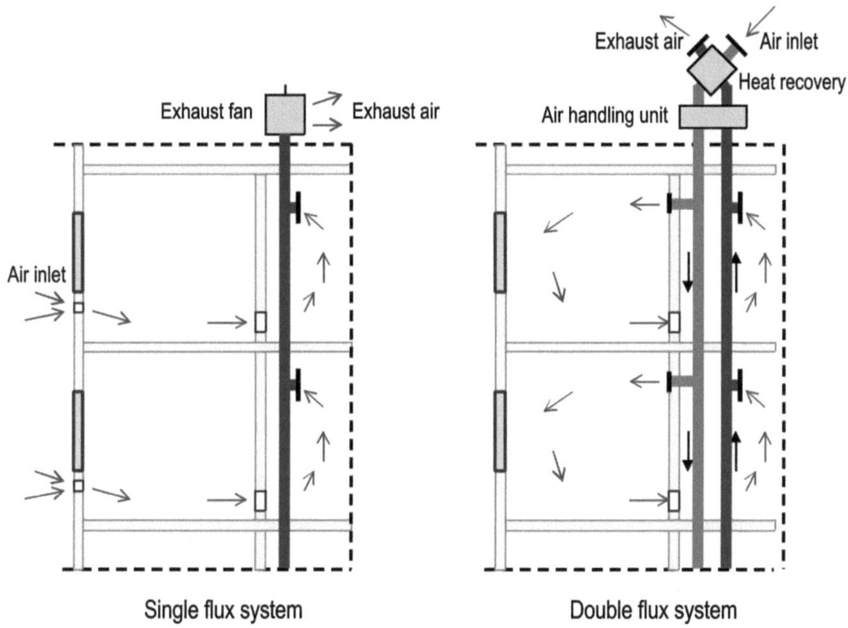

Fig. 13.16 Typical schemes of mechanical ventilation systems

adjust the flow of ventilation through a nozzle, connected on the exhaust duct, which is adjusted as a function of the relative humidity of the environment.

Double flux systems are equipped with two ducts: one for the fresh air supply and one for exhaust air extraction. This plant configuration permits the installation on the top of the building of a heat recovery unit and a significant reduction in the energy consumption for ventilation.

Tips and Warnings

Mechanical ventilation systems, in particular those of double flow with heat recovery, are convenient since they permit a dramatic reduction in the energy use for ventilation. Mechanical ventilation systems also contributes significantly to improving the quality of indoor air. However, their installation in existing buildings becomes convenient only if the building is under major renovation.

13.3.5.2 Replacement of Air Handling Units (2.VE.02)

This measure involves an accurate check of the air handling units with any additional maintenance and/or replacement on the basis of the findings (Table 13.52).

Table 13.52 Replacing of air handling units (2.VE.02): main parameters for the choice

Working C	Obsolete A	New -	O&M -
Saving potential	□■■	Payback	□■
Reliability	□■■	Feasibility	■■■
Environmental effects	**EC**		

General Description

The measure includes an accurate check on the air handling units in order to identify all the inefficiencies. From the examination carried out, if problems which cannot be resolved with routine maintenance operations are encountered, more drastic measures could be scheduled, up to and including the complete replacement of units.

Tips and Warnings

When replacing an air handling unit with a new one, a detailed analytical calculation is necessary in order to define the appropriate technical specifications.

13.3.5.3 Installation of Heat Recovery Systems (2.VE.03)

The measure provides for the installation of heat recovery systems (Table 13.53).

General Description

In existing HVAC systems installed in old buildings often the exhaust air is extracted from an independent circuit and the enthalpy contained in the air is not put to use. In order to increase the efficiency of the HVAC system, this measure provides for the installation of heat recovery systems.

A heat recovery system captures the exhaust air and reuses some of the energy to precondition (preheat in winter and precool in summer) the make-up air before it is supplied to the air-conditioned spaces. Air to air heat recovery systems can be

Table 13.53 Installation of heat recovery systems (2.VE.03): main parameters for the choice

Working C	Obsolete A	New A	O&M -
Saving potential	□■■	Payback	□■
Reliability	□■■	Feasibility	■■■
Environmental effects	**EC**		

in general categorised as *sensible heat* (dry bulb only) systems and *total heat* (wet bulb—sensible heat plus latent heat) systems, also called thermodynamic heat recovery.

Heat recovery units used in ventilation and air-conditioning systems can be classified according to the technology adopted:

- *rotary heat exchangers* in which outlet air heats (or cools) the exchanger when the wheel passes through the outlet air flow, then the energy is transferred to the fresh, make-up air when the wheel passes through the make-up air. Both sensible and latent heat may be transferred; latent heat when moisture in the outlet air condenses on the wheel and in this system moisture may be transferred with heat exchangers by using hygroscopic wheels;
- *air-fluid-air exchangers* where heat is transferred in an heat exchanger from the outlet air to a circulating fluid, that fluid is then pumped to the heat exchanger fitted in the make-up air flow where the heat is transferred to the supply air. Both sensible and latent heat may be transferred; latent heat when moisture in the outlet air condenses on the heat exchanger and this system is used when the point of air exhaust/discharge and the point of air intake are far apart;
- *cross-flow exchangers* heat is transferred directly from the outlet air to the fresh, make-up air through the separating walls of the heat exchanger. Both sensible and latent heat may be transferred: this system is normally integrated in the handling units;
- *heat pumps*: with this system it is possible, with some additional energy, to move more outlet air energy over to the make-up air than can be done in any other system. The energy consumption is approximately 1/3–1/5 of the recovered energy. Both sensible and latent heat may be transferred.

Tips and Warnings

The installation of heat recovery systems in existing buildings can be cost-effective if a global renovation of the HVAC system is scheduled. The installation of heat recovery systems in existing buildings can be no easy undertaking. Nonetheless the significant potential for energy savings means that one should always include for a verification of the technical and economic feasibility of effecting this very interesting measure.

13.3.5.4 Installation of IAQ sensors (2.VE.04)

The measure involves the installation of sensors for controlling the parameters of the air quality in order to manage the ventilation in an efficient manner (Table 13.54).

Table 13.54 Installation of IAQ sensors (2.VE.04): main parameters for the choice

Working	-	Obsolete	-	New	A	O&M	-
Saving potential			□■■	Payback			■■■■
Reliability			■■■	Feasibility			■■■
Environmental effects			EC				

General Description

Mechanical ventilation of indoor spaces is essential to ensure the maximum values of internal air quality (IAQ). However in ventilation systems with a fixed flow rate, it may be that in some periods the air flow rate is excessive compared to the real needs and this gives rise to a considerable wastage of energy.

This measure provides for a network of IAQ sensors that, connected to a control unit, are able to acquire the information necessary to ensure a proper ventilation rate and/or to report promptly alarm situations.

The physical parameters most frequently used to detect the presence of people in closed spaces are the CO_2 (carbon dioxide) and the relative humidity.

Other parameters are used to monitor air quality, including NO_x (nitrogen oxides), CO (carbon monoxide), O_3 (ozone), particulate matter (PM_{10}, $PM_{2.5}$) and volatile organic compounds (VOC).

Once values of air quality have been detected, the flow rate could be modulated using fans with variable speed control (see measure 2.AD.03).

Tips and Warnings

Before implementing a variable flow system, one must check that the minimum air flow is in conformity with the hygiene requirements laid down by the health authorities.

13.3.6 Control and Regulating Systems and Terminals

13.3.6.1 Replacement of Central Control System (2.RE.01)

This measure involves the replacement of a centralised control system which manages a HVAC system (Table 13.55).

General Description

A significant part of the energy wastage of HVAC systems depends upon the inadequacy of the regulating and control system. This measure includes:

Table 13.55 Replacement of central control system (2.RE.01): main parameters for the choice

Working C	Obsolete A	New A	O&M -
Saving potential	□■■	Payback	■■■■
Reliability	■■■	Feasibility	■■■
Environmental effects	EC		

- the analysis of the HVAC control system;
- the identification of the inefficiencies of the regulating and control system;
- the design and the installation of a regulating and control system able to efficiently manage the energy in the building, guaranteeing the best comfort conditions.

Tips and Warnings

The choice of a centralised control and regulating system should take account of the following needs:

- the ability to interface with external networks to facilitate remote management of facilities (remote control);
- the flexibility of the system in managing other plant that may be installed at a later date;
- the ability to interface with local control systems;
- the option to manage the energy accounting.

Provide for periodic check and calibration of the system in order to maintain high performance.

13.3.6.2 Installation of Zone Control Systems (2.RE.02)

The measure provides for installation of local control systems in order to optimise the energy exchange between the terminals of HVAC system and the air-conditioned spaces (Table 13.56).

Table 13.56 Replacement of central control system (2.RE.02): main parameters for the choice

Working -	Obsolete A	New A	O&M -
Saving potential	■	Payback	■■■■
Reliability	□■■	Feasibility	■■■■
Environmental effects	EC		

General Description

The control of thermal energy output (both for heating or cooling) of HVAC system terminals is one of the most cost-effective measures to implement for the reduction in energy consumption and improvement in the thermal comfort of the occupants.

The ability to control the thermal environmental conditions in each room allows the HVAC system to heat and cool the spaces with lower energy consumption, taking into account the free internal gains (presence of people, lighting appliances, equipment etc.), the solar gains as well as the effect of the thermal inertia of the building structures (walls, slabs) itself.

For hydronic HVAC systems the control is made using zone valves or thermostatic valves that modulate the flow rate depending on the air temperature detected by a sensor normally integrated in the valve. The zone valves are controlled by a control unit on the basis of the readings detected by temperature sensors installed inside the controlled spaces.

With thermostatic valves one has two possible solutions: the simplest one consists of the replacement of the manual regulation valve installed at each terminal (e.g., radiator) with a thermostatic valve. The user should adjust the set-point temperature acting on the knob of the valve. In this solution each thermostatic valve act as an independent control system (the temperature sensor is installed inside the valve).

A more complete and functional solution foresees the connection, physically with cables or via wireless, of all the valves to a digital control unit. This solution allows one to exploit many interesting features (e.g., separate set-points for each valve, the definition of different times of operation, the check of the temperature inside each room etc.). The opportunity to use wireless connection is very useful for a retrofits in existing buildings.

For all air HVAC systems or fan-coil systems the local control is done through the installation of thermostats (one for each room) or using more sophisticated local control systems.

Tips and Warnings

So far as hydronic HVAC systems are concerned, local control, which normally acts on the flow rate, allows adjustment of the hydronic circuit malfunctions caused by incorrect sizing or the presence of deposits in some part of the pipes.

It is important to choose a reliable thermostatic valve equipped with a sensor element of quality, characterised by a low inertia: the installation of the valve should avoid the influence of solar radiation on the sensor (some thermostatic valve models allow independent installation of the temperature sensor).

The best performance can be obtained with the choice of a digital control system.

Table 13.57 Installation of energy metering (distributors) (2.RE.03): main parameters for the choice

Working -	Obsolete -	New A	O&M -
Saving potential	□■	Payback	□■■■
Reliability	■■■	Feasibility	■■■■
Environmental effects	EC		

13.3.6.3 Installation of Energy Metering (Distributors) (2.RE.03)

The measure provides for the installation of local metering systems (distributors) for the attribution of the consumptions as a function of the thermal energy actually consumed (Table 13.57).

General Description

The ability for a end-user to be able to account for individual energy consumption (and pay for the energy actually consumed), even if the heating system is centralised, is a strong incentive to use more energy awareness but also to implement energy retrofit actions.

The energy metering is carried out in an indirect way. The devices used for this purpose are installed on each radiator and account for the units of energy directly related to the use of the radiator. The units of energy are then multiplied by a correction factor which accounts for the characteristics of the radiator. A part of the global energy consumption of the entire building is then distributed on the basis of the individual units of energy consumption as recorded and corrected.

The units of energy accounted for each end-user must be read at least twice a year. Modern devices, equipped with a wireless interface, allow the reading to be performed from outside the apartment without interfering with the end-user's privacy.

Tips and Warnings

At the end of the heating season, the total cost for the energy service for heating (including fuel, maintenance, management, repair etc.) will normally be allocated according to the following criterion: a fixed part as a function of the floor surface or the volume of the individual flat (for example 30 %) to account for the overhead costs and the remainder (i.e., 70 %) as a function of the individual consumption detected in an indirect manner through the distribution readings (Fig. 13.17).

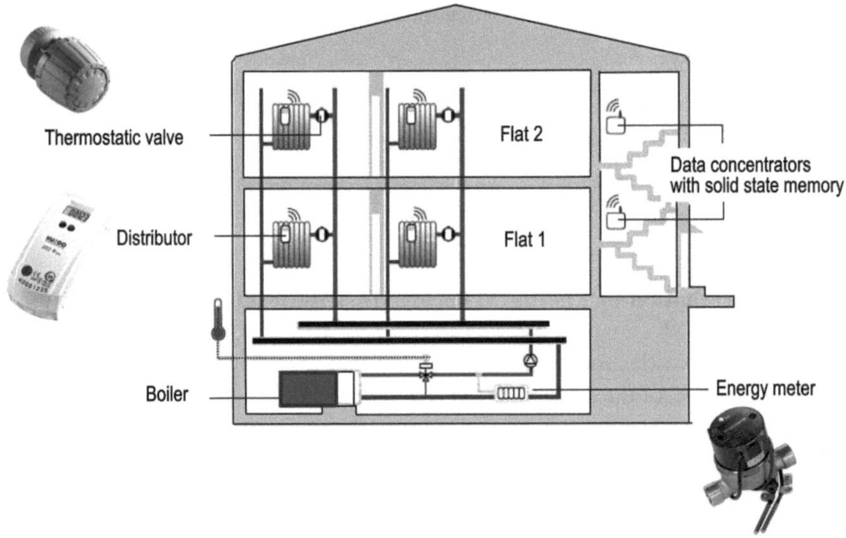

Fig. 13.17 Typical schemes of an individual energy accounting system

13.3.6.4 Installation of Radiant Conditioning Terminals (2.RE.04)

The measure provides for the installation of prefabricated radiant ceiling air-conditioning terminals (Table 13.58).

General Description

This measure involves the installation of prefabricated radiant air-conditioning terminals in order to update existing obsolete HVAC systems or to apply new air-conditioning systems in existing, unair-conditioned buildings under renovation.

The installation of radiant air-conditioning terminals, particularly suitable as a retrofit solution, allows one to procure many advantages. These are improvement in the thermal comfort for occupants, reduction in heating and cooling costs and reduction in the height of the spaces by installing a system with the double function of air-conditioning terminal and false ceiling combined (Fig. 13.18).

Table 13.58 Installation of radiant conditioning terminals (2.RE.04): main parameters for the choice

Working C	Obsolete B	New B	O&M -
Saving potential	■	Payback	□■
Reliability	■■■	Feasibility	■■
Environmental effects	EC		

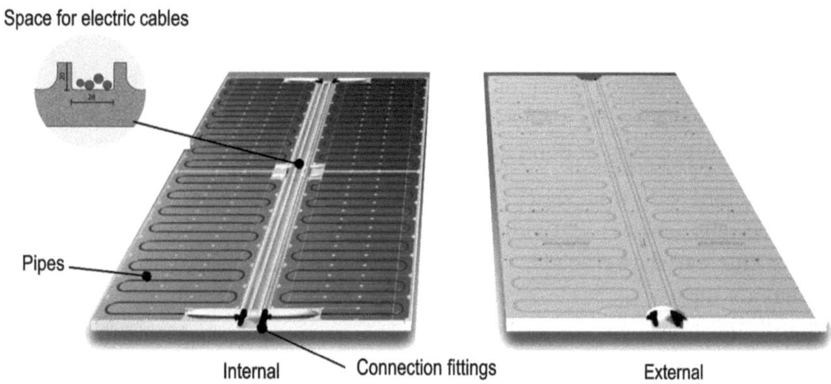

Fig. 13.18 Example of prefabricated radiant panel for ray conditioning applications (courtesy of Messana Co.)

Tips and Warnings

A radiant cooling systems require a constant control of the relative humidity of the air-conditioned spaces. A ventilation system is necessary in order to avoid reaching excessive values of relative humidity inside the spaces. The most reliable solutions are those offered by companies who provide the complete system (radiant panels, ventilation system, regulating and control system).

13.4 Plumbing and DHW Systems

13.4.1 Reduction of Water Consumption

13.4.1.1 Installation of Water Saving Devices for Terminals (2.WR.01)

The measure provides for the installation of devices that contribute to saving water (Table 13.59).

Table 13.59 Installation of water saving devices (2.WR.01): main parameters for the choice

Working	C	Obsolete	A	New	A	O&M	-
Saving potential			□■■	Payback			NA
Reliability			□■■■	Feasibility			■■
Environmental effects			RI				

General Description

This measure provides a check on the plumbing system and on the DHW system, aimed at finding possible further measures to reduce the consumption of potable water. The most common equipment strategies are as follows:-

- lavatory taps with flow restrictions;
- infrared taps sensors;
- water efficient showerheads;
- timers on taps;
- dual-flush toilets.

The reduction in the consumption of water very much depends upon end-user behaviour: any form of communication, education and information to the users is useful to achieve the purpose.

Tips and Warnings

In the general description the potentials on water saving of household appliances such as washing machines and dishwashers were not considered since the choice of these appliances is individual.

Some of the measures listed above are cost-effective in existing buildings as retrofit actions, only if planned in the case of renovation of the building plant.

13.4.1.2 Installation of Potable Water Meters (2.WR.02)

The measure provides for the installation of potable water meters aimed at encouraging a more conscientious use of water (Table 13.60).

General Description

The reduction in the consumption of cold or hot water can be enhanced applying this measure which involves the installation of individual meters. In this way the

Table 13.60 Installation of water saving devices (2.WR.01): main parameters for the choice

Working	B	Obsolete	A	New	-	O&M	-
Saving potential			■	Payback			NA
Reliability			■■■■	Feasibility			■■■■
Environmental effects		**RI**					

end-user pays for the individual consumption of water and thus has incentives to save.

Some water meters are fitted for a remote reading of consumption: this helps the data acquisition activities.

Tips and Warnings

The allocation of the correct consumption depends primarily on the precision with which the measurement is carried out: the meters should be certified and monitoring should be performed over time.

13.4.1.3 Efficient Use of Water for Irrigation Purposes (2.WR.03)

The measure provides for planning and implementing a strategy aimed at an efficient use of water for irrigation purposes (Table 13.61).

General Description

The purpose of this measure is to limit or eliminate the use of potable water for irrigation. The reduction of potable water consumption can be achieved by implementing different strategies:

- installation of efficient irrigation systems capable of irrigating on the basis of the real needs;
- using recycled wastewater (see measure 2.WR.04);
- using rainwater harvesting (see measure 2.WR.05).

The strategies listed above depend upon the types of plants which are related to the climate of the location: the auditor should perform a soil/climate analysis to determine landscape types and design the landscape with indigenous plants to reduce or eliminate irrigation requirements.

Table 13.61 Efficient use of water for irrigation purposes (2.WR.03): main parameters for the choice

Working	**B**	Obsolete	**A**	New	**A**	O&M	-
Saving potential			■■	Payback			NA
Reliability			■■■	Feasibility			■■■■
Environmental effects			**RI**				

Tips and Warnings

The installation choices depend very much on the landscape design: the auditor should consult an expert in landscaping in order to plan the correct measure to adopt.

13.4.1.4 Recovery and Treatment of Grey-Water (2.WR.04)

The measure provides the recovery and the treatment of grey-water for a non-potable usage in the building (Table 13.62).

General Description

With the term *grey-water* one denominates wastewater coming from domestic activities such as showers, lavatories, washing machines and dishwashing which can be recycled on-site for some uses; grey-water is distinguished from black-water, which is wastewater coming from toilets and sinks that contain organic (e.g., human waste) or toxic matter. Grey-water can be collected, filtered and purified from sewage plants available on the market. The treated water can be used for non-potable uses e.g., irrigation, toilets and washing machines.

Tips and Warnings

In the existing buildings normally water waste systems are not distinguish between grey-water and black-water and this is a problem. In existing buildings, moreover, the water distribution system does not distinguish between potable water and non-potable water. All this makes it difficult to implement this measure, which offers a great potential on water saving, in existing buildings for which extensive renovation is not planned. For the possible uses of purified grey-water it is necessary to refer to local hygiene regulations.

13.4.1.5 Rainwater Harvesting (2.WR.05)

The measure provides for the recovery and the reuse of rainwater for irrigation purpose (Table 13.63).

Table 13.62 Recovery and treatment of grey-water (2.WR.04): main parameters for the choice

Working	D	Obsolete	C	New	B	O&M	-
Saving potential			■■	Payback			NA
Reliability			■■	Feasibility			■
Environmental effects		RI					

Table 13.63 Recovery and treatment of grey-water (2.WR.04): main parameters for the choice

Working D	Obsolete B	New -	O&M -
Saving potential	■■	Payback	NA
Reliability	■■	Feasibility	□■■
Environmental effects	RI		

General Description

Rainwater harvesting is the process of intercepting rainwater and putting it to beneficial use. Rainwater is usually collected or harvested from roofs, concrete patios, driveways and other impermeable surfaces. In projects of major renovation, buildings and landscapes can be designed to maximise the amount of catchment area, thereby increasing rainwater harvesting possibilities. Intercepted water then can be collected, captured, retained/stocked and used for non-potable applications as garden irrigation, toilet flushing, indoor plant watering and car washing.

Rainwater harvesting systems vary from the simple and inexpensive to the complex and costly. Typically, these systems are simple, consisting of gutters, downspouts, and underground storage containers. There is a filter which has the function of separating the water from dirt. Through a flexible pipe equipped with a float, the water contained in the storage container is drawn-off by a pump suction and pumped to the users. The whole rainwater harvesting system is operated by an electronic control unit.

Directing rainfall to plants located at the lowest points is the simplest rainwater harvesting system. In this system, the falling rain simply flows to areas with vegetation (Fig. 13.19).

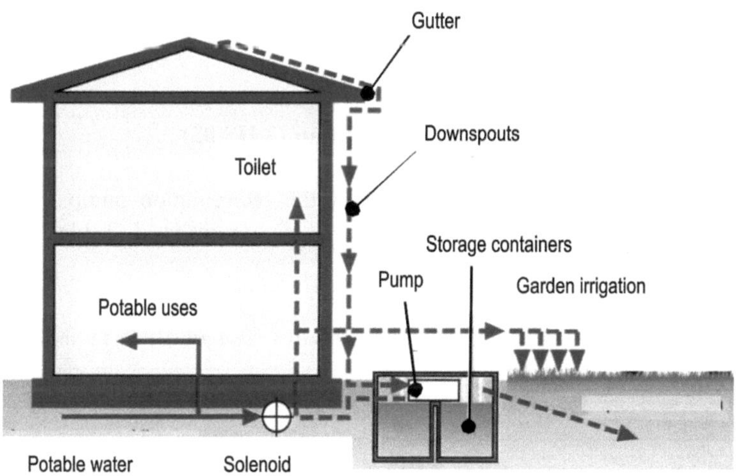

Fig. 13.19 Schematic example of a rainwater harvesting system

Tips and Warnings

For the possible uses of purified rain water it is necessary to refer to local hygiene regulations.

13.4.2 Efficiency of DHW Systems

13.4.2.1 Insulation of DHW Storage Tank (2.HS.01)

The measure covers the thermal insulation of domestic hot water storage tanks in order to increase the energy performance of the DHW system as a whole (Table 13.64).

General Description

An inadequate thermal insulation of the storage tank in a building causes an increase of the heat losses through the walls of its casing. This measure provides for the replacement of poor existing insulation with insulating material of adequate thickness.

The measure is also applicable for inertial storage tanks and cold water tanks.

Tips and Warnings

The inspection of the storage tank could be done using an infrared audit. The existing thermal insulation system should be removed or, if it is inadequate but still in good condition, covered by the new insulation material.

13.4.2.2 Deactivation of Recirculation Pump (2.HS.02)

This measure involves the deactivation of the recirculation pump in order to reduce the use of electricity when DHW systems are not used (Table 13.65).

Table 13.64 Insulation of DHW storage tanks (2.HS.01): main parameters for the choice

Working	C	Obsolete	A	New	-	O&M	-
Saving potential			■	Payback			□□■■
Reliability			■■■	Feasibility			■■■
Environmental effects			EC				

Table 13.65 Deactivation of recirculation pump (2.HS.02): main parameters for the choice

Working A	Obsolete A	New -	O&M A
Saving potential	□■	Payback	■■■■
Reliability	■■■■	Feasibility	■■■■
Environmental effects	EC		

General Description

In a DHW distribution system, the function of the recirculation circuit is to ensure that the hot water is delivered to each usage point/terminal with the design temperature also when the withdrawal of water are very few or null.

On the other hand the activation of the recirculation circuits incurs energy wastage owing to the thermal losses along the distribution piping networks and the electrical consumption of the recirculation pump itself.

If the hot water is not used for long periods, for example in office buildings or in educational buildings during week end or vacations, the deactivation of the recirculation pump could provide a saving in energy.

Tips and Warnings

This measure is cost-effective since the cost is null, for a manual management (switch off end switch off of the pump) system, or very low when a programmable timer is installed. The difficulty consists in scheduling the correct periods of deactivation.

See also measures 2.HD.01 and 2.HD.02.

13.4.2.3 Installation of Centralised DHW System (2.HS.03)

The measure provides the installation of a centralised DHW system in order to increase the energy performance (Table 13.66).

Table 13.66 Installation of centralised DHW system (2.HS.03): main parameters for the choice

Working C	Obsolete B	New -	O&M -
Saving potential	□■■	Payback	□■■
Reliability	■■■	Feasibility	■■
Environmental effects	EC		

General Description

The measure provides for the installation of a centralised DHW system in buildings where DWH production is performed by inefficient independent systems (e.g., a block of flats with joint-ownership of the communal spaces).

The centralisation of DHW production is in some cases more efficient than independent DHW production: this choice permits the use of condensing boilers, heat pumps or thermal solar system integration with a lower installation cost.

Tips and Warnings

This solution is convenient if the users are not very far between. When installing a centralised DHW system, one must provide individual consumption meters (see measure 2.WR.02).

13.4.2.4 Installation of Zone DHW System (2.HS.04)

The measure involves the installation of zone DHW systems in order to increase the energy performance (Table 13.67).

General Description

The measure proposed here is the opposite of measure 2.HS.03: the central DHW system is replaced by individual DHW systems. This choice becomes convenient in those cases in which:

- the users are few and very far between and the heat losses of the distribution circuit are considerable;
- the cost of the thermal insulation of the distribution circuit is high;
- the centralised DHW production system is inadequate for the real needs.

Tips and Warnings

The new DHW production systems should be high-efficiency (see measure 2.HS.06).

Table 13.67 Installation of zone DHW system (2.HS.04): main parameters for the choice

Working	C	Obsolete	B	New	-	O&M	-
Saving potential			□■■	Payback			□■■
Reliability			■■■	Feasibility			■■
Environmental effects			EC				

13.4.2.5 Replacement of Electric Boilers for DHW (2.HS.05)

The measure provides for the replacement of existing electric boilers for DHW with more energy efficiency solutions (Table 13.68).

General Description

The direct thermal use of electricity for the Joule effect is not sustainable and should be avoided. This measure consists of replacing existing electric boilers for DHW with other technical solutions (e.g., condensing boiler or heat pumps).

Tips and Warnings

The new DHW production systems should be high-efficiency, if it is not possible to supply gas, one should consider the opportunity of installing a heat pump (see measure 2.HS.06).

In some cases electric boilers are used for purposes that do not require hot water (e.g., hand washing in offices): in these cases the electric boiler should be removed without any replacement.

13.4.2.6 Installation of DHW Heat Pumps (2.HS.06)

The measure provides for the replacement of an existing water heater for DHW with a heat pump (Table 13.69).

Table 13.68 Replacement of electric boilers for DHW (2.HS.05): main parameters for the choice

Working A	Obsolete A	New -	O&M -
Saving potential	■■■	Payback	□■■
Reliability	■■■	Feasibility	□■
Environmental effects	EC		

Table 13.69 Replacement of electric boilers for DHW (2.HS.05): main parameters for the choice

Working C	Obsolete A	New -	O&M -
Saving potential	□■■■	Payback	□□■■
Reliability	■■■	Feasibility	□□■
Environmental effects	EC		

General Description

Existing inefficient water heaters for DHW (e.g., gas boilers, electric boilers geysers) could be replaced with heat pumps. The market offers compact air to water heat pumps designed in a special way for domestic hot water production.

The use of these equipment items permits the production of hot water with a high energy efficiency. For the operating principle, refer to the description in measure 2.HG.03.

Tips and Warnings

DHW heat pumps could be used as a solution to replace electric boilers in locations where fuels (e.g., gas) are not available: electricity is used but in a more efficient manner (if the COP of the heat pump is 4 the consumption is reduced to ¼ when compared with the previous situation).

13.5 Electrical Systems

Edited by Luca Sarto

13.5.1 Generation, Distribution and Use

13.5.1.1 Power Factor Correction (3.EL.01)

The measure consists of the installation of devices for the power factor correction of electrical loads in order to reduce losses and avoid the payment of penalties (Table 13.70).

General Description

Non-purely resistive loads, for the same power absorbed, cause higher currents in the electrical system: this generates higher losses and the application of a penalty

Table 13.70 Power factor correction (3.EL.01): main parameters for the choice

Working	B	Obsolete	A	New	-	O&M	-
Saving potential			■	Payback			■■■
Reliability			■■■■	Feasibility			■■■
Environmental effects			E				

by the energy supplier. The correction of the power factor, obtained by appropriate devices, cancel almost totally these additional losses, reducing the possibility of overheating.

The retrofit is generally feasible without major difficulties. First one must quantify the necessary compensation through evaluation of the power factor (cos φ) that can be deduced from the electricity bills or tested using a network analyser.

One can choose between localised or centralised power factor correction. In the first option the compensation device, a capacitor for inductive loads, is located on each item of equipment in the network (motors, transformers etc.) The capacity and the number of capacitors should be determined based upon the characteristics of the equipment. Alternatively, or in addition to a network partially compensated locally, an automatic centralised power factor correction unit can be used. The unit consists of a number of capacitors which are switched in, depending upon the power factor of the network. The sizing is performed on the basis of the maximum amount of reactive power to compensate.

A power factor correction system allows energy savings that can reach several percent. It should be emphasised that the savings affect not only the internal network but also the distributor network.

Tips and Warnings

The centralised power factor correction is appropriate in the case of plants with many loads working occasionally, in this case the capacitor bank reactive power (kVAR) is lower the total power that should be provided with distributed power factor correction, bearing in mind also that the cost per kVAR of large power capacitors is less than that of small power capacitors.

The distributed power factor correction, from the technical point of view, is the better solution since it is convenient especially if the majority of the reactive power demand is concentrated in a few high power loads, working many hours a day. With this option the reactive energy reduction affects fully both the utility and the user. Installation is simple and inexpensive, since the capacitor and the load are switched on and off at the same time and can share the same protection against overload and short circuits.

13.5.1.2 Installation of High Efficiency Motors and VSD/Inverters(3.EL.02)

The measure consists of the replacement of electric motors with high efficiency motors and/or the installation of variable speed drives (VSD)/inverters (Table 13.71).

Table 13.71 Installation of high efficiency motors and VSD/inverters (3.EL.02): main parameters for the choice

Working	B	Obsolete	A	New	-	O&M	-
Saving potential			□■	Payback			□■■■
Reliability			■■■■	Feasibility			■■■
Environmental effects			E				

General Description

Technological progress has permitted the production of more efficient electric motors. A motor, especially one of high power, can consume in a single year energy which costs a sum much higher than its own initial cost and this replacement can be very convenient, with limited payback time. For example a 10 kW motor operating 24/24 h consumes in a year about 9,000 euro of energy compared with a purchase cost for the motor itself of less than € 1,000. Obviously where the motor usage is limited to few hours a day, the replacement may not be very convenient because of the extended payback time.

Today specific standards exist for evaluating the energy efficiency of electric motors and manufacturers provide all the information necessary to evaluate the convenience of replacement. For applications where the motor must not always provide maximum power, electronic control of speed and power can be applied. Inverter/VSD controls are useful when the motor runs at variables speeds while they are not convenient to reduce speed permanently because they increase the losses of the motor even if the overall power consumption is reduced. If the motor has excessive power and/or speed for the application it must be replaced with a suitable model. Typical applications are in pumps, fans, air compressors, chillers, conveyors and agitators. To estimate the convenience of this remedial action the energy absorbed by the existing motor should be measured and the load profile, speed and maximum torque should be evaluated.

Based on the information collected, the most suitable model can be selected and energy consumption can be estimated on the basis of the data provided by the manufacturer. The correct sizing is very important since a motor with an efficiency of 95 % at full load may have efficiency falling below 50 % at 10 % partial load.

Losses from electric motors are released into the surrounding space as heat, so reducing motor losses means that the energy consumption for cooling is reduced whereas the energy consumption for heating increases. If the space is unconditioned, costs are not affected.

A new motor can lead to savings that can reach normally 10 % and can even reach 50 % when replacing a motor largely oversized with one of suitable power. The environmental benefits are associated with lower emissions for the production of electricity.

Tips and Warnings

When an high power motor fails, the replacement could be a better alternative to rewinding: even if the cost is, on average, double, savings achieved with a new motor can repay the higher investment in a short time. Often old low power single phase motors are the most inefficient and their replacement can be particularly advantageous.

Thermography can help to identify motors that should be replaced: a motor which is too hot is overloaded, could fail or is inefficient.

For more details on pumps, see measure 2.HD.02 and on fans, see measure 2.AD.03.

13.5.1.3 Loads Management System (3.EL.03)

The measure consists of the installation of an electrical load management system in order to reduce peak power (Table 13.72).

General Description

The implementation of this measure allows the management of electrical loads automatically, with the possibility of limiting the total maximum power required. This limitation avoids the potential problem of disconnection in cases where the supply contract defines a real power limit or the penalties for exceeding the contractual commitment.

In the management of the loads the user has the possibility of defining the priorities, identifying the equipment that must be left active and those items that can be switched-off. A load management system includes a user-programmable control unit and a driver unit for each load to be controlled. The connection between the control unit and driver units can be physically wired (usually using a low voltage BUS connection to) or wireless. This latter solution is useful in installation in existing buildings where the installation of cables can be problematic (e.g., protected historic buildings).

The energy and environmental benefit does not concern the user but rather the local community as a whole because assuring greater available power requires an electrical energy infrastructure of greater capacity. The user still obtains benefits in

Table 13.72 Loads management systems (3.EL.03): main parameters for the choice

Working	C	Obsolete	B	New	-	O&M	-
Saving potential			N/A	Payback			N/A
Reliability			■■■	Feasibility			■■■
Environmental effects			**RI**				

economic terms because penalties are avoided and contractual fixed cost are lower. These systems are normally able to take account of the absorbed energy of each load: in this way the users can be encouraged to optimise their utilisation or to install more efficient appliances.

Tips and Warnings

The choice of loads that can be deactivated in the case of an overload event must be agreed with the client. Load control functionality is usually already implemented in Home Automation systems (see measure 3.DO.01) and in Building Automation Systems (see measure 3.DO.02).

13.5.1.4 Installation of Timers (3.EL.04)

The measure consists of the installation of programmable digital timers for a better management of electrical appliances or plant sections (Table 13.73).

General Description

Devices or plant sections powered-up during periods when their services are not required cause a considerable wastage of electricity. Reliance on manual switching-off is the simplest solution, but there is no guarantee against forgetfulness or laziness. The installation of timers, that enable or disable the power fully automatically, provides far better energy management. The practical solutions may be different depending on technologies, however the digital timers, now available at low cost, are the devices which provide greater flexibility. Programming is usually simple and scheduling can be hourly, daily, weekly or monthly.

The timers can be installed within a switchboard, however, versions that do not require installation as such are available on the market: they must simply be inserted between the plug of the electrical appliance and the socket.

Energy savings and environmental benefits are obvious since the installation of this equipment avoids the running of equipment when it is not needed. The energy savings achieved can be easily calculated by multiplying the value of the electrical power absorbed by the time for which devices are disabled.

Table 13.73 Installation of timers (3.EL.04): main parameters for the choice

Working	C	Obsolete	A	New	-	O&M	-
Saving potential			■	Payback			□■■■
Reliability			■■■■	Feasibility			■■■
Environmental effects		RI					

Tips and Warnings

The installation of timers is part of a more general strategy aimed at drastically reducing or, even better, entirely eliminating wastage. The programming of the periods of switching on or off of the equipment must still be determined with the client, according to the specific situations. The reduction of wastage can sometimes be obtained by devices switching on–off, not on the basis of a time programme, but rather on the basis of other inputs such as presence detectors (see measure 3.LI.03), twilight switches (see measure 3.LI.04) or of daylight sensors (see measure 3.LI.05).

The installation of Home Automation systems in residential buildings (see measure 3.DO.01) or Building Automation systems in tertiary/non-residential buildings, results in (thermal and electrical) power plants being managed with greater efficiency and flexibility .

13.5.2 Lighting

13.5.2.1 Replacement of Lamps For Interior Lighting (3.LI.01)

This measure consists of replacing lamps for interior lightning with higher efficiency ones (Table 13.74).

General Description

The measure consists in the replacement of existing lamp with high-efficiency lamps in order to reduce the consumption of energy for lighting of interior spaces.

Some remedial actions may be limited to the simple replacement of lamps with compatible models but of higher energy efficiency. In other cases it is also necessary to replace the power supply circuit and socket or even the entire *lighting fixture*.

Here are the main features of the most innovative lamps.

Table 13.74 Replacement of lamps for interior lighting (3.LI.01): main parameters for the choice

Working C	Obsolete A	New -	O&M -
Saving potential	□■■■	Payback	□■■■
Reliability	■■■■	Feasibility	■■■■
Environmental effects	EC		

Halogen Lamps

Halogen lamps are incandescent lamps which contain halogen compounds such as iodine or bromine inside the bulb.

Thanks to the presence of these substances the tungsten filament can attain a temperature of 3000–3100 K, higher than conventional bulbs, providing a better performance. Inside the bulb tungsten vapour produced by the sublimation process of the filament are combined with halogen, so a partial regeneration of the filament is made with a consequent increase in the average lamp duration up to about 2,000 h.

The power supply can be 230 V or low voltage (typically 12 V). 230 V lamps are preferable for the diffuse illumination whereas in the case of full lighting, concentrated low voltage models are preferable since they have a compact structure. The start-up is immediate and the lamp can be dimmed just like the traditional incandescent lamps.

Fluorescent Tube Lamps

The operation of the fluorescent lamps is based on the principle of gas-discharge. These lamps consist of a discharge tube glass filled with mercury vapour at low pressure and a small amount of rare inert gas; the inner surface of the tube is covered with fluorescent powders.

Radiation emitted by the mercury vapour is predominantly in ultraviolet range but, thanks to the phosphors present on the surface of the tube, it is transformed into visible radiation with different colours or white tones. Commonly used lamps are hot cathode type, but in some application like signs cold cathode are used.

Fluorescent lamps must use an auxiliary device, a ballast, to regulate the flow of current through the tube and an ignition system starts the discharge. To maximise the luminous effectiveness, a lamp with high output phosphors as well as high efficiency ballast must be chosen.

The ignition is not immediate and, particularly at low temperatures, the maximum luminous flux is obtained after some seconds to minutes. These lamps are only partially dimmable, using special electronic ballasts: this type of lamp is suitable especially for the illumination of interior spaces with diffused light and low glare.

Compact Fluorescent Lamps

Compact fluorescent lamps (CFL) are designed to replace incandescent lamps and, in most cases, fit into light fixtures formerly used for incandescent lamps. These are fluorescent lamps with a curved or folded tube to fit into the space of an incandescent bulb, and a compact electronic ballast integrated in the base of the lamp. Technical specifications are similar to conventional fluorescent tubes.

Usually they are not dimmable but partially dimmable versions with standard regulators for incandescent lamps are also available.

They are particularly suitable for homes and wherever it is not possible to replace the entire light fixture: their lifetime is shorter and therefore involve higher maintenance costs compared to other solutions.

High Pressure Mercury Lamps

The operation of these light sources is analogous to that of fluorescent lamps: the passage of the electrical current inside a tube containing vapours of mercury and other gases emits directly light radiation.

The lamp is composed of a glass ampoule, not a tube, within which is located a little discharge tube containing gas, the internal surface may be phosphor-coated for better colour of emitted light.

Although apparently similar to an incandescent lamp, it requires a ballast for the ignition and control of the discharge, and therefore cannot directly replace these lamps.

The lamp needs 4–5 min to achieve full flux and restart cannot be immediate.

It is suitable in particular for the illumination of external but may also be used indoors. These lamps contain moderate amounts of mercury and will be banned in most countries.

Metal Halide Lamps

These represent an evolution of the mercury vapour lamps and have a similar operating principle, substantially they differ in the kind of gas inside the discharge tube, which contains, in addition to mercury, iodides of sodium, thallium and indium.

In newer lamps versions are also included rare earths such as dysprosium, holmium, thulium and caesium, which give a better spectral distribution and higher luminous effectiveness. For operation they require a special ignition and discharge stabilisation circuit. The metal halide lamps have an obligatory operating position and the time of ignition and reignition are similar to those of mercury vapour lamps.

They can be used both in indoor and outdoor applications when starting and restarting time can be long. They are suitable for producing light beams.

High Pressure Sodium Vapour Lamps

These lamps consist of a discharge tube made of transparent ceramic material, resistant to high temperatures and to the aggressiveness of sodium contained within it. The discharge tube is placed in a glass bulb for protection.

The discharge in vapour at high temperature and pressure generates an emission of golden white light with a continuous spectral distribution.

The high pressure sodium vapour lamps require horizontal positioning with an angle of tolerance of maximum 45°.

For operation they require a special ignition and discharge stabilisation circuit: maximum flux is reached after about 5 min, but after about a minute when the lamp is already hot.

Owing to the high luminous effectiveness and the good quality of the emitted light, this type of lamp is mainly used to illuminate outdoor and industrial premises.

Low Pressure Sodium Vapour Lamps

Similar to the high pressure version they have a glass discharge tube filled with sodium at a very low pressure, protected and thermally insulated by an evacuated glass ampoule. For operation they require a special ignition and discharge stabilisation circuit, maximum flux is reached in a very long time (9–12 min) but restart is immediate. Luminous efficacy is very high but the light is monochromatic yellow, this limits application where colour perception is not necessary.

LED (Light-Emitting Diode) Lamps

These are considered to be the future of lighting and use an operating principle completely different from those of the sources analysed above. The light is produced by a small parallelepiped of semiconductor material, suitably doped, subjected to the flux of an electric current. The light is monochromatic but recently it has become possible to obtain white light by mixing the three primary colours (red, green and blue) or blue-ultraviolet and yellow emitted by fluorescent powders able to convert the ultraviolet light radiation.

A number of semiconductor chips can be used simultaneously to obtain the desired light output but care must be taken to ensure the necessary heat dissipation. Since power must be supplied at low voltage and constant current, an appropriate power supply should be used for the operation.

On the market even more powerful, efficient and low priced led-chip are now available, hence they are in direct competition with traditional lamps even for the most common applications.

The range of colour temperature of the LEDs is very broad and colour quality is very good. In addition to the RGB combination, LEDs can produce any colour in an adjustable manner with continuity. At power-up the luminous flux reaches the maximum value instantly, and on–off switching does not affect the useful life that can attain 150,000 h.

LEDs are easily dimmable from zero to the maximum flux and for a few seconds they can generate flux much higher than the nominal value (flash applications).

Thanks to the small size of the chip which is possible to make light sources of any shape, so they can replace both linear light sources or spotlight and can be used for both diffuse or concentrated lighting. Their flexibility means that they can be applied in any context, both for the illumination of interiors and exteriors. To date, their cost is still quite high but the long life and thus the low maintenance costs repay the initial investment, indeed disposable light fittings, for which no provision for lamp replacement is necessary, are now commercially available.

The replacement of existing lamps with high efficiency lamps brings a significant reduction, even of 80 %, in the electricity consumption for lighting. This is therefore a particularly convenient measure and, if it is not necessary to replace the lighting body, is also inexpensive.

Table 13.75 compares the technical characteristics of the different types of light sources (models not described are shown for comparison). The evaluation of energy saving can be made by comparing the efficiency of the old lamp with the efficiency of the replacement lamp and also the longer lifetime of the latter that can lead to reductions in maintenance costs.

In the assessment of energy benefits it should be considered that less power means a reduction in heat emitted into the environment. During the summer this gives rise to additional energy saving in the air-conditioning system.

Tips and Warnings

In the replacement of an existing lamp, energy efficiency is an important parameter but not the only one to be considered. Qualitative aspects such as the colour rendering index, colour temperature and the starting time are also very important.

Replacing light sources should be included within a more complete lighting project aimed not only at energy saving but also at the improvement of the lighting comfort. The analysis of the "status quo" may lead not only to the replacement of the lamps but also to the replacement, addition or relocation of the light fittings. A complete lighting project will also evaluate lighting management (see sheets 3.LI.03, 3.LI.05, 3.LI.07 and 3.EL.04).

High-efficiency lamps contain harmful substances and, therefore, cannot be disposed of as common wastes but must be brought to designated collection points.

13.5.2.2 Replacement of Lamps for Outdoor Lighting (3.LI.02)

The measure involves the replacement of lamps for outdoor lightning with models of higher efficiency (Table 13.76).

Table 13.75 Technical characteristics of the main types of lamps

Type of lamp	Efficiency (lm/W)	Color rendering index (Ra)	Light flux (min–max) (klm)	Mean duration (hours)	Color temperature (K)	Start time (min.)
Incandescent	10÷20	100	0.1÷20	1,000	2,700÷3,000	0
Halogen high voltage	13÷22	100	1÷44	2,000	3,000	0
Halogen low voltage	17÷18	100	0.35÷1.35	3,000	3.000	0
Fluorescent tube	50÷95	85÷96	0.1÷15.3	5,000÷16,000	3,000÷6,300	0[a]
Compact fluorescent	50÷70	85÷98	0.2÷4.2	5,000÷20,000	2,700÷5,400	0[a]
High pressure mercury lamps	30÷60	<65	1.8÷5.8	10,000÷12,000	3,000÷4,200	4÷6
Mercury vapour mixed light	30÷50	45÷75	3÷32	5,000÷10,000	3,000÷4,200	4÷6
Metal halide	80÷95	65÷95	3÷30	6,000	2,700÷3,000	4÷6
High pressure sodium vapour	90÷120	20÷80	12.5÷38	5,000÷12,000	2,000÷2,500	1÷5
Low pressure sodium vapour	180	0	1.8÷26	8,000÷10,000	1,700	9÷12
LED	50÷120	80÷96	0.1÷6	50,000÷15,0000	3,000÷6,300	0

[a] Start is within seconds, but at low temperature full flux is reached in 1–2 min

Table 13.76 Replacement of lamps for outdoor lighting (3.LI.02): main parameters for the choice

Working	B	Obsolete	A	New	-	O&M	-
Saving potential			□■■■	Payback			□■■■
Reliability			■■■■	Feasibility			■■■■
Environmental effects			EI				

General Description

The measure consists of the replacement of existing lamps in outdoor spaces with high-efficiency lamps in order to reduce the consumption of energy.

The lamps that are used for outdoor lighting are practically the same as those used for lighting internally, for their description and associated comments therefore, please refer to the information already provided in the sheet 3.LI.01.

The replacement of existing lamps with high efficiency lamps permits a significant reduction, even of 80 %, in the electricity consumption for lighting.

This is therefore a particularly convenient measure and, if it is not necessary to replace the lighting body, is also inexpensive.

The evaluation of energy saving can be made by comparing the efficiency of the old lamp with the efficiency of the replacement lamp and also the longer lifetime of the latter that can lead to reductions in maintenance costs.

Tips and Warnings

In the replacement of an existing lamp, energy efficiency is an important parameter but not the only one to be considered. Qualitative aspects such as the colour rendering index, colour temperature and the starting time are also very important. Replacing light sources should be included within a more complete lighting project aimed not only at energy saving but also at the improvement of the lighting comfort. The analysis of the "status quo" may lead not only to the replacement of the lamps but also to the replacement, addition or relocation of light fittings. A complete lighting project will also evaluate lighting management (see measures 3.LI.03, 3.LI.05, 3.LI.07 and 3.EL.04).

High-efficiency lamps contain harmful substances and, therefore, cannot be disposed of as common wastes but must be brought designated collection points.

13.5.2.3 Installation of Presence Detectors (3.LI.03)

The measure consists of the installation of presence detectors capable of automatically switching off artificial lighting in rooms when there is no one present (Table 13.77).

General Description

The objective of this measure is to handle automatically the switching on or off of artificial lights within the premises as a function of the presence of the people. The technologies used are three: passive infrared sensors, ultrasonic sensors and dual technology sensors.

Table 13.77 Installation of presence detectors (3.LI.03): main parameters for the choice

Working	C	Obsolete	A	New	A	O&M	-
Saving potential			□■	Payback			■■■
Reliability			■■■	Feasibility			■■■
Environmental effects			EC				

The *passive infrared sensors* are activated by the presence of thermal sources in the infrared range such as the human body. Analysing the difference between the energy emitted by bodies present and that from the surrounding environment, the sensors can detect the presence of persons and activate the load. In order to function properly and effectively, the passive infrared sensors require an unobstructed view of the environment in which they must operate.

Ultrasonic sensors emit non audible sound waves that hit the objects in the volume of space in which they operate, then measuring the time the wave requires to return. When there is movement, the sound waves change in wavelength and sensors can thus detect the presence of persons and if necessary activate the load. Ultrasonic sensors are ideal in environments with obstacles and moving people.

The *dual-technology sensors* utilise both of the technologies described above in an AND/OR mode for maximum flexibility and detection sensitivity.

Detectors are only the sensor elements of the system and must be connected to the devices set-up for the management of the activation signals. The sensors can be mounted on the ceiling or installed on the walls in the same way as a normal burglar alarm (this solution is less valid aesthetically but functional and suitable for retrofit).

Avoiding the (unnecessary) lighting of spaces when not in use by people provides an energy saving which is the product of the lighting power and the switch off time. Reducing working time lengthens the operating time of the lamps with a valuable economic saving. There are also the environmental benefits of less energy, pollution and wastes.

Tips and Warnings

Presence detectors are usually installed in transit areas, such as hallways. The initial darkness of spaces can be uncomfortable or generate panic. In such cases it may be useful to install the sensor so that the detection is timely or, better yet, provide two levels of light: in addition to normal lights activated by the presence of people, a permanent security low-level illumination.

Placement of sensors must be carefully planned considering the different points of entry into the zone. To obtain a greater saving the installation of the presence detectors can also be provided in other areas than hallways, choosing areas where disabling lighting is acceptable to the user customer.

In the case of areas which are not of transit or passageways but rather areas where people stay without moving, the dual-technology sensors must be chosen.

Only lamps that require very short ignition timing must be chosen, especially if switched-on time is very short.

Lighting management with detection of the presence of people is part of a more general strategy that aims at reducing the energy consumption without decreasing service quality.

These sensors can be installed alone or paired with other technologies, such as timers (see measure 3.EL.04), daylight sensors (see measure 3.LI.05) or dimmers (see measure 3.LI.07).

13.5.2.4 Installation of Twilight Switches (3.EL.04)

The measure consists of the installation of twilight switches to manage lighting automatically as a function of natural light (Table 13.78).

General Description

The twilight switch is a device equipped with a light sensor that, when properly adjusted, is automatically activated when it becomes dark (at dusk) and deactivated when it becomes light (at dawn). Hence lights are switched on automatically whenever the lighting of the place in which it is installed becomes low, regardless of the season and/or time of day.

The switch's most common format is that of a white globe which contains the sensor and in some cases also the control switch, which can be electronic or electromechanical depending upon the model. If control unit is separate, it can be mounted in the electrical control panel so that calibration is easier.

The twilight switches help to reduce the energy consumption for lighting with little investment. Especially for outdoor lighting this solution is better than the most sophisticated time switch. (see measure 3.EL.04) The lamps come on when natural light falls below a certain threshold level, a circumstance which not only varies from day to day but also as a function of the cloud cover. The twilight switch, apart from its energy saving role, ensures regular illumination levels in all conditions.

Tips and Warnings

For proper operation of the twilight switch, the device must not be affected by the light from the lamps that it controls. The twilight switch can be easily integrated in existing plants without requiring substantial changes.

Table 13.78 Installation of twilight switches (3.EL.04): main parameters for the choice

Working	B	Obsolete	A	New	-	O&M	-
Saving potential			□■	Payback			■■■
Reliability			■■■	Feasibility			■■■
Environmental effects			E				

13.5.2.5 Installation of Daylight Sensors (3.EL.05)

The measure consists of the installation of daylight sensors to adjust artificial lighting as a function of available natural light (Table 13.79).

General Description

The objective of this measure is to manage artificial lighting according to the availability of natural light, avoiding artificial illumination of rooms where or when natural lighting is sufficient.

Using daylight sensors the adjustment of artificial lights can happen in two ways: either on–off or gradually. The gradual adjustment allows better integration of natural light with artificial support, improving visual comfort and energy savings.

Maintaining an artificial lighting system switched on when natural lighting is sufficient to ensure the correct illumination of the premises is a primary cause of energy wastage in lighting systems.

This measure is effective and the related energy benefits are considerable, in particular for those premises in which the contribution of natural light is high.

Reduced use lengthens the operating time of the lamps with a valuable economic saving. There are also the environmental benefits of less energy, pollution and wastes.

Tips and Warnings

To render this measure fully effective it is necessary that the lamps used in the lighting fixtures are compatible with this kind of control. It is important to position properly the sensors to obtain correct measurement.

Lighting management based on natural light levels is part of a more general strategy that aims at reducing the energy consumption for lighting without sacrificing the quality of service.

These sensors can be installed alone or paired with other technologies, such as timers (see measure 3.EL.04), presence detectors (see measure 3.LI.03) or dimmers (see measure 3.LI.07).

Table 13.79 Installation of daylight sensors (3.EL.05): main parameters for the choice

Working	C	Obsolete	A	New	-	O&M	-
Saving potential			■	Payback			■■■
Reliability			■■■	Feasibility			■■■
Environmental effects			EC				

13.5.2.6 Installation of High Frequency Electronic Ballasts (3.EL.06)

The measure consists of replacing magnetic ballasts with high frequency electronic ballasts (Table 13.80).

General Description

Discharge lamps (fluorescent, metal halide, sodium etc.) require a special transformer for their ignition and operation, commonly known as "ballast". This component causes dissipation of energy which is lower when using high frequency electronic ballasts. The new ballasts allow faster ignition, no flicker and increased lamp life.

Often, the ballast installed inside the light fixture can be easily replaced with a new one, but sometimes the replacement is not feasible or not convenient, and in such cases it is necessary to replace the entire fixture.

The new electronic ballasts allow energy savings of up to 20 %. The environmental benefits are linked to lower emissions for the electricity generation and the increased useful life of the lamps.

Tips and Warnings

First and foremost a check on the state of the light fixtures is required in order to establish which are in a good state or those for which a complete replacement would be preferable.

When replacing the ballasts, the cleaning of the light fixtures should also be performed and, where possible, the replacement of lamps with more efficient compatible models should be made. Electronic ballasts which allow the adjustment of the luminous flux are available on the market, see measure 3.LI.06.

13.5.2.7 Installation of Dimmers (3.EL.07)

The measure consists of the installation of dimmers so that the users can adjust lighting levels to meet their preferences (Table 13.81).

Table 13.80 Installation of high frequency electronic ballasts (3.EL.05): main parameters for the choice

Working	C	Obsolete	A	New	-	O&M	-
Saving potential			■	Payback			■■■
Reliability			■■■■	Feasibility			■■■
Environmental effects			EC				

Table 13.81 Installation of dimmers (3.EL.07): main parameters for the choice

Working	C	Obsolete	A	New	-	O&M	-
Saving potential			■	Payback			■■■
Reliability			■■■	Feasibility			■■■
Environmental effects			EC				

General Description

This measure consists of the installation of devices that allow the user to adjust the light level as a function of the visual task to be performed.

Dimmers control the power transferred to the load, limiting it at will. Older technology used a rheostat for voltage regulation, with substantial losses. The current electronic dimmers act by varying the time in which the power is supplied to the load (duty cycle modulation), with little energy loss.

There are dimmers suitable for low and high power equipment and control can be effected directly by means of a button, knob, touch sensor, or remotely via a control signal that may originate, for example, from a centralised management system, home automation or from a light sensor.

Common lighting dimmers are used for adjusting the light intensity of incandescent lamps or halogen lamps, but cannot be used for the adjustment of discharge lamps unless they are equipped with special ballast that permits the variation of the supply voltage. The best solution is to use appropriate dimmable ballast with input for adjusting the luminous flux because this is more efficient. The technology can also be applied to the lighting of outdoor spaces, to operate at reduced capacity during the middle of the night, when sites are rarely visited.

The use of low loss electronic dimmers result in a lower consumption of energy and, therefore, in environmental benefits.

Tips and Warnings

The use of this technology brings benefits from an energy point of view, even though, most often, the installation of dimmers is mainly aimed at managing lighting quality. In the past particularly, dimmers were used only with certain types of lamp that are among the least efficient. If the goal is to save energy in lighting then it is perhaps appropriate to review the entire project, first changing the light sources, even if dimmer installation alone could achieve energy savings.

The lighting management using the dimmer technology is part of a more general strategy that aims at reducing the energy consumption for lighting without sacrificing the quality of the service.

The dimmer can be installed alone or combined with other technologies, such as that of timers (see measure 3.EL.04), the daylight sensors (see measure 3.LI.05) or presence detectors (see measure 3.LI.03).

13.5.3 Building Management Systems and Remote Controls

13.5.3.1 Installation of a Home Automation System (3.DO.01)

The measure provides for the installation of a home automation system in order to reduce the energy consumption by optimising the use of facilities and improving comfort for occupants (Table 13.82).

General Description

Home automation, also denominated domotics, is the interdisciplinary science that deals with the study of technologies to improve the quality of life of end-users. It can be considered to be a field of building automation specialised in the specific automation requirements of private homes and in application of automation techniques for the comfort and security of the resident.

The evolution of automation technologies makes it difficult to distinguish the boundary between home and building automation which often use the same equipment and products: the main difference is, however, the human interface. In home automation ergonomics is a particularly important factor since the control should be largely image-based and self-explanatory.

The objectives of a home automation systems can be summarised into four categories:

- improvement of the quality of life;
- improvement in safety and security;
- more efficient energy usage;
- reduction in operating costs.

A home equipped with a home automation system can be controlled by the user via user-interfaces such as buttons, remote controls, touch-screens and keyboards that allow him to send commands and receive information. The initial programming of the system is done by the contractor who performs the.

installation or indeed by the vendor's factory, but the user can change some parameters, adapting the software to the actual needs.

Table 13.82 Installation of a home automation system (3.DO.01): main parameters for the choice

Working	B	Obsolete	A	New	-	O&M	-
Saving potential			■	Payback			NA
Reliability			■■■	Feasibility			■■■
Environmental effects			**ECIC**				

The elements of a home automation system are: controllers, sensors and actuators. The interconnections between the elements can be wired (coaxial and twisted-pair cables, optical fibre) or wireless (Wi-Fi, Bluetooth, DECT, GPRS UMTS).

A home automation system plays an important role in making the building and the activities more sustainable. Here are some examples of features that a home automation system is able to manage:

- automatic operation based upon the recognition of the presence of persons;
- adjustment of the indoor environmental parameters (air temperature, relative humidity, air flow rate);
- automatic shut-off of HVAC terminals if windows are open;
- energy accounting functions;
- appliance management;
- isolation and automatic protection during thunder storms;
- lighting control;
- management of electrical loads;
- automatic opening and closing of blinds and outdoor awnings.

An important feature of the home automation systems is its ability to communicate with the outside world by exploiting the available technologies (e.g., Internet and mobile 'phone network), in this way it is possible to display the status also remotely.

Tips and Warnings

A home automation system must be flexible in order to meet current but also possible future needs. It must also be simple since the end-user normally has not advanced technical skill and it must be reliable.

13.5.3.2 Installation of a Building Automation System (3.DO.02)

The measure provides for the installation of a building automation system in order to reduce the energy consumption by optimising the use of facilities and improving comfort for occupants (Table 13.83).

Table 13.83 Installation of a building automation system (3.DO.02): main parameters for the choice

Working	B	Obsolete	A	New	-	O&M	-
Saving potential			■	Payback			NA
Reliability			■■■	Feasibility			■■■
Environmental effects			ERIC				

General Description

Functional integration is the true innovation introduced by building automation: the different areas (e.g., security, safety, HVAC, lighting, communication etc.), previously considered independent, interact, communicate and create synergies.

A building automation system is an example of a distributed control system. The control system is a computerised, intelligent network of electronic devices designed to monitor and control the mechanical, electronic, and lighting systems in a building.

The architecture of a building automation system can be expanded in the field to match the technological and functional evolution of a building. A building automation system plays an important role in making the building and the activities carried out therein more sustainable (see measure 3.DO.01 for a list of typical features).

Tips and Warnings

The advantages offered by building automation are many: increased safety and security, better indoor environmental comfort, improved sustainability of the building and reduction in operating costs. However the correct design of a building automation system is not particularly easy: the difficulty is not the available of technology but the ability to put together all that technology offers in order to define a flexible, simple and reliable system.

Even before defining the structure of the system it is essential to address a correct and in-depth analysis to the real needs of the user, finding the best solution to meet them.

A building automation system is a perfect tool to support the implementation of an energy management system model in accordance with the ISO 50001 standard.

13.5.3.3 Installation of a Remote Control System (3.DO.03)

The measure provides for the installation of a remote control system in order to reduce the energy consumption by optimising the use of facilities and improving comfort for occupants (Table 13.84).

Table 13.84 Installation of a remote control system (3.DO.03): main parameters for the choice

Working	B	Obsolete	A	New	-	O&M	-
Saving potential			■	Payback			NA
Reliability			■■■	Feasibility			■■■
Environmental effects			EC				

General Description

A remote control system could be considered to be an extension of the building automation system which allows for an external permanent service of control and monitoring of the building's facilities.

By using a remote control system, an external service company is able to manage the building facilities also for clients who do not have an internal staff (e.g., public buildings, schools etc.).

Tips and Warnings

A remote control system combines two important elements: the technology of the system (hardware and software) and the human factor, intended as the technical competences of the external personnel charged of the supervision. Professionalism and skill on the part of the said personnel is therefore a key element for the success of the remote control service.

13.6 Renewable Energy Sources

13.6.1 Solar Energy Systems

13.6.1.1 Installation of a Thermal Solar System for Low Temperature Uses (4.RS.01)

The measure provides for the installation of a thermal solar system for hot water production compatible with low temperature uses (Table 13.85).

General Description

Solar energy can be converted into thermal energy using thermal solar system. A typical solar system essentially consist of:

Table 13.85 Installation of e thermal solar system for low temperature uses (4.RS.01): main parameters for the choice

Working -	Obsolete -	New A	O&M -
Saving potential	■■	Payback	■■■
Reliability	■■■	Feasibility	■■■■
Environmental effects	EC		

- *solar collectors*, devices whose function is to convert radiant energy from the sun into thermal energy that is transferred to the heat transfer medium (normally water);
- a *solar storage tank*, whose function is to accumulate thermal energy within a mass of water in order to guarantee to the users availability of heat energy also when solar radiation is not available (for a period depending upon the volume of water). The device is equipped with a heat exchanger in order to separate the primary solar circuit from the user circuit;
- a *primary solar circuit* equipped with one or more pumps, whose function is to guarantee the circulation of the heat medium between the solar collectors and the solar storage tank;
- a *control system*, equipped with two sensors (a sensor installed in the pilot collector and a sensor installer in the solar storage tank) whose function is to activate the primary solar circuit (i.e., switch on the pump) only if the thermal energy output from the solar collectors is effectively useful for the hot water system.

As regard the solar collectors, the most diffused are the flat plate collectors and the vacuum solar collectors. In addition to the components listed above, which permit sizing of a thermal solar system on the basis of the energy demand, the market also offers solar water heaters, i.e., compact solar systems in which the circulation of the heat medium between solar collector and storage tank is by natural convection i.e., without the need for a pump.

Thermal solar systems are cost-effective for all the applications that require hot water at a relatively low temperature, for example:

- domestic hot water;
- integration for heating systems;
- reheat for HVAC systems;
- swimming pools.

Thermal solar plants may also be used for high temperature applications, such as summer cooling (see measure 4.RS.02) (Fig. 13.20).

Tips and Warnings

The main constraint to the installation of a solar thermal pant is the space available in which to install the solar collectors: a well-oriented surface (possibly in the direction of the equator) and tilted according to the latitude of the location. Solar collectors normally are installed on pitched roofs or flat roof (in this case a support is necessary in order to maintain the correct orientation and tilt angle).

A not perfect orientation (i.e., ±45° over East or West) is nonetheless acceptable and the performance is little affected. Other orientations, up to 90° over East or West, are still acceptable but require an integration of the solar surface. the general rule is to make some effort to integrate the solar collectors with the architecture of the buildings.

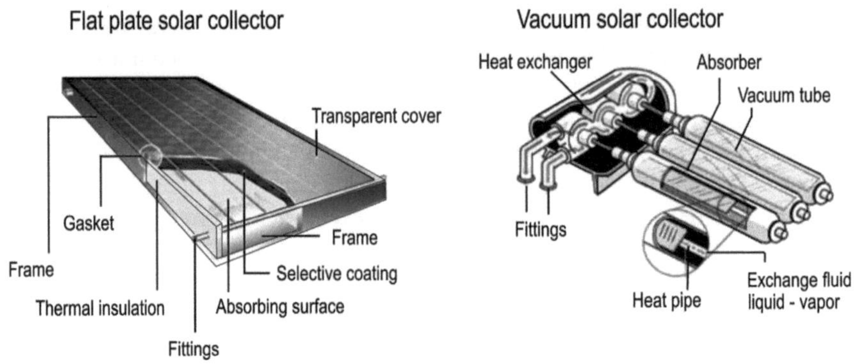

Fig. 13.20 Example of solar collectors: flat plate collector (on the *left*), vacuum solar collector (on the *right*)

Solar thermal systems need an energy integration with a conventional system (e.g., boiler): for average latitudes (around 30–60°) it is not cost-effective to size the thermal solar system for more than 70 % of annual integration.

In order to guarantee the performance, and above all the durability, of a thermal solar system over time a maintenance programme should be drawn-up and scheduled.

13.6.1.2 Installation of a Solar Heating and Cooling System (4.RS.02)

The measure provides for the installation of a thermal solar system that, by means of an absorption chiller, is able to supply energy for cooling demand. The same system is also used for heating in the winter season (Table 13.86).

General Description

During the summer the demand for electricity increases because of the ever more extensive use HVAC systems (even in more temperate climates), which increase the peak electric load, causing major problems in electric supply.

Table 13.86 Installation of a solar heating and cooling system (4.RS.02): main parameters for the choice

Working	C	Obsolete	B	New	B	O&M	-
Saving potential			■	Payback			□■■■
Reliability			■■■	Feasibility			■■■■
Environmental effects		EC					

The use of solar energy to drive cooling cycles for space air-conditioning of most buildings is an attractive concept, since the (peak) cooling load coincides generally with the (greatest) solar energy availability and therefore cooling requirements of a building are approximately in phase with the solar incidence.

Solar cooling systems have the advantage of using absolutely harmless working fluids such as water, or aqueous solutions of certain salts. They are both energy efficient and environmentally safe.

The principle of operation of a solar cooling system is quite simple: solar collectors heat the mass of water contained in the solar storage tank which has the function of a thermal buffer. When the water reaches the right temperature to activate the process (higher than 80–85 °C) the hot water coming from the storage tank feeds the generator of the absorption chiller which in turn produces chilled water for the HVAC terminals. A solar cooling system needs also a cooling tower, it is also possible to provide the installation of an integration boiler to use for feeding the absorption chiller in the periods in which the temperature of the water inside the solar storage tank does not attain the minimum temperature required. For solar cooling other solutions are available, e.g., desiccant cooling, however the scheme described above is that most common.

Tips and Warnings

Solar heating and cooling systems are the best way to control internal climate using solar energy as energy source. The possibility of using solar energy in summer for cooling, the coincidence of the cooling loads with the availability of solar energy and the potential of the solar system are key factors for the success of these systems, at least from a technological point of view (Fig. 13.21).

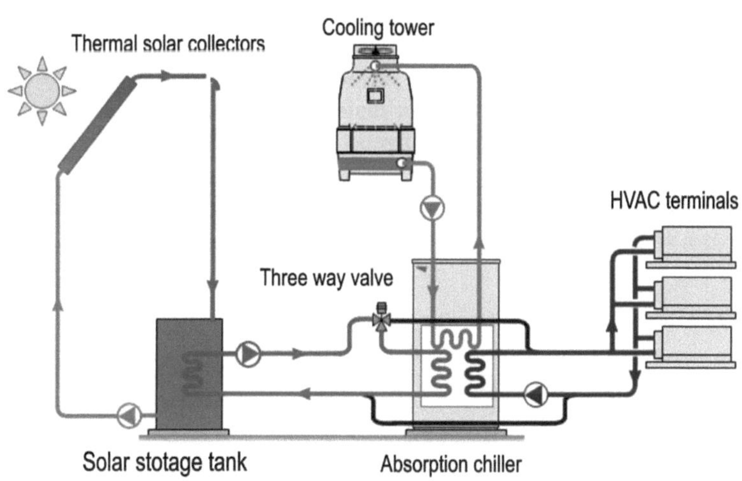

Fig. 13.21 Typical scheme of a solar cooling system (courtesy of Yazaki Co.)

Another point to consider is that these systems, although not yet very common, have been applied since the 1970s, so the technology is mature and reliable.

However, the first cost of these systems is still high when compared with conventional HVAC systems. Another critical aspect is the limited availability on the market of models of absorption chillers suitable to operate with the supply temperatures compatible with solar technologies.

For these reasons, before planning the installation of this interesting application of solar energy it is necessary to make some effort to reduce the thermal loads of the building in order to reduce the heating and cooling capacity of the system. A second suggestion is to design the cooling capacity of the solar system covering 60 % of the peak load (this means to cover 90 % of the annual energy demand). So far as integration is concerned, two solutions are possible: the first is to install a boiler to heat the absorption chiller, the second one is to install a compression cycle reversible chiller (able to work as heat pump for winter heating) in parallel with the absorption chiller. The second solution is the most convenient as the compression chiller guarantees a complete cooling back-up/reserve.

In order to guarantee the performance, and above all the durability, of a thermal solar system over time a maintenance programme should be drawn-up and scheduled.

13.6.1.3 Installation of a Photovoltaic (PS) Solar System (4.RS.03)

The measure provides the installation of a PV solar system aimed at producing electrical energy (Table 13.87).

General Description

Photovoltaic (PV) technology permits the conversion of solar energy directly into electrical energy that can be used for various purposes. The PV effect is based on the proprieties of some semiconductor materials (most commonly silicon) which, if appropriately treated, are capable of generating electricity when exposed to solar radiation.

The basic device able to effect a conversion of solar energy into electrical energy is a PV cell, the set of multiple cells, on average 50–60, constitute a PV

Table 13.87 Installation of a photovoltaic (PV) solar system (4.RS.03): main parameters for the choice

Working	A	Obsolete	A	New	-	O&M	-
Saving potential			□■■■	Payback			■■■
Reliability			■■■■	Feasibility			□□■■
Environmental effects			E				

module that has a gross surface of about 1 m^2. A PV field is a set of PV modules, connected in series and in parallel, so as to achieve the desired operating conditions. The most diffuse modules are crystalline silicon, either mono-crystalline with an efficiency of about 13–17 % or polycrystalline with efficiency approx. 11–15 %. The amorphous silicon modules, which are less expensive, have a lower efficiency, approx. 6–8 %.

PV systems can be stand-alone or grid-connected. In grid-connected systems, a frequency inverter is installed with the function to transform DC current into AC current.

Tips and Warnings

The ability of grid-connected PV solar systems to exchange energy with the grid and then use the grid as an energy storage vehicle, together with the incentives that many countries have made available to users, are the key factors for the diffusion of these systems all over the world. The enlargement of the solar PV market has contributed to a dramatic reduction in the prices of the equipment and it is expected that in a few years time grid parity[2] can be achieved and financial support of solar PV systems will no longer be required.

For solar PV plants the main constraint to the installation is the available space in which to install the solar modules: on this issue the principles are the same as those for solar thermal collectors (refer to the tips and warnings of measure 4.RS.01).

Solar PV system technology is better suited to an integration with the architecture of the buildings and this will undoubtedly contribute to these systems becoming widespread all over the world.

13.6.2 Biomass

13.6.2.1 Installation of a Biomass Boiler (4.RS.04)

The measure covers the installation of a boiler powered with biomass originating from vegetation (Table 13.88).

General Description

Biomass heating systems generate heat, which can be used for non-industrial applications (e.g., winter heating and DHW) from biomass, a renewable energy

[2] Grid parity occurs when an alternative energy source can generate electricity at a cost that is less than or equal to the price of purchasing power from the electricity grid. Reaching grid parity is considered to be an important point in the development of new sources of power.

Table 13.88 Installation of a biomass boiler (4.RS.04): main parameters for the choice

Working C	Obsolete B	New B	O&M -
Saving potential	■■	Payback	□■■
Reliability	■■	Feasibility	■■■
Environmental effects	E		

source. This type of energy production has a limited long term effect on the environment because the carbon in biomass is part of the natural carbon cycle; while the carbon in fossil fuels is not, and permanently adds carbon-dioxide to the environment (atmosphere) when burned for fuel. An interesting feature of biomass is that, unlike other renewable sources, can be stored for being used when required.

For non-industrial application biomass of plant origin is used, made available on the market in various forms: firewood logs, wood pellets and wood chips.

Modern biomass boilers using wood pellets or wood chips are equipped with automatic feeding systems which draw the fuel from the buffer vessel and burn it respecting environmental standard. These boilers have emissions of SO_2 (Sulphur Dioxide) similar to, or lower than those of conventional systems. Values for NO_X (Nitrogen Oxides) and CO (Carbon Monoxide) are slightly higher than conventional, whilst the dust emissions are higher.

Tips and Warnings

The main critical aspects of woody biomass systems are as follow:

- a greater space required for the placement of the boiler, of the buffer tank/storage vessel and to allow the access of the vehicles for refueling of the wood pellet or the wood chips;
- the supply of the wood fuel is not always available locally (indeed it is ecologically sound to ensure locally derived fuel to avoid negative environmental effects from significant road transport of the fuel to site);
- the more complex maintenance for the boiler;
- necessity of indoor storage
- emission of dust and other polluting substances which involve additional safety considerations and complicate handling requirements.

13.6.3 Green Energy

13.6.3.1 Purchase of Certified Energy from Renewable Sources (4.RS.05)

The measure is for the activation of a contract for the supply of certified energy from renewable sources (green energy) (Table 13.89).

Table 13.89 Purchase of certified energy from renewable sources (4.RS.05): main parameters for the choice

Working	C	Obsolete	A	New	-	O&M	A
Saving potential			NA	Payback			NA
Reliability			NA	Feasibility			■■■■
Environmental effects			E				

General Description

The production of energy from renewable energy sources sometimes cannot be done locally (production on-site) owing to technical constraints (e.g., lack of space for installation of solar PV modules or solar collectors) or economics constraints (e.g., lack of available funds for the investment).

Renewable energy, however, could be purchased from external producers (production off-site), thus contributing to promote the sustainability of the territory.

Purpose of this measure is to activate contracts for the supply of certified green energy. The LEEDS protocols reward this measure if certain requirements are met (at least 35 % of the electricity needs of the building must be satisfied with energy produced from renewable sources by means of certified supply contracts for a period of at least 2 years).

Tips and Warnings

The energy must be certified by e.g., RECSRenewable Energy Certificate System(RECS) or other similar certification systems recognised internationally.

13.7 Improvement in Management of the Building

13.7.1 Reduction in Operation Times for HVAC Systems (5.MI.01)

The measure consists of checking the periods of use of HVAC systems and a possible new scheduling of the hours of operation (Table 13.90).

13.7.1.1 General Description

The checks carried out by the audit on-site and especially the results of the monitoring campaign could highlight inconsistencies between the hours of use of the indoor spaces and the times of activation of HVAC systems.

Table 13.90 Reduction of operation times for HVAC systems (5.MI.01): main parameters for the choice

Working	-	Obsolete	-	New	-	O&M	B
Saving potential			□■	Payback			□□■■
Reliability			NA	Feasibility			■■■■
Environmental effects			**EC**				

The objective of this measure is to remove these critical aspects by rescheduling, where possible, the times of activation of HVAC systems according to the actual hours of usage of the spaces concerned, thus avoiding any unnecessary air-conditioning.

13.7.1.2 Tips and Warnings

The new schedule of the times of activation of HVAC systems should be agreed with the client. This measure represents no extra cost if a building management system is available (see measures 3.DO.01, 3.DO.02, 2.DO03) or if it is possible to operate, modifying the time schedules on the control systems. See also measure 5.MI.02.

13.7.2 Control of Indoor Environmental Conditions (5.MI.02)

The measure consist of scheduling a check on the indoor environmental conditions in order to save energy, still maintaining high levels of thermal comfort (Table 13.91).

13.7.2.1 General Description

A regular monitoring of the environmental conditions of the indoor spaces (e.g., air temperature, relative humidity, CO_2 concentration) permits the detection of potential problems and implementation of action to restore the optimum situation.

Table 13.91 Checks on indoor environmental conditions (5.MI.02): main parameters for the choice

Working	-	Obsolete	-	New	-	O&M	A
Saving potential			■	Payback			□■■■
Reliability			NA	Feasibility			■■■■
Environmental effects			**EC**				

This measure consists of programming monitoring campaigns to verify at least twice a year (once in the winter and once in the summer) the indoor environmental conditions.

13.7.2.2 Tips and Warnings

The effect of this measure is a reduction of possible energy wastage but above all the maintenance of the best indoor environmental conditions in terms of the thermal comfort of occupants.

The scheduling of the monitoring campaigns should be discussed and agreed with the client.

13.7.3 Deactivating Standby (5.MI.03)

The measure is the adoption of all the strategies to deactivating standby of the electrical equipment, when not used, to save energy (Table 13.92).

13.7.3.1 General Description

The number of electronic devices in homes and offices, in recent years has grown immensely. Almost all devices are equipped with standby, introduced to speed-up the activating phase and in some case to permit the use of electronic remote controls.

The energy consumption of equipment, particularly older items, with standby activated is high if one considers the significant length of time during which the devices are in this state (in some cases printers and personal computers remain on standby throughout the entire weekend).

This measure consists of adopting all the strategies to deactivate standby of the electrical equipment, when not used, to save energy.

13.7.3.2 Tips and Warnings

The strategies of standby deactivation should be discussed and agreed with the client.

Table 13.92 Deactivating standby (5.MI.03): main parameters for the choice

Working -	Obsolete -	New -	O&M A
Saving potential	■	Payback	□□■■
Reliability	NA	Feasibility	■■■■
Environmental effects	EC		

13.7.4 Control of DHW Temperature (5.MI.04)

The measure consists of the reduction of the set-point temperature of the DHW in order to reduce the distribution heat losses (Table 13.93).

13.7.4.1 General Description

The cases in which the domestic hot water is distributed to the user at a temperature higher than the required value are quite frequent: this situation then requires the users to mix that hot water with cold water in order to obtain the right temperature.

The reduction of the temperature of DHW through a simple adjustment of the thermostat or of the control system brings a significant reduction in thermal losses along the distribution circuit and thus saves energy.

13.7.4.2 Tips and Warnings

Verify that, by adjusting the DHW temperature in the boiler room/central hot water plant room, the hot water reaches the users furthest away (normally the most unfavorable situation) at an acceptable temperature. If the temperature difference between the starting point in the plant room and the point corresponding to the farthermost user(s) is relevant, the thermal insulation of the distribution circuit should be checked-over (see measure 2.HD.01).

13.7.5 Maintenance of Lighting Fixtures (5.MI.05)

The measure consists of the periodic maintenance of the lighting fixtures in order to maintain the high performance of the lighting system, thereby reducing energy consumption (Table 13.94).

Table 13.93 Control of DHW temperature (5.MI.04): main parameters for the choice

Working	-	Obsolete	-	New	-	O&M	A
Saving potential			■	Payback			□□■■
Reliability			NA	Feasibility			■■■■
Environmental effects			EC				

Table 13.94 Maintenance of lighting fixtures (5.MI.05): main parameters for the choice

Working	-	Obsolete	-	New	-	O&M	A
Saving potential			NA	Payback			NA
Reliability			NA	Feasibility			■■■■
Environmental effects			**EC**				

13.7.5.1 General Description

A scheduled maintenance of the lighting fixtures in one of the activities of the building maintenance plan. The objective of this measure is to ensure the continued high performance of these equipment items, so reducing energy consumption but also maintaining lighting comfort for occupants over time.

Maintenance operations, to be carried out on a three monthly or six monthly basis, depending upon the activities, are as follows:

- removal of the lamps;
- verification of the integrity of the appliances;
- cleaning of the fixture, in particular of the transparent screens, grids and reflecting elements;
- verification of the lamps and the eventual need for their replacement;
- reinstallation of the lamps.

Instrumental checks should be made periodically to ensure that the lighting requirements are met.

13.7.5.2 Tips and Warnings

Maintaining the quality of the lighting equipment improves lighting comfort and, in a workplace, improves productivity.

13.7.6 Drafting of an Instruction Manual for the Users (5.MI.06)

The measure consists of the drafting of an instruction manual for the users in order to facilitate the understanding of how to operate the building facilities efficiently (Table 13.95).

Table 13.95 Drafting instruction manual for users (5.MI.06): main parameters for the choice

Working	-	Obsolete	-	New	-	O&M	B
Saving potential			■	Payback			□□■■
Reliability			NA	Feasibility			■■■■
Environmental effects		EC					

13.7.6.1 General Description

End-users rarely have enough technical and practical knowledge to understand how to use properly the equipment of the building facilities. This is sometimes one of the root causes of energy wastage. This lack of knowledge can be overcome by the drafting of a small user manual which contains the necessary information.

13.7.6.2 Tips and Warnings

The manual should be clear, concise and easy to read. In writing the manual it is important to take fully account of the fact that the intended readers will not be experts.

13.7.7 Activation of Reward Strategies (5.MI.07)

The measure consists on the activation of reward strategies in order to encourage the end-users to use energy more efficiently (Table 13.96).

13.7.7.1 General Description

A lot of energy wastage, both thermal and electrical, can be avoided if the end-user is more involved in and responsible for the results of energy usage. Reward strategies, already tested with positive results in offices and schools, may be implemented through agreements whereby the economic benefits thanks to energy savings are shared with the employees, or the pupils/students in the case of (high) schools.

Table 13.96 Activation of reward strategies (5.MI.07): main parameters for the choice

Working	-	Obsolete	-	New	-	O&M	B
Saving potential			■	Payback			□□■■
Reliability			NA	Feasibility			■■■■
Environmental effects		EC					

13.7.7.2 Tips and Warnings

This measure should be accompanied by a training phase and a user manual (see measure 5.MI.06).

13.7.8 Scheduling of Monitoring Procedures (5.MI.08)

The measure involves the programming of monitoring procedures in order to verify the management process of the building facilities (Table 13.97).

13.7.8.1 General Description

The scheduling of monitoring procedures is one of the measures that allow control of the energy management of the facilities. Monitoring can be performed periodically by checking the indoor environmental conditions and comparing these with the design conditions.

13.7.8.2 Tips and Warnings

For a comprehensive approach to efficient energy management in a building it is always useful to implement an energy management system model in accordance with the ISO 50001 standard.

13.7.9 Scheduling of Energy Accounting Procedures (5.MI.09)

The measure consists of the scheduling of energy accounting procedures in order to verify the management process of the facilities (Table 13.98).

Table 13.97 Scheduling of monitoring procedures (5.MI.08): main parameters for the choice

Working	-	Obsolete	-	New	-	O&M	B
Saving potential			□■	Payback			□□■■
Reliability			NA	Feasibility			■■■■
Environmental effects			**E**				

Table 13.98 Scheduling of energy accounting procedures (5.MI.08): main parameters for the choice

Working	-	Obsolete	-	New	-	O&M	B
Saving potential			□■	Payback			□□■■
Reliability			NA	Feasibility			■■■■
Environmental effects			E				

13.7.9.1 General Description

The scheduling of energy accounting procedures is one of the measures that allow control of the energy management of the facilities. Energy accounting can be performed periodically by checking the energy consumption of the various users.

Internal benchmark indicators are useful for this measure.

13.7.9.2 Tips and Warnings

For a comprehensive approach to efficient energy management in a building it is always useful to implement an energy management system model in accordance with the ISO 50001 standard.

Glossary

Absorption Chiller A refrigeration machine which uses heat as the power input to generate chilled water. The heat can be supplied indirectly, through a heat medium (e.g., hot water, high pressure hot water, steam) or directly using a fuel burner.

Aeraulic System A heating and/or cooling system that uses air as the medium for heat transfer.

Air Handling Unit (AHU) Takes return air from the conditioned space, mixes it with fresh air, filters it, provides or removes heat and returns it to the conditioned space. Normally comprised of a fan, cooling/heating coils, humidifiers and filters.

Air Changes per Hour (ACH) Is a measure of how many times the air within a defined space is replaced.

Air sealing Air-sealing is the practice of filling gaps, holes or cracks in a building's envelope in an effort to tighten the home's pressure boundary.

ASHRAE American Society of Heating, Refrigerating and Air-Conditioning Engineers.

Audit report A document that informs the client about what has been done during the audit and the measures that can be implemented to make the building or facility more energy efficient and more sustainable.

Average Occupancy Normally is the average number of people in a building or area over a 24-hour period (it is possible to specify also different periods).

Ballast An electrical device used for starting and regulating fluorescent and high-intensity discharge (HID) lamps.

G. Dall'O', *Green Energy Audit of Buildings*, Green Energy and Technology,
DOI: 10.1007/978-1-4471-5064-0,

Baseline Energy consumption or energy cost for a specific period of time in which one will have to compare the costs or future use after the retrofit measures will be implemented.

Benchmark Also called reference parameters, are indices commonly used to compare the performance of a system. Benchmarks can also be used to assess the improvement of the performance of a building or a system after the implementation of an energy saving strategy.

Blower door test A diagnostic tool designed to measure the air-tightness of buildings and to help locate air leakage sites.

Boiler Capacity Rate of heat output measured at the boiler outlet at the design pressure and/or temperature.

BREEAM (Building Research Establishment Environmental Assessment Method) An environmental assessment method that uses recognised measures of performance, which are set against established benchmarks, to evaluate a building's specification, design, construction and use. The measures used represent a broad range of categories and criteria from energy to ecology. They include aspects related to energy and water use, the internal environment (health and well-being), pollution, transport, materials, waste, ecology and management processes.

Building Envelope The physical separator, including the roof, walls, windows, floor, between the interior and the exterior environments of a building.

Building facilities All the plants that supplies services in a building (e.g., HVAC systems, DHW systems, electrical systems, lighting systems, etc.).

Building Shell See Building Envelope.

Capital Cost The total investment needed to complete a project of implementation of retrofit measures and bring it to an operable status.

Chiller A central plant refrigeration device that produces chilled water for use in cooling coils of HVAC systems or for other uses (see also absorption chiller).

CHP (Combined heat and power) The cogeneration systems allow to self-produce electricity using different fuels supplied to endothermic engines or turbines and, at the same time, to recover thermal energy from the transformation process.

Cogeneration See CHP.

Combustion Efficiency Ratio of heat obtained from the combustion of a fuel to the theoretical heat content of the fuel.

Commissioning (Cx) The basic purpose of building commissioning is to provide a quality-based process with documented confirmation that building systems are planned, designed, installed, tested, operated and maintained in compliance with the Owner's Project Requirements (OPR) (Definition of BCA-Building Commissioning Association).

Cooling Tower The condensing unit of a central chiller plant which uses evaporation and air movement to provide cooling.

COP (Coefficient of Performance) Ratio of heat produced (including circulating fan heat but not supplemental or backup heat) divided by the total electric energy input including condenser fan and defrost.

Condensing boiler Are water heaters in which a high efficiency (typically greater than 90%) is achieved by using the waste heat in the flue gases to pre-heat the cold water entering the boiler.

Daylighting Using natural light through windows and skylights. Ideally used in conjunction with dimming controls to reduce amount of electrical light input to maintain constant lighting levels.

Degree Days The difference between the average daily temperature (°C) and a standard temperature (e.g., 20 °C in winter or 26 °C in summer). Degree days are used to indicate patterns of deviation from a given temperature standard.

Demand Limiting A technique to reduce demand by measuring incoming electrical power and turning off specified loads to keep the rate of electrical usage under a preset level.

DHW Domestic hot water systems.

Distributed generation Distributed generation is the generation of electricity at some point or points other than a central station power energy conservation measures definitonlant. The source of the electricity can be diesel or gas-fired generator sets, gas turbines, small gas turbines, photovoltaic plants, fuel cells, wind or biomass. Advantages: reduction in peak demand and thereby in the demand for adding more generating capacity, reduction in need for more transmission and distribution infrastructure, more efficient use of renewable energy sources (the electric grid play a virtual rule as energy storage). Disadvantage: initial costs, if compared with centralised generation, and requirement for developing expertise.

ECM (Energy Conservation Measure) A building modification or equipment change of an installation in a facility which is primarily intended to reduce energy consumption or allow the use of an alternative energy source.

Economizer Cycle A method of operating a ventilation system to reduce refrigeration load. Whenever the outdoor air conditions are more favourable (lower heat content) that return air conditions, outdoor air quantity is increased to provide "free" cooling.

Emissions Waste substances released into the air or water.

End-Use Sectors The residential, commercial, industrial and transportation sectors of the economy.

Energy Is the ability of a physical system to do work on other physical systems. The laws of thermodynamics demonstrate that not all energy can perform work. Depending on the boundaries of the physical system in question, energy as understood in the above definition may sometimes be better described by concepts such as exergy, and thermodynamic free energy.

Energy Accounting A formal process of providing long-term organisation that can describe using energy usage data, such as that from utility bills or regular meter readings, to create a usage profile. A usage profile can provide insight into energy usage (electricity, gas, water, etc.) such as baseline usage, peak usage and usage trends.

Energy Audit Systematic inspection and analysis of energy use and energy consumption of a system or organisation with the objective of identifying energy flows and the potential for energy efficiency improvements.

Energy Auditor Individual, group or body undertaking the energy audit.

Energy Balance The accounting of energy inputs and outputs in any process.

Energy Star Is a joint program of the U.S. Environmental Protection Agency and the U.S. Department of Energy helping us all save money and protect the environment through energy efficient products and practices.

Energy Management System Defined by ISO 50001 is based on the management system model of continual improvement also used for other well-known standards such as ISO 9001 or ISO 14001. This makes it easier for organisations to integrate energy management into their overall efforts to improve quality and environmental management.

ESCO (Energy Service Company) A company that offers to reduce a client's energy consumption with the cost savings being split with the company to pay for installation costs of energy conservation measures.

Exhaust Air Air removed from the conditioned space to the outdoors by a dedicated exhaust fan or by the ventilation system.

Fenestration An architectural term referring any opening in a building's envelope including windows, doors and skylights. It also refers to the design and placement of windows in buildings.

Fixture A complete lighting unit, or luminaire, consisting of one or more lamps, ballast if needed, and elements necessary to position and protect lamps, distribute light and connect to a power supply.

Glare Can be defined as the contrast lowering effect of stray light in a visual scene. Any excessive brightness from a direct or reflected source can annoys, distracts or reduces visibility.

Green Building is the practice of increasing the efficiency with which buildings use resources—energy, water and materials—while reducing building impacts on human health and the environment, through better siting, design, construction, operation, maintenance and removal—the complete building life cycle.

Green Energy Audit An assessment of the energy and resources flows in a building, with a view to identify opportunities to increase the sustainability.

Green Energy Plan Operational plan that defines the strategies and then implement actions to improve the environmental sustainability of a building or a system.

Gross Calorific Value (GCV) is the amount of heat available from the complete combustion, at constant pressure and the density of the fuel, when the products of combustion are returned to the initial temperature of the fuel and the combustion air. Gross calorific value can be obtained approximately by calculation, based on elemental analysis of the fuel or directly through the use of special calorimetric instruments.

HDD (Heating Degree Day) The difference of the mean daily temperature below a base temperature (e.g., 20 °C). A relative measure of how weather imposes a heating load on a building which assumes that the building will not require heating.

Heat Pump A cooling system that can operate in the reverse mode: it can move thermal energy in a direction which is opposite to the direction of spontaneous heat flow. It can be used as a heating unit as well as a cooling unit.

Home Energy Audit Comprehensive set of procedures built to facilitate the investigation of the inefficiencies of the residential buildings.

HVAC (Heating, Ventilation, and Air-Conditioning) A system that provides the process of comfort heating, ventilating and/or air conditioning within a building.

Hydronic System A heating and/or cooling system that uses a liquid, usually water, as the medium for heat transfer.

IES Illuminating Engineering Society.

Illuminance Lighting level measured in foot-candles or lux on a working surface such as a desktop or floor.

Indoor Air Quality (IAQ) IAQ deals with the content of interior air that could affect the health and comfort of building occupants. The IAQ may be compromised by microbial contaminants (mold, bacteria), chemicals (such as carbon monoxide, radon), allergens, or any mass or energy stressor that can induce health effects.

Infrared thermography Is a type of infrared imaging science. Thermographic cameras detect radiation in the infrared range of the electromagnetic spectrum and produce images of that radiation.

Infiltration The process by which outdoor air leaks into a building through cracks and holes in the building envelope.

Installation Buildings, equipment, systems, energy-using processes or services which are the subject of the energy audit.

IPMPV The International Performance Measurement and Verification Protocol (IPMVP)is an international protocol that provides an overview of current best practice techniques available for verifying results of energy efficiency, water efficiency and renewable energy projects. It may also be used by facility operators to assess and improve facility performance.

IRR (Internal Rate of Return) A measure of the return percentage to be expected on a capital investment.

Latent Heat The amount of energy required to cause a liquid to change its physical state to a vapour. When a vapour condenses back into a liquid, it releases the same amount of energy without any change in temperature.

LEED (Leadership in Energy and Environmental Design) An environmental rating and certification system for residential and commercial buildings from the U.S. Green Building Council. LEED provides benchmarks for the design, construction and operation of a property and covers site makeup, building materials, water and energy efficiency as well as indoor environmental quality. It also provides certification for people who demonstrate an understanding of green building practices.

Life Cycle Cost (LCC) The cost of owning, operating and maintaining a piece of equipment over its entire useful life.

Life Cycle Assessment (LCA) Is a tool used to analyse environmental consequences (from cradle to grave) of the construction and use of products or services (including energy services).

Load Factor The relationship between the peak rate of consumption to the total consumption for the period. For electricity a load factor ideal for a user who operates in the 24 hours is 1, while decreases to 0.5 for a final user who operates in a period of 12 hours.

Low-E-glass A common term used to refer to glass which has low emissivity due to a film or metallic coating on the glass or suspended between the two lights of glass to restrict the passage of radiant heat.

Lower Calorific Value (LCV) is the gross calorific value decreased by the heat of condensation of water–vapour during the combustion process. This is the value to which one usually refers when generally discussing the calorific value of fuel and the efficiency of a heat engine or a heating device.

Luminaire See fixture.

Make-up Air Outdoor air supplied to a building to compensate for air exhausted from the building.

Make-up Water Water supplied to a system to replace water lost by blow-down, leakage and evaporation.

Monitoring Is a systematic procedure that, also through the use of instrumentation (data-logger) is able to acquire the values of physical environmental parameters (e.g., luminance, temperature, relative humidity, etc.) with defined time rate. The purpose of monitoring is to acquire information boobies actual mode of use of the building and facilities.

Night Cycle A cycle of an HVAC control system that distinguishes between occupied and unoccupied operation.

O&Ms (Operation and Maintenance Measures) Low cost or no cost energy efficiency opportunities involving changes in the operation and maintenance practices taken to improve facilities or building efficiency.

Occupied Hours Period during which the buildings or some building areas are occupied by people involved in their work.

Organization Person or body who owns, operates, uses or manages the installations being audited and who in normal circumstances benefits from the audit.

Outside Air Air taken from the outdoors and therefore not previously circulated through the HVAC system.

Payback Period The length of time necessary to recover the initial investment of a project through energy or maintenance savings.

Peak Load or Peak Demand The electric load that corresponds to a maximum level of electric demand in a specified time period.

Predictive Maintenance Those activities involving continuous or periodic monitoring and diagnosis in order to forecast component degradations so that as-needed, planned maintenance can be performed prior to equipment failure.

Present Value The present worth of a dollar saved or spent at a determined point of time in the future. This concept reflects the time value of money.

Preventive Maintenance The care and servicing by personnel for the purpose of pre-scheduling adjustment, cleaning, calibration, lubrication, component replacement, repairs or whatever is necessary to eliminate minor equipment problems before they become major.

Radiant (Ray) conditioning system A climatisation system which controls the climate of a space using radiant terminals both for winter heating and summer cooling.

Radiation The transfer of heat from one body to another by heat waves without heating the air between the bodies.

Return Air Air that is drawn back into the ventilation system from the conditioned space.

Sensible heat The heat which, when added or subtracted, causes a temperature change.

Simple Payback The length of time required for an investment to pay for itself determined by dividing the initial investment by the annual savings.

Simulation Audit Has the same characteristics of the Standard Audit with the difference that the calculations are done using dynamic simulation models.

Solar cooling A solar thermal system that, by means of an absorbption chiller, is able to produce cold water that can be used for cooling systems.

Standard Audit Defines a standard approach to energy audits: a comprehensive approach that provides surveys, monitoring and calculations to evaluate the effects of the proposed retrofit measures. The result is a detailed report including a description of the current situation (building and plants), pinpointing inefficiencies in the systems or their management; definition and description of actions to be taken and works to be carried out, economics.

Supply Air Conditioned air going to a conditioned space. The end product of the HVAC system.

Task Lighting Lighting directed to a specific area to provide illumination for the performance of an individual visual task.

Thermal insulation Thermal insulation in buildings is an important factor to achieving thermal comfort for its occupants. Insulation reduces unwanted heat loss or gain and can decrease the energy demands of heating and cooling systems. It does not necessarily deal with issues of adequate ventilation and may or may not affect the level of sound insulation.

Thermal comfort Human thermal comfort is defined by ASHRAE as the state of mind that expresses satisfaction with the surrounding environment. Maintaining thermal comfort for occupants of buildings or other enclosures is one of the important goals of HVAC design engineers.

Trigeneration defines the combined production of electricity, heat and heat and chilled water. For the production of chilled water using an absorption chiller which is fed with the heat produced by the CHP system.

U-Value Rate of heat flow-value through the complete heat barrier, from room air to outside air. The lower the U-value, the better the insulating value.

Unoccupied Hours The time when a building is normally empty of people, except for a few attendants or maintenance personnel.

Useful Life That period of time for which a modification used under specific conditions is able to fulfil its intended function and which does not exceed the period of remaining use of the building being modified.

VAV (Variable Air Volume) Air flow is varied to match the heating or cooling loads.

Visual Task Those details and objects which must be seen for the performance of a given activity, including the immediate background of details or objects.

Walkthrough audit Also called a One-day Audit or Preliminary Audit is the first level of approach to energy audit. The field survey phase is normally limited to a single visit. The result is a brief report pinpointing inefficiencies in the systems or their management; rough list of actions to be taken and works to be carried out; suggestions for further in-depth surveys.

MIX
Papier aus verantwortungsvollen Quellen
Paper from responsible sources
FSC® C105338
FSC
www.fsc.org

If you have any concerns about our products,
you can contact us on
ProductSafety@springernature.com

In case Publisher is established outside the EU,
the EU authorized representative is:
Springer Nature Customer Service Center GmbH
Europaplatz 3, 69115 Heidelberg, Germany

Printed by Libri Plureos GmbH
in Hamburg, Germany